D1393665

Grundlehren der mathematischen Wissenschaften 267

A Series of Comprehensive Studies in Mathematics

Grundlehren der mathematischen Wissenschaften

A Series of Comprehensive Studies in Mathematics

A Selection

180. Landkof: Foundations of Modern Potential Theory
181. Lions/Magenes: Non-Homogeneous Boundary Value Problems and Applications I
182. Lions/Magenes: Non-Homogeneous Boundary Value Problems and Applications II
183. Lions/Magenes: Non-Homogeneous Boundary Value Problems and Applications III
184. Rosenblatt: Markov Processes, Structure and Asymptotic Behavior
185. Rubinowicz: Sommerfeldsche Polynommethode
186. Handbook for Automatic Computation. Vol. 2. Wilkinson/Reinsch: Linear Algebra
187. Siegel/Moser: Lectures on Celestial Mechanics
188. Warner: Harmonic Analysis on Semi-Simple Lie Groups I
189. Warner: Harmonic Analysis on Semi-Simple Lie Groups II
190. Faith: Algebra: Rings, Modules, and Categories I
191. Faith: Algebra II, Ring Theory
192. Mallcev: Algebraic Systems
193. Pólya/Szegö: Problems and Theorems in Analysis I
194. Igusa: Theta Functions
195. Berberian: Baer*-Rings
196. Athreya/Ney: Branching Processes
197. Benz: Vorlesungen über Geometric der Algebren
198. Gaal: Linear Analysis and Representation Theory
199. Nitsche: Vorlesungen über Minimalflächen
200. Dold: Lectures on Algebraic Topology
201. Beck: Continuous Flows in the Plane
202. Schmetterer: Introduction to Mathematical Statistics
203. Schoeneberg: Elliptic Modular Functions
204. Popov: Hyperstability of Control Systems
205. Nikollskii: Approximation of Functions of Several Variables and Imbedding Theorems
206. André: Homologie des Algébres Commutatives
207. Donoghue: Monotone Matrix Functions and Analytic Continuation
208. Lacey: The Isometric Theory of Classical Banach Spaces
209. Ringel: Map Color Theorem
210. Gihman/Skorohod: The Theory of Stochastic Processes I
211. Comfort/Negrepontis: The Theory of Ultrafilters
212. Switzer: Algebraic Topology—Homotopy and Homology
213. Shafarevich: Basic Algebraic Geometry
214. van der Waerden: Group Theory and Quantum Mechanics
215. Schaefer: Banach Lattices and Positive Operators
216. Pólya/Szegö: Problems and Theorems in Analysis II
217. Stenström: Rings of Quotients
218. Gihman/Skorohod: The Theory of Stochastic Process II
219. Duvant/Lions: Inequalities in Mechanics and Physics
220. Kirillov: Elements of the Theory of Representations
221. Mumford: Algebraic Geometry I: Complex Projective Varieties
222. Lang: Introduction to Modular Forms
223. Bergh/Löfström: Interpolation Spaces. An Introduction
224. Gilbarg/Trudinger: Elliptic Partial Differential Equations of Second Order

Continued after Index

E. Arbarello
M. Cornalba
P. A. Griffiths
J. Harris

Geometry of Algebraic Curves

Volume I

With 10 Illustrations

Springer-Verlag
New York Berlin Heidelberg Tokyo

E. Arbarello
Dipartimento di Matematica
Istituto "Guido Castelnuovo"
Università di Roma "La Sapienza"
00187 Roma
Italia

M. Cornalba
Dipartimento di Matematica
Università di Pavia
27100 Pavia
Italia

P. A. Griffiths
Office of the Provost
Duke University
Durham, NC 27760
U.S.A.

J. Harris
Department of Mathematics
Brown University
Providence, RI 02912
U.S.A.

AMS Subject Classification: 14-XX

Library of Congress Cataloging in Publication Data
Main entry under title:
Geometry of algebraic curves.
(Grundlehren der mathematischen Wissenschaften; 267)
Bibliography: v. 1, p.
Includes index.
1. Curves, Algebraic. I. Arbarello, E. II. Series.
QA565.G46 1984 512′.33 84-5373

Typeset by Composition House Ltd., Salisbury, England.
Printed and bound by R. R. Donnelley & Sons, Harrisonburg, Virginia.
Printed in the United States of America.

9 8 7 6 5 4 3 2 1

ISBN 0-387-90997-4 Springer-Verlag New York Berlin Heidelberg Tokyo
ISBN 3-540-90997-4 Springer-Verlag Berlin Heidelberg New York Tokyo

To the memory of Aldo Andreotti

Preface

In recent years there has been enormous activity in the theory of algebraic curves. Many long-standing problems have been solved using the general techniques developed in algebraic geometry during the 1950's and 1960's. Additionally, unexpected and deep connections between algebraic curves and differential equations have been uncovered, and these in turn shed light on other classical problems in curve theory. It seems fair to say that the theory of algebraic curves looks completely different now from how it appeared 15 years ago; in particular, our current state of knowledge represents a significant advance beyond the legacy left by the classical geometers such as Noether, Castelnuovo, Enriques, and Severi.

These books give a presentation of one of the central areas of this recent activity; namely, the study of linear series on both a fixed curve (Volume I) and on a variable curve (Volume II). Our goal is to give a comprehensive and self-contained account of the extrinsic geometry of algebraic curves, which in our opinion constitutes the main geometric core of the recent advances in curve theory. Along the way we shall, of course, discuss applications of the theory of linear series to a number of classical topics (e.g., the geometry of the Riemann theta divisor) as well as to some of the current research (e.g., the Kodaira dimension of the moduli space of curves).

A brief description of the contents of the various chapters is given in the Guide for the Reader. Here we remark that these volumes are written in the spirit of the classical treatises on the geometry of curves, such as Enriques–Chisini, rather than in the style of the theory of compact Riemann surfaces, or of the theory of algebraic functions of one variable. Of course, we hope that we have made our subject understandable and attractive to interested mathematicians who have studied, in whatever manner, the basics of curve theory and who have some familiarity with the terminology of modern algebraic geometry.

We now would like to say a few words about that material which is *not* included. Naturally, moduli of curves play an essential role in the geometry of algebraic curves, and we have attempted to give a useful and concrete discussion of those aspects of the theory of moduli that enter into our work. However, we do not give a general account of moduli of curves, and we say nothing about the theory of Teichmüller spaces. Secondly, it is obvious that theory of abelian varieties and theta functions is closely intertwined with algebraic curve theory, and we have also tried to give a self-contained presentation of those aspects that are directly relevant to our work, there being no pretense to discuss the general theory of abelian varieties and theta functions. Thirdly, again the general theory of algebraic varieties clearly underlies this study; we have attempted to utilize the general methods in a concrete and practical manner, while having nothing to add to the theory beyond the satisfaction of seeing how it applies to specific geometric problems. Finally, arithmetical questions, as well as the recent beautiful connections between algebraic curves and differential equations including the ramifications with the Schottky problem, are not discussed.

These books are dedicated to Aldo Andreotti. He was a man of tremendous mathematical insight and personal wisdom, and it is fair to say that Andreotti's view of our subject as it appears, for example, in his classic paper "On a theorem of Torelli," set the tone for our view of the theory of algebraic curves. Moreover, his influence on the four of us, both individually and collectively, was enormous.

It is a pleasure to acknowledge the help we have received from numerous colleagues. Specifically, we would like to thank Corrado De Concini, David Eisenbud, Bill Fulton, Mark Green, Steve Kleiman, and Edoardo Sernesi for many valuable comments and suggestions. We would also like to express our appreciation to Roy Smith and Harsh Pittie, who organized a conference on Brill–Noether theory in February, 1979 at Athens, Georgia, the notes of which formed the earliest (and by now totally unrecognizable) version of this work.

Also, it is a pleasure to thank Steve Diaz, Ed Griffin, and Francesco Scattone for excellent proofreading.

Finally, our warmest appreciation goes to Laura Schlesinger, Carol Ferreira, and Kathy Jacques for skillfully typing the various successive versions of the manuscript.

Contents

Guide for the Reader xii
List of Symbols xv

CHAPTER I
Preliminaries 1
§1. Divisors and Line Bundles on Curves 1
§2. The Riemann–Roch and Duality Theorems 6
§3. Abel's Theorem 15
§4. Abelian Varieties and the Theta Function 20
§5. Poincaré's Formula and Riemann's Theorem 25
§6. A Few Words About Moduli 28
Bibliographical Notes 30
Exercises 31
A. Elementary Exercises on Plane Curves 31
B. Projections 35
C. Ramification and Plücker Formulas 37
D. Miscellaneous Exercises on Linear Systems 40
E. Weierstrass Points 41
F. Automorphisms 44
G. Period Matrices 48
H. Elementary Properties of Abelian Varieties 48

APPENDIX A
The Riemann–Roch Theorem, Hodge Theorem, and Adjoint Linear Systems 50
§1. Applications of the Discussion About Plane Curves with Nodes 56
§2. Adjoint Conditions in General 57

CHAPTER II
Determinantal Varieties 61
§1. Tangent Cones to Analytic Spaces 61
§2. Generic Determinantal Varieties: Geometric Description 67
§3. The Ideal of a Generic Determinantal Variety 70
§4. Determinantal Varieties and Porteous' Formula 83
(i) Sylvester's Determinant 87
(ii) The Top Chern Class of a Tensor Product 89
(iii) Porteous' Formula 90
(iv) What Has Been Proved 92

§5. A Few Applications and Examples 93
Bibliographical Notes 100
Exercises 100
A. Symmetric Bilinear Maps 100
B. Quadrics 102
C. Applications of Porteous' Formula 104
D. Chern Numbers of Kernel Bundles 105

CHAPTER III
Introduction to Special Divisors 107
§1. Clifford's Theorem and the General Position Theorem 107
§2. Castelnuovo's Bound, Noether's Theorem, and Extremal Curves 113
§3. The Enriques–Babbage Theorem and Petri's Analysis of the Canonical Ideal 123
Bibliographical Notes 135
Exercises 136
A. Symmetric Products of $\mathbb{P}^1$ 136
B. Refinements of Clifford's Theorem 137
C. Complete Intersections 138
D. Projective Normality (I) 140
E. Castelnuovo's Bound on k-Normality 141
F. Intersections of Quadrics 142
G. Space Curves of Maximum Genus 143
H. G. Gherardelli's Theorem 147
I. Extremal Curves 147
J. Nearly Castelnuovo Curves 149
K. Castelnuovo's Theorem 151
L. Secant Planes 152

CHAPTER IV
The Varieties of Special Linear Series on a Curve 153
§1. The Brill–Noether Matrix and the Variety C_d^r 154
§2. The Universal Divisor and the Poincaré Line Bundles 164
§3. The Varieties $W_d^r(C)$ and $G_d^r(C)$ Parametrizing Special Linear Series on a Curve 176
§4. The Zariski Tangent Spaces to $G_d^r(C)$ and $W_d^r(C)$ 185
§5. First Consequences of the Infinitesimal Study of $G_d^r(C)$ and $W_d^r(C)$ 191
Biographical Notes 195
Exercises 196
A. Elementary Exercises on μ_0 196
B. An Interesting Identification 197
C. Tangent Spaces to $W_1(C)$ 197
D. Mumford's Theorem for g_d^2's 198
E. Martens–Mumford Theorem for Birational Morphisms 198
F. Linear Series on Some Complete Intersections 199
G. Keem's Theorems 200

CHAPTER V
The Basic Results of the Brill–Noether Theory 203
Bibliographical Notes 217
Exercises 218

A. $W_4^1(C)$ on a Curve C of Genus 6 218
B. Embeddings of Small Degree 220
C. Projective Normality (II) 221
D. The Difference Map $\phi_d: C_d \times C_d \to J(C)$ (I) 223

CHAPTER VI
The Geometric Theory of Riemann's Theta Function 225
§1. The Riemann Singularity Theorem 225
§2. Kempf's Generalization of the Riemann Singularity Theorem 239
§3. The Torelli Theorem 245
§4. The Theory of Andreotti and Mayer 249
Bibliographical Notes 261
Exercises 262
A. The Difference Map ϕ_d (II) 262
B. Refined Torelli Theorems 263
C. Translates of W_{g-1}, Their Intersections, and the Torelli Theorem 265
D. Prill's Problem 268
E. Another Proof of the Torelli Theorem 268
F. Curves of Genus 5 270
G. Accola's Theorem 275
H. The Difference Map ϕ_d (III) 276
I. Geometry of the Abelian Sum Map u in Low Genera 278

APPENDIX B
Theta Characteristics 281
§1. Norm Maps 281
§2. The Weil Pairing 282
§3. Theta Characteristics 287
§4. Quadratic Forms Over $\mathbb{Z}/2$ 292

APPENDIX C
Prym Varieties 295
Exercises 303

CHAPTER VII
The Existence and Connectedness Theorems for $W_d^r(C)$ 304
§1. Ample Vector Bundles 304
§2. The Existence Theorem 308
§3. The Connectedness Theorem 311
§4. The Class of $W_d^r(C)$ 316
§5. The Class of C_d^r 321
Bibliographical Notes 326
Exercises 326
A. The Connectedness Theorem 326
B. Analytic Cohomology of C_d, $d \leq 2g - 2$ 328
C. Excess Linear Series 329

CHAPTER VIII
Enumerative Geometry of Curves 330
§1. The Grothendieck–Riemann–Roch Formula 330

§2. Three Applications of the Grothendieck–Riemann–Roch Formula 333
§3. The Secant Plane Formula: Special Cases 340
§4. The General Secant Plane Formula 345
§5. Diagonals in the Symmetric Product 358
Bibliographical Notes 364
Exercises 364
A. Secant Planes to Canonical Curves 364
B. Weierstrass Pairs 365
C. Miscellany 366
D. Push–Pull Formulas for Symmetric Products 367
E. Reducibility of $W_{g-1} \cap (W_{g-1} + u)$ (II) 370
F. Every Curve Has a Base-Point-Free g^1_{g-1} 372

Bibliography 375

Index 383

Guide for the Reader

This book is not an introduction to the theory of algebraic curves. Rather, it addresses itself to those who have mastered the basics of curve theory and wish to venture beyond them into more recently explored ground. However, we felt that it would be useful to provide the reader with a condensed account of elementary curve theory that would serve the dual purpose of establishing our viewpoint and notation, and furnish a handy reference for those results which are more frequently used in the main body of our work. This is done in the first chapter; quite naturally, few proofs are given in full or even sketched. One notable exception is provided by the theorem of Riemann, describing the theta divisor of the Jacobian of a curve of genus g as a translate of the image under the Abel–Jacobi map of the $(g - 1)$-fold symmetric product of the curve.

The reader is assumed to have a working knowledge of basic algebraic geometry such as is given, for example, in the first chapter of Hartshorne's book *Algebraic Geometry*. Occasionally, however, we have been compelled to make use of relatively more advanced results, such as the theory of base change or the Grothendieck–Riemann–Roch theorem. Our policy, in this situation, has been to give complete statements and adequate references to the existing literature when the results are first used. The main exception to this rule is provided by Chapter II, which contains a down-to-earth and utilitarian presentation, with complete proofs, of the first and second fundamental theorems of invariant theory for the general linear group, the local structure of determinantal varieties, and their global enumerative properties such as Porteous' formula. The main reason for this exception is, of course, that the varieties of special divisors, which form one of the main objects of study in this book, have a natural determinantal structure and many of their properties are essentially direct consequences of general facts about determinant varieties. We also felt that most readers would prefer a unified account of the results rather than a sequence of references to scattered sources in the literature.

The first two chapters are thus of a preliminary nature, and most readers will probably want to use them primarily for reference purposes. The main theme of the book, that is, the study of special divisors and the extrinsic geometry of curves, is introduced in Chapter III. Here will be found, beside elementary facts such as Clifford's theorem (which are discussed here and not in Chapter I simply because they are close in nature to some of the results that will be encountered in later chapters), Castelnuovo's description of extremal curves, Noether's theorem, and the theorems of Enriques–Babbage and Petri on the canonical ideal.

Chapter IV is of a foundational nature. In it the varieties of special divisors and linear series on a fixed curve—the main characters of this book—are defined, and the functors they represent are identified. This is where the results of Chapter II are first applied in a systematic way. Although containing no major results of independent interest, except for Martens' improvement of Clifford's theorem and its subsequent refinement by Mumford, this chapter is, in a sense, the cornerstone on which most of the later chapters rest.

In Chapter V the main theorems of Brill–Noether theory are stated and illustrated by means of examples drawn from low-genus cases. In fact most of the theorems are proved by *ad hoc* arguments, for genus up to six.

Chapter VI is probably the most geometric in nature and collects many of the central results about the geometry of the theta divisor of a Jacobian. The main topics touched upon are Riemann's singularity theorem and its generalization by Kempf, Andreotti's proof of the Torelli theorem for curves, and Andreotti and Mayer's approach to the Schottky problem via the heat equation for the theta function.

Chapter VII contains the proofs of some of the results stated in Chapter V, notably those of the existence and connectedness theorems, and of the enumerative formulas for the classes of the varieties of special linear series and divisors. The enumerative geometry of these varieties, and related ones, is further investigated in the eighth and final chapter of this volume.

The second volume will contain an exposition of the fundamentals of deformation theory and of the main properties of the moduli space of curves, the proof of the remaining results of Brill–Noether theory, a presentation of the basic properties of the varieties of special linear series on a moving curve with special attention to series of dimension one and two (that is, to Hurwitz spaces and varieties of plane curves), and a proof of the theorem that the moduli space of curves of sufficiently high genus is of general type.

List of Symbols

χ 1
$p_a(C)$ 1
$g(C)$ 1
$h^i(V, \mathscr{F})$ 1
$\mathrm{mult}_p(D)$ 2
$\deg(D)$ 2
$\mathrm{Div}(C)$ 2
μ_p 2
(ϕ) 2
Res_p 2
$\mathscr{L}(D)$ 3
$D \geq 0$ 3
$D \sim 0$ 4
$|D|$ 4
g_d^r 4
$\phi_{\mathscr{D}}$ 5
ϕ_L 6
$\phi_{|D|}$ 6
K_C 7
Θ_C 7
$l(D)$ 7
$i(D)$ 7
$r(D)$ 7
$f^*(D)$ 8
$\overline{\phi(D)}$ 12
$\bar{D}$ 12
$\overline{p_1, \ldots, p_d}$ 12
$J(C)$ 17
u 17, 18
$\mathrm{Div}^d(C)$ 18
C_d 18
$\mathrm{Pic}(C)$ 18
$\mathscr{H}_g$ 22
$\mathscr{A}_g$ 22
Γ_g 22
θ 23
Θ 23
w_d 25
c_d 25
θ 25
κ 27
$j(C)$ 28
$\mathscr{M}_g$ 28
Δ_i 29
$\mathscr{C}$ 29
$\phi^{(k)}$ 37
$\mathscr{M}(C)$ 41
$T_p(Y)$ 61
$\mathscr{T}_p(X)$ 61
$\mathbb{P}\mathscr{T}_p(X)$ 62
$X_k(\phi)$ 83
$-F$ 85
$E - F$ 85
$\Delta_{p,q}(a)$ 86
μ_0 108, 159
$\pi(d, r)$ 116
C_d^r 153, 177
$W_d^r(C)$ 153, 177
$G_d^r(C)$ 153
ρ 159
$\mu_{0,W}$ 187
ϕ_d 223
V_d 223
$\mathscr{J}_g$ 249
Nm 281
$\mathbb{P}E$ 304
$\mathscr{O}_{\mathbb{P}E}(1)$ 305
$K(X)$ 331

$\mathrm{ch}(E)$ 331
$\pi_!$ 331
$\mathrm{td}(E)$ 332
$\binom{n}{i}$ 341
$(g(t))_{t_1^{a_1} \cdots t_n^{a_n}}$ 341
$\Gamma_d(\mathscr{D})$ 341
V_d^r 345
$\mu(a, b, m, i, \beta)$ 352
ϕ_a 358

Note. Throughout this book, if V is a vector space (resp. if E is a vector bundle) we will denote by $\mathbb{P}V$ (resp. $\mathbb{P}E$) the space of one-dimensional subspaces of V (resp. of the fibers of E); thus

$$\mathbb{P}V = \mathrm{Proj}(\bigoplus \mathrm{Sym}^n V^*).$$

More generally, if C is a cone, $\mathbb{P}C$ will stand for its projectivization. Similarly, by $G(k, V)$ (resp. $G(k, E)$) we will mean the space of k-dimensional subspaces of V (resp. of the fibers of E).

Chapter I

Preliminaries

This book will be concerned with geometric properties of algebraic curves. Our central problem is to study the various projective manifestations of a given abstract curve. In this chapter we shall collect various definitions, notations, and background facts that are required for our work.

§1. Divisors and Line Bundles on Curves

By *curve* we shall mean a complete reduced algebraic curve over $\mathbb{C}$; it may be singular or reducible. When speaking of a *smooth curve*, we shall always implicitly assume it to be irreducible. Sometimes, when no confusion is possible we shall drop the adjective "smooth"; this will only be done in sections where exclusively smooth curves are being considered.

We shall assume known the basic properties of sheaves and line bundles on algebraic varieties and analytic spaces, and shall make the usual identification of invertible sheaves with line bundles and of locally free sheaves with vector bundles. When tensoring with line bundles we shall often drop the tensor product symbol. If $\mathscr{F}$ is a sheaf of $\mathbb{C}$-vector spaces over a topological space V, we shall set, as customary

$$h^i(V, \mathscr{F}) = \dim_{\mathbb{C}} H^i(V, \mathscr{F}),$$

$$\chi(\mathscr{F}) = \sum (-1)^i h^i(V, \mathscr{F}).$$

The basic invariant of a curve C is its genus. To be more precise, we shall use the words *arithmetic genus* of C to denote the integer

$$p_a(C) = 1 - \chi(\mathcal{O}_C).$$

Of course, when C is connected, the arithmetic genus of C equals $h^1(C, \mathcal{O}_C)$. On the other hand, when C is irreducible, we shall denote by $g(C)$ its *geometric genus*, which is defined to be the (arithmetic) genus of its normalization. In this book we shall usually talk about the genus of a curve without further specification; hopefully, it will always be clear from the context which genus we are referring to. A very basic but non-elementary fact, which can be proved via potential theory, is that the genus of a smooth curve C is one-half of the first Betti number of the underlying topological surface; in symbols

$$g(C) = \tfrac{1}{2}\operatorname{rank}(H^1(C, \mathbb{Z})).$$

Throughout this chapter we shall fix a smooth curve C.

To understand the geometry of C it is essential to study its meromorphic functions; this is best done in the language of divisors and line bundles. A *divisor*

$$D = \sum_i n_i p_i, \qquad n_i \in \mathbb{Z} \quad \text{and} \quad p_i \in C,$$

is a formal linear combination of points on C. We may assume that the p_i are distinct, and then n_i is the *multiplicity* $\operatorname{mult}_{p_i}(D)$ of D at p_i. The divisors form a group $\operatorname{Div}(C)$, and the *degree* homomorphism

$$\deg\colon \operatorname{Div}(C) \to \mathbb{Z}$$

is defined by

$$\deg(D) = \sum_i n_i.$$

The group of divisors of degree zero is denoted by $\operatorname{Div}^0(C)$.

If ϕ is a meromorphic function (resp., a meromorphic differential) on C, then in terms of a local holomorphic coordinate z on C,

$$\phi = f(z) \qquad (\text{resp., } \phi = f(z)\,dz),$$

where $f(z)$ is a meromorphic function. If the point $p \in C$ corresponds to the origin in the z-plane, and if we write

$$f(z) = z^{\mu} g(z), \qquad g(0) \neq 0, \infty,$$

then the *order* $\mu_p(\phi) = \mu$ of ϕ at p is well defined. The divisor (ϕ) associated to ϕ is defined to be

$$(\phi) = \sum_{p \in C} \mu_p(\phi) p.$$

In case ϕ is a meromorphic differential, its *residue* at p is

$$\operatorname{Res}_p(\phi) = \frac{1}{2\pi\sqrt{-1}} \int_{\gamma} \phi,$$

where γ is any curve homotopic to $\{|z| = \varepsilon\}$ in a small punctured neighborhood of p. A simple but basic fact is the residue theorem

$$\sum_{p \in C} \operatorname{Res}_p(\phi) = 0.$$

This is a straightforward consequence of Stokes' theorem. In fact, suppose ϕ has poles at $p_1, \ldots, p_d$ and let U_i be a small parametric disc around p_i such that $U_i \cap U_j = \emptyset$ if $i \neq j$. Setting $C^* = C - \bigcup_i U_i$, we obtain

$$\begin{aligned}\sum_i \operatorname{Res}_{p_i}(\phi) &= \sum_i \frac{1}{2\pi\sqrt{-1}} \int_{\partial U_i} \phi \\ &= -\frac{1}{2\pi\sqrt{-1}} \int_{C^*} d\phi \\ &= 0.\end{aligned}$$

When applied to the logarithmic differential $\phi = df/f$ of a meromorphic function f, the residue theorem gives

$$\deg((f)) = 0.$$

If we view f as a holomorphic map

$$f\colon C \to \mathbb{P}^1,$$

this expresses the well-known topological fact that the degree of the divisor $f^{-1}(q)$ is independent of $q \in \mathbb{P}^1$, and agrees with the degree (also called sheet number) of f.

A divisor D is said to be *effective*, and we write $D \geq 0$, if all points of D appear with non-negative multiplicity. We shall write $D \geq D'$ to mean $D - D'$ is effective.

To any divisor D one can attach the sheaf $\mathcal{O}(D)$ defined by the prescription

$$\Gamma(U, \mathcal{O}(D)) = \left\{\begin{matrix}\text{meromorphic functions on } U \\ \text{that satisfy } (f) + D|_U \geq 0.\end{matrix}\right\}$$

Actually, $\mathcal{O}(D)$ turns out to be a line bundle, since it is generated, over any sufficiently small open set, by $1/g$, where g is a local defining equation for D. It is customary to write

$$\mathcal{L}(D) = H^0(C, \mathcal{O}(D)).$$

Conversely, given a line bundle L and a non-zero meromorphic section s, we may define, in complete analogy with the case of meromorphic functions, the divisor $D = (s)$ of s, and division by s yields an isomorphism

$$L \cong \mathcal{O}(D).$$

It will be an easy consequence of the Riemann–Roch theorem that any line bundle on C has a non-zero meromorphic section, thus showing that every line bundle on C is of the form $\mathcal{O}(D)$, up to isomorphism.

The following formal rules are clear from the definitions:

$$\mathcal{O}(D) \otimes \mathcal{O}(D') \cong \mathcal{O}(D + D'),$$
$$\mathcal{O}(D)^{-1} \cong \mathcal{O}(-D).$$

A basic notion in the study of divisors is the one of linear equivalence. A divisor D is *linearly equivalent to zero*, and we write $D \sim 0$, if

$$D = (f)$$

for some meromorphic function f. Two divisors D and D' are *linearly equivalent* if

$$D - D' \sim 0,$$

and the linear equivalence class of a divisor D is called the divisor class of D and denoted by $[D]$. Clearly, two divisors D and D' are linearly equivalent if and only if there is an isomorphism between $\mathcal{O}(D)$ and $\mathcal{O}(D')$.

To each projective manifestation of C there is attached a linear series of divisors on the curve itself. First of all, given a divisor D, the *complete linear series* (or *system*) $|D|$ is the set of effective divisors linearly equivalent to D. Given two meromorphic functions f and g, notice that $(f) = (g)$ if and only if there is a non-zero constant λ such that $f = \lambda g$. We then have an identification

$$|D| = \mathbb{P}\mathcal{L}(D)$$

obtained by associating to each non-zero $f \in \mathcal{L}(D)$ the divisor $(f) + D$. A complete linear series is therefore a projective space. More generally, any linear subspace of a complete linear series is called a *linear series* (or *system*). A linear series $\mathcal{D} = \mathbb{P}V$, where V is a vector subspace of $\mathcal{L}(D)$, is said to be a g^r_d if

$$\deg(D) = d; \qquad \dim(V) = r + 1.$$

A g^1_d is called a *pencil*, a g^2_d a *net*, and a g^3_d a *web*. By a base point of a linear series $\mathcal{D}$ we mean a point common to all divisors of $\mathcal{D}$. If there are none we say that the linear series is *base-point-free*. The *base locus* of $\mathscr{D}$ is defined to be the divisor

$$B = \sum_{p \in C} n_p \cdot p$$

where

$$n_p = \min\{\operatorname{mult}_p(D): D \in \mathscr{D}\}.$$

As we mentioned before, there is a fundamental relation between linear series on C and maps of C to projective spaces. This is best expressed in the language of line bundles. As we already said, given a non-zero holomorphic section s of a line bundle L, and denoting by D the divisor of s, there is an isomorphism between L and $\mathscr{O}(D)$. It is therefore natural to use the symbol $|L|$ to denote the complete linear series

$$|D| = \mathbb{P}H^0(C, L)$$

of all divisors of section of L. Now, given a g_d^r $\mathscr{D} = \mathbb{P}V$, where V is a subspace of $H^0(C, L)$, notice that a base point of $\mathscr{D}$ is a point of C where all sections in V vanish. If $\mathscr{D}$ is base-point-free, we can then define a map

$$\phi_{\mathscr{D}}: C \to \mathbb{P}V^* = \mathbb{P}^r$$

by

$$\phi_{\mathscr{D}}(p) = \{s \in V: s(p) = 0\}.$$

When, on the other hand, $\mathscr{D}$ has a non-empty base locus B, defining, as is customary

$$L(-B) = L \otimes \mathscr{O}(-B),$$

we may view V as a subspace of $H^0(C, L(-B))$ with the property that the corresponding linear series $\mathscr{D}(-B)$ is base-point-free; we may then set

$$\phi_{\mathscr{D}} = \phi_{\mathscr{D}(-B)}.$$

It is useful to express $\phi_{\mathscr{D}}$ in homogeneous coordinates on $\mathbb{P}^r$. If we choose a basis $s_0, s_1, \ldots, s_r$ of V, we have

$$\phi_{\mathscr{D}}(p) = [s_0(p), s_1(p), \ldots, s_r(p)].$$

The notation means this. Given a point $q \in C$, choose a local coordinate z centered at q and a local trivialization of L near q. In this trivialization we can write

$$s_i = z^{\mu} g_i(z),$$

where μ is the multiplicity of q in B (of course, $\mu = 0$ if q is not a base point) and the g_i's are holomorphic functions such that $g_i(0) \neq 0$ for at least one value of i. We then set

$$[s_0(p), \ldots, s_r(p)] = [g_0(z(p)), \ldots, g_r(z(p))].$$

In view of the local description of $\phi_{\mathscr{D}}$ we have isomorphisms

$$\phi_{\mathscr{D}}^*(\mathcal{O}_{\mathbb{P}^r}(1)) \cong L(-B),$$

$$\phi_{\mathscr{D}}^*: H^0(\mathbb{P}^r, \mathcal{O}_{\mathbb{P}^r}(1)) \xrightarrow{\sim} V.$$

We may also notice that the image of $\phi_{\mathscr{D}}$ is a *non-degenerate* curve C' in $\mathbb{P}^r$. This means that C' is not contained in any hyperplane. Moreover, it follows from the above formulas that

$$d = \deg(L) = \deg(C') \cdot \deg(f) + \deg(B),$$

where f is just $\phi_{\mathscr{D}}$ viewed as a map from C to C'.

If D is a divisor and L is a line bundle we shall write ϕ_D instead of $\phi_{|D|}$ and ϕ_L instead of $\phi_{|L|}$.

A fundamental example is provided by the so-called *rational normal curve* of degree n, which by definition is the image of $\mathbb{P}^1$ under the embedding

$$v_n = \phi_L: \mathbb{P}^1 \to \mathbb{P}^n, \qquad L = \mathcal{O}_{\mathbb{P}^1}(n).$$

In terms of an affine coordinate z on $\mathbb{P}^1$, the mapping v_n is given by

$$z \to [1, z, \ldots, z^n].$$

The symbol v_n is suggestive of the fact that such a map is a special case of a Veronese map.

§2. The Riemann–Roch and Duality Theorems

A basic problem concerning linear series on curves is to compute the dimension of $\mathscr{L}(D)$ and for this the main tool is the Riemann–Roch theorem. To state it we need a few definitions. Recall that the group of isomorphism classes of line bundles on C can be identified with $H^1(C, \mathcal{O}_C^*)$. From the fundamental *exponential sequence*

$$0 \to \mathbb{Z} \to \mathcal{O}_C \to \mathcal{O}_C^* \to 0$$

we get the coboundary map

$$\deg: H^1(C, \mathcal{O}_C^*) \to H^2(C, \mathbb{Z}) \cong \mathbb{Z}.$$

If L is a line bundle the integer $\deg(L)$ is called the *degree* of L. It is an easy consequence of Stokes' theorem that this definition agrees with the one previously given for divisors, that is

$$\deg(\mathcal{O}(D)) = \deg(D).$$

We also recall that the *canonical line bundle* K_C is defined to be the dual of the tangent bundle Θ_C or, which is the same, the sheaf whose sections over an open set U are the holomorphic differentials on U. Thus the vector space $H^0(C, K_C)$ is the space of holomorphic differentials on C; these are also called *abelian* differentials. For brevity we shall often write K and Θ instead of K_C and Θ_C. We can now state the

Riemann–Roch Theorem. *For any line bundle L on a smooth curve C of genus g*

$$h^0(C, L) - h^0(C, KL^{-1}) = \deg(L) - g + 1.$$

From a cohomological point of view the main ingredient of the proof, in one way or another, is the duality theorem. Recalling that each class in $H^1(C, K)$ is represented in Dolbeault cohomology by a $(1, 1)$-form, this can be stated as follows.

Duality Theorem. *For any line bundle L on a smooth curve C the pairing*

$$\langle\ ,\ \rangle : H^1(C, L) \times H^0(C, KL^{-1}) \to \mathbb{C}$$

given by $\langle \xi, \eta \rangle = \int_C \xi \wedge \eta$ is a duality.

The Riemann–Roch theorem can in fact be gotten by combining the duality theorem with the easily proved formula

$$h^0(C, L) - h^1(C, L) = \deg(L) - g + 1.$$

As a first consequence we find that for any point p on C and any large enough integer n the line bundle $L(np)$ has a non-zero section, thus showing that L has a non-zero meromorphic section and hence is of the form $\mathcal{O}(D)$ for suitable D. Without losing any information we can therefore state the Riemann–Roch theorem in terms of divisors.

Historically this is the form in which the theorem was originally stated (cf., for instance, Enriques–Chisini [1, Vol. III]) and is also the form in which we will most often use it. A proof of it is sketched in the exercises. It is customary to write

$$l(D) = \dim \mathcal{L}(D) = h^0(C, \mathcal{O}(D)),$$

$$i(D) = h^0(C, K(-D)),$$

$$r(D) = l(D) - 1.$$

Of course, if $|D|$ is non-empty

$$r(D) = \dim|D|.$$

The Riemann–Roch formula can now be rephrased in either one of the following two ways:

$$l(D) - i(D) = \deg(D) - g + 1,$$
$$r(D) - i(D) = \deg(D) - g.$$

The integer $i(D)$ is called the *index of speciality* of the divisor D and is equal to the number of linearly independent abelian differentials ω such that

$$(\omega) \geq D.$$

The divisor D is said to be *special* if it is effective and $i(D) > 0$. In a sense it may be said that the entire book is devoted to the theory of special divisors.

The simplest case of the Riemann–Roch formula arises when D is the trivial divisor, in which case we obtain

$$h^0(C, K) = g.$$

This remarkable equality due to Riemann is the first instance of Hodge theory on an algebraic variety (cf., for instance, Griffiths–Harris [1]). Next, applying the Riemann–Roch formula to the canonical bundle, we find

$$\deg(K) = 2g - 2.$$

Since K is the dual of the tangent bundle to C, this formula is nothing but the Hopf index theorem for compact Riemann surfaces. As a corollary we can prove the classical Riemann–Hurwitz formula. Let

$$f: C \to C'$$

be a non-constant (and hence surjective) holomorphic map between smooth curves C and C' of genera g and g'. If D is any divisor on C, we take the inverse image $f^*(D)$ of D to be the divisor on C whose local defining equations are obtained by composing with f the local defining equations of D. For any q on C and $p = f(q)$ on C' we choose local coordinates z and w centered at q and p such that f has the standard form $w = z^{\nu(q)}$. Then, for any p on C',

$$f^*(p) = \sum_{q \in f^{-1}(p)} \nu(q)q.$$

The degree n of the divisor $f^*(p)$ is well known to be independent of p and is called the *degree* or *sheet number* of the map f. For general p it is simply the number of points in the preimage of p. We define the *ramification divisor* R on C by

$$R = \sum_{q \in C} (\nu(q) - 1)q.$$

The integer $\nu(q) - 1$ is called the *ramification index* of f at q. Since, around q, the map f is of the form $w = z^{\nu(q)}$, for any meromorphic differential ϕ on C' we have

$$\mu_q(f^*\phi) = \mu_{f(q)}(\phi)\nu(q) + \nu(q) - 1.$$

Thus

$$(f^*(\phi)) = f^*((\phi)) + R.$$

Counting degrees and recalling that $\deg K_C = 2g - 2$ we obtain the *Riemann–Hurwitz formula*

$$2g - 2 = n(2g' - 2) + \deg R.$$

As a further straightforward application of the Riemann–Roch formula we may compute $l(D)$ in extreme degree ranges. It suffices to notice that if $d = \deg(D) < 0$ then $l(D) = 0$, and similarly $i(D) = 0$ when $d > \deg(K) = 2g - 2$, to conclude that

$$l(D) = \begin{cases} 0 & \text{if } d < 0, \\ d - g + 1 & \text{if } d > 2g - 2. \end{cases}$$

For divisors in the intermediate range

$$0 \leq d \leq 2g - 2$$

the situation is more interesting. It will immediately follow from the geometric version of the Riemann–Roch theorem, to be discussed later, that if d is in the above degree range and D consists of d general points of C then

$$r(D) = \begin{cases} 0 & \text{if } d \leq g, \\ d - g & \text{if } d \geq g. \end{cases}$$

We may call an effective divisor D, for which the above fails to hold, an *exceptional* special divisor.

The admissible values (d, r) for a complete g_d^r on C all lie above the heavy lines in the diagram below. The exceptional special divisors are those for which (d, r) is strictly above the heavy lines. It is, of course, understood that when $d > 2g - 2$ the dimension r is forced to equal $d - g$.

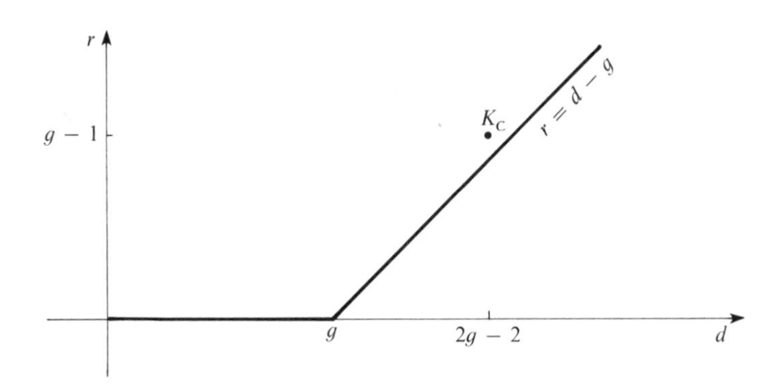

We may note in passing that the only g_{2g-2}^{g-1} is the canonical series. This follows, by the Riemann–Roch theorem, from the remark that the only line bundle of degree zero with non-zero sections is the trivial one.

A useful concept that emphasizes the duality implicit in the Riemann–Roch formula is the concept of residuation. A divisor D' is said to be *residual* to an effective divisor D in case

$$D' \in |K(-D)|.$$

We note that, conversely, D is residual to D' and the relations

$$i(D) = l(D'),$$
$$l(D) = i(D')$$

hold. Using these, and the fact that the degree of the canonical series is $2g - 2$, it follows that the Riemann–Roch formula for D is equivalent to that for D'.

As a first geometric application of the Riemann–Roch theorem we are going to study the canonical map

$$\phi_K: C \to \mathbb{P}H^0(C, K)^*,$$

and this in turn will provide us with a geometrical interpretation of the theorem itself.

We first need to recall that a smooth curve C of genus $g > 1$ is said to be *hyperelliptic* if it carries a g_2^1 (since $g > 0$ this has to be base-point-free).

Equivalently, there should be a degree two map

$$f: C \to \mathbb{P}^1.$$

The corresponding involution of C is called the *hyperelliptic involution*. By the Riemann–Hurwitz formula f has $2g + 2$ ramification points and hence, being of degree two, has $2g + 2$ branch points $a_1, \ldots, a_{2g+2}$. In other words, the curve C can be thought of as the Riemann surface of the algebraic function

$$w = \sqrt{(z - a_1) \cdots (z - a_{2g+2})}.$$

A distinguishing feature of such a curve is that one can be totally explicit in describing the abelian differentials on it. If suffices to notice that, in case $p(z)$ is a polynomial of degree not greater than $g - 1$,

$$p(z)\frac{dz}{w}$$

represents a holomorphic differential on C, and therefore an explicit basis for $H^0(C, K)$ is

$$\omega_1 = \frac{dz}{w}, \omega_2 = z\frac{dz}{w}, \ldots, \omega_g = z^{g-1}\frac{dz}{w}.$$

Historically the integrals of these differentials, that is the abelian integrals, were the subject of extensive study and it was only much later that their investigation led to the very notion of the Riemann surface of an algebraic curve.

It is apparent that, with the choice of coordinates in $\mathbb{P}^{g-1} = \mathbb{P}H^0(C, K)^*$ given by $\omega_1, \ldots, \omega_g$, the canonical map ϕ_K is simply the composition of the two-sheeted covering f with the Veronese embedding v_{g-1} which we described at the end of Section 1:

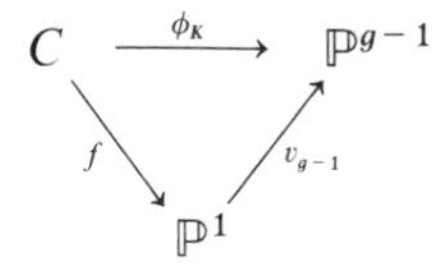

Thus ϕ_K represents the hyperelliptic curve C as a ramified two-sheeted covering of a rational normal curve in $\mathbb{P}^{g-1}$.

On the other hand we are now going to show that, when C is a non-hyperelliptic curve of genus $g > 2$ (any curve of genus 2 is hyperelliptic), the canonical map ϕ_K is an *embedding*. First of all notice that $|K|$ has no base point. If it had one, call it p, we would have $h^0(C, K(-p)) = g$ and therefore, by the Riemann–Roch theorem, there would be a meromorphic function having a simple pole at p and no other singularity, implying that $g = 0$. This shows, incidentally, that for any p on C, $h^0(C, K(-p)) = g - 1$. Now notice that, for any choice of p and q on C, $l(p + q) = 1$ since there are no g_2^1's on C. By the Riemann–Roch theorem this means that

$$h^0(C, K(-p - q)) = g - 2.$$

When $p \neq q$ this means that there are abelian differentials vanishing at p but not at q, showing that ϕ_K is one-to-one. When $p = q$ we find that there are abelian differentials vanishing simply at p and this exactly means that the differential of ϕ_K at p is not zero.

The image of ϕ_K, in case C is non-hyperelliptic, is called a *canonical curve*. It is a smooth non-degenerate curve in $\mathbb{P}^{g-1}$ of degree $2g - 2$. Conversely, any non-degenerate, irreducible curve in $\mathbb{P}^{g-1}$ of geometric genus g and degree $2g - 2$ is a canonical curve. In fact, we have already observed that the only g_{2g-2}^{g-1} on a genus g curve is the canonical series.

Using the canonical image of a (possibly hyperelliptic) curve we want to give a geometric interpretation of the index of speciality of an effective divisor on a smooth curve of genus $g > 1$ and then use it to restate the Riemann–Roch theorem in a geometric form. For this we need a notation that will be very often used throughout this book.

Let C be a smooth curve and let $\phi\colon C \to \mathbb{P}^r$ be a holomorphic mapping. For any effective divisor D on C we denote by $\overline{\phi(D)}$ the intersection of the hyperplanes H such that either $\phi(C) \subset H$ or $\phi^*(H) \geq D$.

If $D = p_1 + \cdots + p_d$ and in case ϕ is an embedding we shall often write $\bar{D}$ or $\overline{p_1, \ldots, p_d}$ for $\overline{\phi(D)}$. If we make the further assumption that the p_i are distinct, then $\bar{D}$ is the ordinary linear span in $\mathbb{P}^r$ of these points; in general, if we write $D = \sum \mu_\alpha q_\alpha$, where the q_α are distinct, then D is the linear span of the appropriate osculating spaces to C at the points q_α.

With this notation it is clear that

$$i(D) = g - 1 - \dim \overline{\phi_K(D)}.$$

Substituting this in the Riemann–Roch formula gives the

Geometric Version of the Riemann–Roch Theorem. *For D an effective divisor of degree d on a smooth curve of genus $g > 1$*

$$r(D) = d - 1 - \dim \overline{\phi_K(D)}.$$

When the points of $D = p_1 + \cdots + p_d$ are distinct we may restate this as saying that $r(D)$ is the number of independent linear relations among the points $\phi_K(p_i)$ on the canonical curve. The geometric version of the Riemann–Roch theorem implies that the exceptional special divisors are those divisors whose points (counted with multiplicities) fail to be independent on the canonical image. This, together with the fact that the canonical image is non-degenerate, gives, as previously announced, that for a general effective divisor D of degree d

$$r(D) = \begin{cases} 0 & \text{if } d \leq g, \\ d - g & \text{if } d \geq g. \end{cases}$$

Residuation may be interpreted as follows. If D is a special effective divisor on C, and D' is residual to D, the complete linear series $|D|$ is the moving part of the series cut out on C by the hyperplanes through $\overline{\phi_K(D')}$. In other words, $|D|$ consists of all divisors D_H such that

$$D_H + D' = \phi_K^*(H),$$

where H is a hyperplane through $\overline{\phi_K(D')}$.

In passing we note that canonical curves have the remarkable property that if they admit one d-secant $(d - r - 1)$-plane then there are ∞^r such planes. For example, if a canonical curve $C \subset \mathbb{P}^3$ of genus 4 has one trisecant line, then there are ∞^1 such trisecant lines. In a sense that is difficult to make precise, the above seems to be the characteristic property of the canonical curve.

A useful consequence of the geometric version of the Riemann–Roch theorem is the following. Let C be a hyperelliptic curve of genus g. Let d be an integer such that $0 \leq d \leq g$. Then any complete g_d^r on C is of the form

$$rg_2^1 + p_1 + \cdots + p_{d-2r},$$

where no two of the p_i's are conjugate under the hyperelliptic involution. In fact, the Riemann–Roch theorem implies that any series of this form is complete, by the well-known fact that any effective divisor of degree $\delta \leq g$ on a rational normal curve in $\mathbb{P}^{g-1}$ spans a $\mathbb{P}^{\delta-1}$. (When the points of the divisor are distinct, this amounts to the non-vanishing of the Vandermonde determinant.) On the other hand, any degree d effective divisor on C belongs to one of these series.

To conclude this section we want to deduce from the duality theorem a constructive version of the Riemann–Roch theorem for effective divisors. Let D be such a divisor; the *Mittag–Leffler sequence* attached to D is

$$0 \to \mathcal{O}_C \to \mathcal{O}_C(D) \to \mathcal{O}_D(D) \to 0.$$

Setting

$$D = \sum_{\alpha=1}^{m} n_\alpha q_\alpha,$$

where $n_\alpha > 0$ and the q_α are distinct, and choosing a local coordinate z_α centered at p_α, a section of $\mathcal{O}_D(D)$ can be written as

$$\eta = \sum \eta_\alpha,$$

where

$$\eta_\alpha = \sum_{k=-n_\alpha}^{-1} a_k z_\alpha^k$$

is a *Laurent tail* (or *principal part*) centered at q_α. The Mittag–Leffler problem consists in deciding which collections of Laurent tails come from a global meromorphic function on C. This is equivalent to solving the equation $\delta\eta = 0$, where δ is the coboundary map

$$\delta\colon H^0(C, \mathcal{O}_D(D)) \to H^1(C, \mathcal{O}_C)$$

gotten from the Mittag–Leffler sequence. In view of the duality theorem this amounts to solving the system of linear equations

$$\langle \delta\eta, \omega \rangle = 0 \quad \text{for every} \quad \omega \in H^0(C, K).$$

To make this pairing computable we will show that

$$\langle \delta\eta, \omega \rangle = 2\pi\sqrt{-1} \sum_\alpha \operatorname{Res}_{q_\alpha}(\eta_\alpha \omega). \tag{2.1}$$

By definition

$$\langle \delta\eta, \omega \rangle = \int_C \phi,$$

where ϕ is a Dolbeault representative of the cup product $\delta\eta \cdot \omega$. To compute ϕ choose, for each α, a small disc U_α centered at q_α and set

$$U_0 = C - \{q_1, \ldots, q_m\}, \qquad \eta_0 = 0.$$

We may find C^∞ (1, 0)-forms ξ_α on U_α, $\alpha = 0, \ldots, m$, such that

$$\xi_\beta - \xi_\alpha = \eta_\beta\omega - \eta_\alpha\omega;$$

then ϕ is the global (1, 1) form defined by

$$\phi \,|\, U_\alpha = \bar{\partial}\xi_\alpha = d\xi_\alpha.$$

Denoting by $T_\alpha(\varepsilon)$ a disc of radius ε centered at q_α, we have

$$\begin{aligned}
\langle \delta\eta, \omega \rangle &= \int_C \phi \\
&= \lim_{\varepsilon \to 0} \int_{C - (\bigcup T_\alpha(\varepsilon))} \phi \\
&= \lim \int_{C - (\bigcup T_\alpha(\varepsilon))} d\xi_0 \\
&= -\lim \sum \int_{\partial T_\alpha(\varepsilon)} \xi_0 \\
&= \lim \sum \int_{\partial T_\alpha(\varepsilon)} \xi_\alpha - \xi_0 \\
&= \lim \sum \int_{\partial T_\alpha(\varepsilon)} \eta_\alpha \omega \\
&= 2\pi\sqrt{-1} \sum_\alpha \operatorname{Res}_{q_\alpha}(\eta_\alpha \omega),
\end{aligned}$$

as desired. Summarizing we can say that *a collection of Laurent tails* η *comes from a global meromorphic function* $f \in \mathscr{L}(D)$ *if and only if*

$$\sum_\alpha \operatorname{Res}_{q_\alpha}(\eta_\alpha \omega) = 0 \quad \text{for every} \quad \omega \in H^0(C, K).$$

As a system of linear equations for η this is a system in d unknowns consisting of g equations, among which those corresponding to differentials ω in $H^0(C, K(-D))$ are trivial. A moment of reflection will convince the reader that, in fact, there are exactly $g - i(D)$ independent equations in the above system. Since two meromorphic functions having the same principal parts differ by a constant we recover the Riemann–Roch theorem.

Finally, formula (2.1), when applied to the divisor consisting of a single point p on C, tells us that the image of

$$\delta\colon H^0(C, \mathcal{O}_p(p)) \to H^1(C, \mathcal{O}) = H^0(C, K)^*$$

represents the point $\phi_K(p)$ on the canonical curve, thus providing another description of the canonical map.

§3. Abel's Theorem

We have already remarked on the central role played by the abelian differentials in the study of linear systems of divisors on an algebraic curve. These differentials also define cohomology classes, as well as systems of differential equations on the curve, and we shall now recount the relevant definitions and facts in this development.

Let ω and ω' be holomorphic differentials on a smooth curve C. Locally $\omega = f(z)\,dz$, where $f(z)$ is a holomorphic function, and consequently

$$\omega \wedge \omega' = 0,$$
$$\sqrt{-1}\,\omega \wedge \bar{\omega} = |f(z)|^2(\sqrt{-1}\,dz \wedge d\bar{z}),$$

which implies

$$\int_C \omega \wedge \omega' = 0,$$
$$\sqrt{-1}\int_C \omega \wedge \bar{\omega} > 0 \quad \text{if} \quad \omega \neq 0.$$

Since $d\omega = 0$, the form ω defines a cohomology class in the deRham cohomology group $H^1_{DR}(C) \approx H^1(C, \mathbb{C})$. We shall now define the period matrix of the curve, and translate the above relations into the famous Riemann bilinear relations. For this we let $\omega_1, \ldots, \omega_g$ be a basis of $H^0(C, K)$, and $\gamma_1, \ldots, \gamma_{2g}$ a basis for $H_1(C, \mathbb{Z})$. The *period matrix* Ω is then the $g \times 2g$ matrix whose jth column Ω_j is given by

$$\Omega_j = {}^t\!\left(\int_{\gamma_j} \omega_1, \ldots, \int_{\gamma_j} \omega_g\right).$$

Using the standard fact that cup product in cohomology is Poincaré dual to intersection in homology we obtain the *Riemann bilinear relations*

$$\Omega Q\,{}^t\Omega = 0,$$
$$\sqrt{-1}\,\Omega Q\,{}^t\bar{\Omega} > 0,$$

where Q is an integral skew-symmetric matrix whose transposed inverse is equal to the intersection matrix (γ_i, γ_j). It is an elementary fact that we may always choose the basis $\gamma_1, \ldots, \gamma_{2g}$ to be symplectic, meaning that the intersection matrix is of the form

$$\begin{pmatrix} 0 & I_g \\ -I_g & 0 \end{pmatrix}.$$

The Riemann bilinear relations that one encounters in the classical literature are written in terms of such a basis.

A well-known consequence of these relations is that the $2g$ column vectors of the period matrix generate a lattice Λ in $\mathbb{C}^g$, so that the quotient

$\mathbb{C}^g/\Lambda$ is a complex torus. This torus is called the *Jacobian variety* of C and denoted by the symbol $J(C)$.

To say that Λ is a lattice is equivalent to saying that integration over cycles defines an injection

$$H_1(C, \mathbb{Z}) \to H^0(C, K)^*.$$

This gives rise to a coordinate-free descripton of the Jacobian variety as

$$J(C) = H^0(C, K)^*/H_1(C, \mathbb{Z}).$$

The Riemann bilinear relations also imply that the Jacobian variety of C has a principal polarization. This aspect, which is unimportant here, will be dealt with extensively in the next section.

The geometry of the curve C interacts with the geometry of $J(C)$ via a basic group homomorphism

$$u\colon \mathrm{Div}(C) \to J(C), \tag{3.1}$$

where, as is customary, $\mathrm{Div}(C)$ denotes the group of all divisors on C. This homomorphism depends on the choice of a base point p_0. Having fixed this, we first define

$$u\colon C \to J(C)$$

by setting

$$u(p) = \left(\int_{p_0}^{p} \omega_1, \ldots, \int_{p_0}^{p} \omega_g\right).$$

The right-hand side of this equality depends on the choice of a path from p_0 to p, but the ambiguity involved in this choice disappears when we pass to the quotient modulo Λ. The general mapping (3.1) is then obtained by linearity: explicitly, for

$$D = \sum p_i - \sum q_j$$

we set

$$u(D) = \sum_i u(p_i) - \sum_j u(q_j).$$

The individual components

$$u_\alpha(D) = \sum_i \int_{p_0}^{p_i} \omega_\alpha - \sum_j \int_{p_0}^{q_j} \omega_\alpha$$

of $u(D)$ are frequently called *abelian sums.*

As we already remarked, the definition of u depends on the choice of a base point. However, if we denote by $\mathrm{Div}^d(C)$ the set of all divisors of degree d on C, the restriction of u to the *subgroup* $\mathrm{Div}^0(C)$ is independent of p_0, as is apparent from the definitions.

The version of the basic mapping u which we will find most useful in the sequel involves the symmetric products of the curve. The *d-fold symmetric product* of C is, by definition, the set of effective divisors of degree d on C, and is denoted by C_d. Clearly, C_d is the quotient of the ordinary d-fold product C^d by the natural action of the symmetric group S_d; as such, C_d is a projective variety. Moreover, the map

$$u: C_d \to J(C) \tag{3.2}$$

is holomorphic, since its composition with the quotient map from C^d to C_d is both holomorphic and invariant.

Let U be a coordinate open subset of C. We take, as coordinates on the d-fold product of U, the compositions of the projections onto the factors with the coordinate on U. Recall then that any S_d-invariant holomorphic function on U^d depends holomorphically on the d elementary symmetric functions of the local coordinates. A consequence is that C_d is smooth.

The main purpose of this section is to discuss Abel's theorem, which is the basic result relating the algebro-geometric concept of linear equivalence and the transcendental concept of abelian sum.

Abel's Theorem. *Let D and D' be effective divisors of degree d on a smooth curve C. Then D is linearly equivalent to D' if and only if $u(D) = u(D')$.*

Abel's theorem may be rephrased as saying that the (set-theoretic) fibres of the mapping u from C_d to $J(C)$ are complete linear series; more precisely, for each $D \in C_d$, we have

$$u^{-1}(u(D)) = |D|.$$

Another useful interpretation of Abel's theorem involves the *Picard group*

$$\mathrm{Pic}(C) = H^1(C, \mathcal{O}^*).$$

In more detail, we shall denote by $\mathrm{Pic}^d(C)$ the subset of $\mathrm{Pic}(C)$ consisting of isomorphism classes of degree d line bundles. As we already observed in Section 1, $\mathrm{Pic}^d(C)$ is just the quotient of $\mathrm{Div}^d(C)$ modulo linear equivalence. The exponential sequence provides an identification

$$\mathrm{Pic}^0(C) \cong H^1(C, \mathcal{O})/H^1(C, \mathbb{Z}),$$

thus exibiting $\mathrm{Pic}^0(C)$ as a complex torus.

Abel's theorem then states, first of all, that for each d there is a factorization

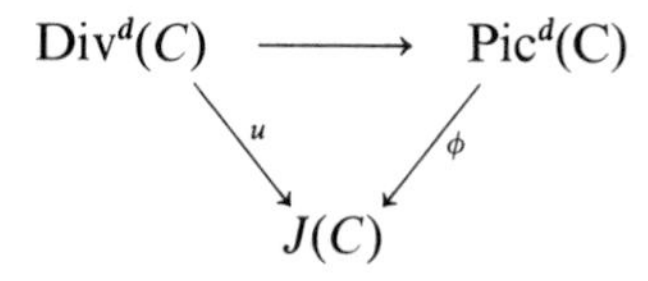

and, secondly, that ϕ is injective.

Indeed, even more is true, namely that ϕ is a bijection. The statement that ϕ is surjective is equivalent to the *Jacobi inversion theorem*, stating that if g is the genus of C,

$$u: C_g \to J(C)$$

is onto. To see this, since C_g and $J(C)$ have the same dimension, it suffices to show that u is generically finite. By Abel's theorem, it then suffices to produce a divisor $D \in C_g$ such that $r(D) = 0$. Since the image of C under the canonical mapping is non-degenerate, the geometric version of the Riemann–Roch theorem tells us that a general D will do.

As for the proof of Abel's theorem, here we shall only sketch the easy part, namely that, for any $D \in \mathrm{Div}^d(C)$, u is constant on $|D|$. Identify $J(C)$ with $\mathbb{C}^g/\Lambda$ and denote by $w_1, \ldots, w_g$ linear coordinates on $\mathbb{C}^g$; then dw_α descends to a holomorphic differential on $J(C)$. Since $|D|$ is a projective space, $u^*(dw_\alpha)$ vanishes on $|D|$ for every α. This proves that u is constant on $|D|$.

An argument based on the "easy part" of Abel's theorem provides one of the classical proofs of the Riemann–Roch theorem, which we now sketch.

Let $D = p_1 + \cdots + p_d$ be an effective divisor of degree d; for simplicity of notation we shall assume that D consists of d distinct points. We set $r = r(D)$ and let $\lambda_1, \ldots, \lambda_r$ be local coordinates in the projective space $|D|$ near D; we shall write $D_\lambda = p_1(\lambda) + \cdots + p_d(\lambda)$ to denote the point of $|D|$ corresponding to the value λ of the coordinates. Further, for each i we let z_i be a local coordinate on C around p_i and write

$$z_i(p_i(\lambda)) = z_i(\lambda),$$
$$\omega_\alpha = f_{\alpha i}(z_i)\, dz_i, \qquad \alpha = 1, \ldots, g.$$

Since the points $p_i(\lambda)$ move with r degrees of freedom, for general λ we have that

$$\operatorname{rank} \frac{\partial(z_1, \ldots, z_d)}{\partial(\lambda_1, \ldots, \lambda_r)}(\lambda) = r.$$

Now, the classical way of writing the easy part of Abel's theorem is

$$\sum_i \int_{p_0}^{z_i(\lambda)} f_{\alpha i}(z_i)\, dz_i \equiv \text{constant} \quad \text{(modulo periods)},$$

and differentiating with respect to λ_ν we obtain

$$\sum_i f_{\alpha i}(z_i(\lambda)) \frac{\partial z_i}{\partial \lambda_\nu}(\lambda) = 0.$$

This relation, together with the assertion on the rank of the matrix $\partial z_i/\partial \lambda_\nu(\lambda)$, tells us that

$$\dim \overline{\phi_K(D)} \leq d - 1 - r. \tag{3.3}$$

To show that, in this relation, we have an equality, we proceed by contradiction. Suppose (3.3) is a strict inequality. Then, given $D' \in |K(-D)|$, we would have

$$\dim |D'| > g - d + r - 1.$$

Now, applying (3.3) to D' we would get

$$\begin{aligned} \dim \overline{\phi_K(D')} &< 2g - 2 - d - 1 - (g - d + r - 1) \\ &= g - 2 - r. \end{aligned}$$

But then, again by residuation, we would deduce that $\dim |D| > r$, which is absurd. We therefore have equality in (3.3), proving the Riemann–Roch theorem in its geometric version.

§4. Abelian Varieties and the Theta Function

In this section we shall briefly discuss principally polarized abelian varieties and Riemann's theta function. The reason for doing this is that the theta divisor on the Jacobian of a curve C very closely reflects the geometric properties of the curve itself; many beautiful and classical results such as Riemann's theorem, the Riemann singularity theorem, and Torelli's theorem deal with this interplay.

Given a general complex torus

$$A = \mathbb{C}^g/\Lambda,$$

then, upon making the natural identification

$$\Lambda \cong H_1(A, \mathbb{Z})$$

we may view a skew-symmetric bilinear form

$$E: \Lambda \times \Lambda \to \mathbb{Z}$$

as an element

$$\xi \in \operatorname{Hom}(\Lambda^2 H_1(A, \mathbb{Z}), \mathbb{Z}) \cong H^2(A, \mathbb{Z}).$$

Now we choose generators $e_1, \ldots, e_{2g}$ for Λ and, regarding each e_i as a column vector, we define the *period matrix* to be

$$\Omega = (e_1, \ldots, e_{2g}).$$

This terminology is justified by the fact that

$$e_{j\alpha} = \int_{\gamma_j} du_\alpha,$$

where $\gamma_j \in H_1(A, \mathbb{Z})$ corresponds to e_j and $u_1, \ldots, u_g$ are the coordinates on $\mathbb{C}^g$. Let Q be the $2g \times 2g$ matrix representing E with respect to the basis $e_1, \ldots, e_{2g}$. To say that ξ is a (1, 1)-form is equivalent to saying that Ω and Q satisfy the *first Riemann bilinear relation*

$$\Omega Q\, {}^t\Omega = 0.$$

On the other hand, the *second Riemann bilinear relation*

$$\sqrt{-1}\,\Omega Q\, {}^t\overline{\Omega} > 0$$

is equivalent to the assertion that ξ is a positive (1, 1)-form. Now any such ξ is the first Chern class of an ample line bundle L on A; moreover, this line bundle is uniquely determined up to translation (see the bibliographical notes at the end of the chapter). The *Riemann–Roch theorem* for L states that

$$\begin{aligned} h^0(A, L) &= \sqrt{\det Q} \\ &= \frac{1}{g!} \int_A c_1(L)^g. \end{aligned}$$

A *principally polarized abelian variety* is a pair consisting of a complex torus A together with the Chern class ξ of an ample line bundle on A such that

$$\frac{1}{g!} \int_A \xi^g = 1.$$

This simply means that the skew-symmetric matrix Q representing ξ is unimodular. Thus a principal polarization ξ on A is the fundamental class of a divisor

$$\Theta \subset A,$$

which is unique up to translation and is called the *theta divisor*. We may observe, in passing, that the Riemann bilinear relations satisfied by the period matrix of a smooth curve C mean that the intersection pairing defines a canonical principal polarization on $J(C)$.

We will now explicitly construct the theta divisor of a principally polarized abelian variety. In particular, this construction applies to the Jacobian variety of a smooth curve.

To begin with, it is a standard linear algebra lemma that, since the pairing

$$E\colon \Lambda \times \Lambda \to \mathbb{Z}$$

is unimodular, we may choose generators for Λ such that the matrix of E has the standard form

$$Q = \begin{pmatrix} 0 & I_g \\ -I_g & 0 \end{pmatrix}.$$

Writing the period matrix Ω in block form

$$\Omega = (B, B'),$$

where B and B' are $g \times g$ matrices, the Riemann bilinear relations imply that B and B' are both non-singular. If we change our basis for $\mathbb{C}^g$ by B^{-1}, then Ω becomes

$$\Omega = (I_g, Z),$$

and the Riemann bilinear relations are now, respectively

$$Z = {}^tZ,$$

$$\operatorname{Im} Z > 0.$$

We shall refer to such a period matrix as a *normalized period matrix*. The *Siegel generalized upper half plane* is defined to be the set of all $g \times g$ symmetric matrices with positive definite imaginary part, and is denoted by the symbol $\mathscr{H}_g$. We also let $\mathscr{A}_g$ denote the set of isomorphism classes of principally polarized abelian varieties of dimension g. Clearly there is a surjective map

$$\mathscr{H}_g \to \mathscr{A}_g.$$

For $g = 1$ this is the well-known parametrization of smooth elliptic curves by the upper half plane in $\mathbb{C}$. In general there is an identification

$$\mathscr{H}_g/\Gamma_g \xrightarrow{\cong} \mathscr{A}_g,$$

where

$$\Gamma_g \cong Sp(2g, \mathbb{Z})/\pm I_{2g}$$

is the *Siegel modular group* operating on $\mathscr{H}_g$ by

$$T(Z) = (AZ + B)(CZ + D)^{-1},$$

where

$$T = \begin{pmatrix} A & B \\ C & D \end{pmatrix} \in Sp(2g, \mathbb{Z}).$$

On the product $\mathbb{C}^g \times \mathscr{H}_g$ we define *Riemann's theta function*

$$\theta(u, Z) = \sum_{m \in \mathbb{Z}^g} e^{\pi\sqrt{-1}\langle m, Zm\rangle + 2\pi\sqrt{-1}\langle m, u\rangle},$$

where the bracket denotes the standard duality pairing $\langle m, v\rangle = \sum_\alpha m_\alpha v_\alpha$ on $\mathbb{C}^g$. Since $\operatorname{Im} Z > 0$, the above series converges normally, and hence uniformly on compact sets, on $\mathbb{C}^g \times \mathscr{H}_g$ to a holomorphic function. One easily checks that the theta function satisfies the functional equations

$$\theta(u + e_\alpha, Z) = \theta(u, Z),$$
$$\theta(u + e_{g+\alpha}, Z) = e^{-2\pi\sqrt{-1}(u_\alpha + Z_{\alpha\alpha})}\theta(u, Z),$$

where $Z = (Z_{\alpha\beta})$ and $e_1, \ldots, e_{2g}$ are the $2g$ column vectors in the period matrix Ω. It follows from these equations that for each fixed $Z \in \mathscr{H}_g$, the divisor (θ) of Riemann's theta function is invariant under translation by elements of the lattice Λ generated by $e_1, \ldots, e_{2g}$, and therefore induces a divisor Θ on $A = \mathbb{C}^g/\Lambda$. It is clear from its definition that the theta function is even in u and therefore Θ is symmetric, i.e.

$$-\Theta = \Theta.$$

We now want to verify that Θ is the principal polarization of A. This amounts to showing that the fundamental class of Θ is represented by the differential form

$$\xi = \sum_{\alpha=1}^{g} dx_\alpha \wedge dx_{g+\alpha},$$

where $x_1, \ldots, x_{2g}$ are the real linear coordinates on $\mathbb{C}^g$ corresponding to the basis $e_1, \ldots, e_{2g}$. Since this is a purely topological statement and since $\mathscr{H}_g$

is connected (in fact, convex), it is enough to give the proof for a particular Z in $\mathscr{H}_g$. It will be convenient to choose Z to be a diagonal matrix

$$Z = \begin{pmatrix} \tau_1 & & 0 \\ & \ddots & \\ 0 & & \tau_g \end{pmatrix}, \qquad \operatorname{Im} \tau_i > 0.$$

Geometrically Z corresponds to a product of g elliptic curves. It is apparent from the definition of the theta function that for such a Z

$$\theta(u, Z) = \prod_{i=1}^{g} \theta(u_i, \tau_i).$$

We are thus reduced to the one variable case, where all we have to show is that the divisor of θ consists of a single point on the elliptic curve. In fact, using the functional equation for θ, we find

$$\begin{aligned} \deg(\theta) &= \frac{1}{2\pi\sqrt{-1}} \int d \log \theta \\ &= \frac{1}{2\pi\sqrt{-1}} \left(\int_0^1 d \log \theta + \int_{1+\tau}^{\tau} d \log \theta \right) \\ &\quad + \frac{1}{2\pi\sqrt{-1}} \left(\int_1^{1+\tau} d \log \theta + \int_{1+\tau}^{0} d \log \theta \right) \\ &= 1, \end{aligned}$$

where the first integral is taken around the fundamental parallelogram

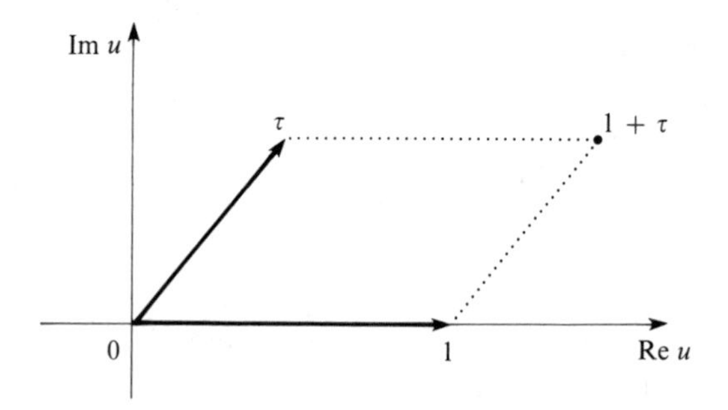

We conclude this section by observing that term-wise differentiation of the theta series shows that Riemann's theta function is a solution of the *heat equation*

$$\frac{\partial^2 \theta}{\partial u_\alpha \, \partial u_\beta}(u, Z) = 2\pi\sqrt{-1}(1 + \delta_{\alpha\beta}) \frac{\partial \theta}{\partial Z_{\alpha\beta}},$$

where $Z = (Z_{\alpha\beta})$ and $\delta_{\alpha\beta}$ is the Kronecker symbol. This equation, as we shall see in Chapter VI, provides a tight link between the intrinsic geometry and the infinitesimal deformations of an algebraic curve.

§5. Poincaré's Formula and Riemann's Theorem

We now return to the abelian sum mapping

$$u: C_d \to J(C)$$

defined in Section 3. We shall denote by $W_d(C)$ the image of C_d under u. As usual we let Θ be the theta divisor on $J(C)$. A natural problem is to try and relate the subvarieties Θ and $W_d(C)$. We shall use the symbols θ, w_d, and c_d to denote the fundamental classes of Θ, $W_d(C)$, and C_d, respectively. The fundamental relation between θ and w_d is

Poincaré's Formula. *When* $0 \leq d \leq g$,

$$w_d = u_*(c_d) = \frac{1}{(g-d)!}\,\theta^{g-d}.$$

Proof. If we can prove the right-hand side equality it follows, in particular, that $W_d(C)$ is d-dimensional (this, by the way, is just the content of the Jacobi inversion theorem). Therefore, by Abel's theorem, u is generically 1–1, which proves the left-hand side equality as well. Clearly $d!\,u_*(c_d)$ is the image of the fundamental class of the d-fold cartesian product C^d via the mapping

$$v: C^d \to J(C)$$

obtained by composing u with the $d!$-sheeted covering of C_d by C^d.

We now choose a symplectic basis $\gamma_1, \ldots, \gamma_{2g}$ of $H_1(C, \mathbb{Z}) = \Lambda$, think of $J(C)$ as $\mathbb{C}^g/\Lambda$, and let $x_1, \ldots, x_{2g}$ be the real coordinates on $\mathbb{C}^g$ corresponding to the basis $\gamma_1, \ldots, \gamma_{2g}$. Denoting by

$$dx_I = dx_{i_1} \wedge \cdots \wedge dx_{i_{2d}}, \qquad I = (i_1, \ldots, i_{2d}),$$

the exterior monomials that generate

$$H^{2d}(J(C), \mathbb{C}) \cong \Lambda^{2d} H^1(J(C), \mathbb{C}),$$

we have to show that

$$\text{(5.1)} \qquad \frac{1}{d!}\int_{C^d} v^*(dx_I) = \frac{1}{(g-d)!}\int_{J(C)} \theta^{g-d} \wedge dx_I$$

for any index set I. Since

$$\int_{\gamma_i} u^*(dx_j) = \delta_{ij},$$

$u^*(dx_i)$ is Poincaré dual to γ_i. Therefore

$$\begin{aligned}\int_C u^*(dx_\alpha \wedge dx_{g+\alpha}) &= -\int_C u^*(dx_{g+\alpha} \wedge dx_\alpha) \\ &= (\gamma_\alpha, \gamma_{g+\alpha}) = 1,\end{aligned}$$

and

$$\int u^*(dx_i \wedge dx_j) = (\gamma_i, \gamma_j) = 0 \quad \text{if} \quad |i - j| \neq g.$$

Thus, by Fubini's theorem, the left-hand side of (5.1) can be non-zero only when dx_I is a product of monomials

$$\eta_\beta = dx_\beta \wedge dx_{g+\beta}.$$

The same is true for the right-hand side, since the intersection matrix is in standard form, i.e.

$$\theta = \sum_{\alpha=1}^{g} \eta_\alpha.$$

For every index set $B = (\beta_1, \ldots, \beta_d)$, where $1 \leq \beta_1 < \cdots < \beta_d \leq g$, we set

$$\eta_B = \eta_{\beta_1} \wedge \cdots \wedge \eta_{\beta_d},$$

so that

$$\theta^{g-d} = (g-d)! \sum_{|A|=g-d} \eta_A,$$

and therefore

$$\int_{J(C)} \theta^{g-d} \wedge \eta_B = (g-d)!.$$

On the other hand, denoting by u_λ the composition of the projection of C^d on the λth factor with $u\colon C \to J(C)$, we have

$$v^*(dx_i) = \sum_\lambda u_\lambda^*(dx_i),$$

and

$$\begin{aligned}\int_{C^d} v^*(\eta_B) &= \int_{C^d} \bigwedge_{h=1}^{d} \sum_{\lambda} u_\lambda^*(\eta_{\beta_h}) \\ &= \sum_{\sigma} \int_{C^d} \bigwedge_{h=1}^{d} u_h^*(\eta_{\beta_{\sigma(h)}}) \\ &= d!,\end{aligned}$$

where the last sum is over all permutations of $\{1, \ldots, d\}$. Q.E.D.

The first and most important corollary of Poincaré's formula is

Riemann's Theorem. *For a suitable point κ of $J(C)$*

$$W_{g-1}(C) = \Theta - \kappa.$$

By Poincaré's formula, $w_{g-1} = \theta$. Since, as we already observed, Θ is determined by θ up to translation, Riemann's theorem follows.

The point $\kappa \in J(C)$ is called *Riemann's constant*. Of course, it depends on the base point introduced to define the map u. A simple residue computation (cf. Lewittes [1], for example) shows that

$$-2\kappa = u(K),$$

where K is a canonical divisor. It is often convenient to define a new map

$$\pi\colon C_{g-1} \to \Theta \subset J(C)$$

by setting $\pi(D) = u(D) + \kappa$. This has the property that

$$-\pi(D) = \pi(K - D),$$

which gives a geometrical content to the symmetry property of the theta divisor.

When $g = 2$ the map π provides an isomorphism between C and the theta divisor. This is the first instance of Torelli's theorem, which states that the curve C can be recovered from the pair $(J(C), \Theta)$, i.e., from the knowledge of $J(C)$ as a principally polarized abelian variety. Torelli's theorem will be proved in Chapter VI.

In the course of the proof of Poincaré's formula we have remarked that part of the statement really boils down to the Jacobi inversion theorem. In fact, using Poincaré's formula, one may describe Jacobi inversion explicitly as follows (see Lewittes [1] for details). Consider the map

$$u\colon C \to J(C).$$

For any $\lambda \in J(C)$ denote by Θ_λ the translate of the theta divisor by λ. Poincaré's formula implies that, when $u(C)$ is not contained in Θ_λ, the pull-back $u^*(\Theta_\lambda)$ is an effective divisor of degree g. It can be proved that the rational map

$$\psi: J(C) \dashrightarrow C_g$$

defined by

$$\psi(\lambda) = u^*(\Theta_{\lambda+\kappa}),$$

where κ is Riemann's constant, is an inverse of

$$u: C_g \to J(C).$$

§6. A Few Words About Moduli

In this work we shall deal primarily not with a fixed curve, but with one which is allowed to "vary with moduli." By definition, the *moduli space* $\mathcal{M}_g$ of curves of genus g is the set of isomorphism classes of smooth, genus g curves.

For genus one, it is well known that any smooth elliptic curve has a plane model

$$y^2 = 4x^3 - g_2 x - g_3.$$

Denoting this curve by C we set

$$j(C) = 1728 g_2^3/(g_2^3 - 27g_3^2).$$

It is classical that $j(C)$ depends only on the isomorphism class of C and induces a 1–1 map

$$j: \mathcal{M}_1 \to \mathbb{C}.$$

The set $\mathcal{M}_1$ is thus endowed with a complex structure. That this is true for higher genera as well was long taken for granted, and Riemann already indicated how to compute the dimension of $\mathcal{M}_g$, which he showed to be $3g - 3$ (when $g > 1$, of course). In a transcendental setting all this can be rigorously established via Teichmüller theory.

It was only much later that Baily showed that $\mathcal{M}_g$ has a natural structure of quasi-projective normal variety of dimension $3g - 3$. The most natural way of proving this is probably by means of geometric invariant theory: this is due to Mumford. It turns out that $\mathcal{M}_g$ is also irreducible. This was already observed by Klein and follows from results of Lüroth and Clebsch.

Following a classical idea, Mumford–Deligne and Mayer proposed a natural compactification $\overline{\mathcal{M}}_g$ of $\mathcal{M}_g$. This consists of the isomorphism classes of the so-called stable curves of (arithmetic) genus g. A *stable curve* is a curve whose only singularities are nodes and whose smooth rational components contain at least three singular points of the curve. It was later proved by Knudsen that $\overline{\mathcal{M}}_g$ is in fact a projective variety. The complement of $\mathcal{M}_g$ in $\overline{\mathcal{M}}_g$ turns out to be a codimension one subvariety Δ which is the union

$$\Delta = \Delta_0 \cup \Delta_1 \cup \cdots \cup \Delta_{[g/2]}$$

of $[g/2] + 1$ irreducible subvarieties. The general point of Δ_0 is an irreducible stable curve of genus g with one node, while the general point of Δ_i, $i > 0$, consists of a smooth curve of genus i joined at one point to a smooth curve of genus $g - i$.

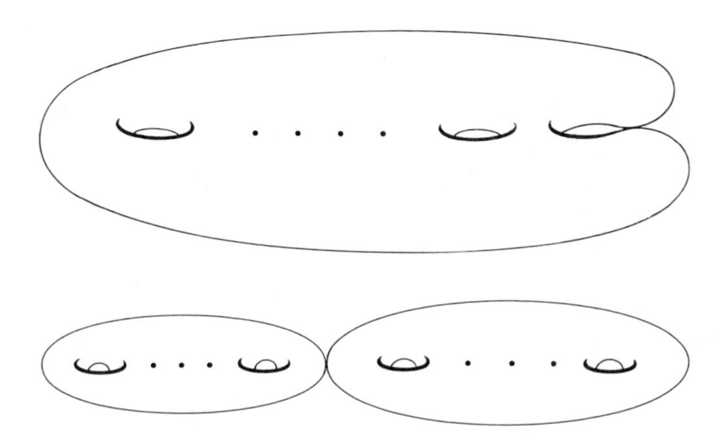

In general, both $\mathcal{M}_g$ and $\overline{\mathcal{M}}_g$ are singular. All the singularities arise from curves which have a non-trivial group of automorphisms, and we shall denote by $\overline{\mathcal{M}}_g^0$ the smooth open subset of $\overline{\mathcal{M}}_g$ consisting of automorphism-free curves. Over $\overline{\mathcal{M}}_g^0$ there is a *universal family*

$$f: \mathscr{C} \to \overline{\mathcal{M}}_g^0$$

with the property that, for each $p \in \overline{\mathcal{M}}_g^0$, the fiber $f^{-1}(p)$ is a stable curve whose isomorphism class is p, and, most important, that each flat family of stable, automorphism-free, genus g curves parametrized by a variety S is induced by a unique morphism from S to $\overline{\mathcal{M}}_g^0$. No such object can exist on $\overline{\mathcal{M}}_g$ (or on $\mathcal{M}_g$). However, the structure of $\overline{\mathcal{M}}_g$ is natural in the sense that, for each flat family

$$g: X \to S$$

of genus g stable curves, there is assigned a morphism from S to $\bar{\mathcal{M}}_g$ which is compatible with base change and is such that, when S is a point, the image of S is the isomorphism class of X.

It was shown by Severi that $\mathcal{M}_g$ is a unirational variety when $g \leq 10$, and it was for a long time an open problem whether this held true for higher genera as well. Sernesi showed that $\mathcal{M}_{12}$ is in fact unirational. The problem was finally given a negative answer by Mumford and Harris who proved that $\mathcal{M}_g$ is of general type whenever g is sufficiently large.

In this volume we shall make free use of those few basic properties of $\mathcal{M}_g$ that we will occasionally need. A more complete discussion of moduli space will appear in the first chapters of the second volume. In the same volume proofs of the irreducibility of $\mathcal{M}_g$, of Severi's unirationality result, and of the theorem of Harris and Mumford will also be given.

Bibliographical Notes

We give a few sources for background reading, from various points of view, in the theory of algebraic curves.

For a classical Riemann surface viewpoint we suggest Farkas and Kra [1], Siegel [1], Springer [1], and Weyl [1].

For a modern Riemann surface point of view; i.e., the study of one-dimensional compact, complex manifolds, the reader may consult Gunning [1].

Algebraic curves are treated in Coolidge [1], Enriques–Chisini [1], Fulton [1], Severi [2], and Walker [1].

The theory of curves and abelian varieties is discussed in Fay [1], Lang [1], Mumford [1], [3], [5], and [6].

Curves, as part of general algebraic geometry, are treated in Griffiths–Harris [1], Hartshorne [2], and Shafarevich [1], among many other places.

Deformation theory and moduli appear in Kodaira–Spencer [1] (local theory) and Mumford [3] and [4].

The quasi-projectivity of $\mathcal{M}_g$ is proved in Baily [1]. For the invariant-theoretic approach we refer to Mumford–Fogarty [1]. Klein's remark on the irreducibility of $\mathcal{M}_g$ appears in Klein [1, no. 19]: his argument is based on earlier work by Lüroth [1] and Clebsch [1]. The compactification of $\mathcal{M}_g$ by stable curves was introduced in Deligne–Mumford [1]. The projectivity of $\bar{\mathcal{M}}_g$ was established by Knudsen [1]. The main results concerning the unirationality (or non-rationality) of $\mathcal{M}_g$ can be found in Severi [2], Sernesi [1], Harris–Mumford [1], and Harris [3].

A more complete discussion of the literature on the moduli space of curves will be found in the bibliographical notes to the second volume.

Exercises

A. Elementary Exercises on Plane Curves

Algebraic curves are frequently given by irreducible equations

$$f(x, y) = 0.$$

By this we shall always mean the *irreducible affine algebraic curve*

$$\Gamma_0 \subset \mathbb{C}^2$$

given by the complex solutions to the above equation. Associated to Γ_0 are the following objects:

(i) The *projective curve*

$$\Gamma \subset \mathbb{P}^2$$

given by the zeros

$$F(X_0, X_1, X_2) = 0$$

of the homogeneous form

$$F(X_0, X_1, X_2) = X_0^d f(X_1/X_0, X_2/X_0), \qquad d = \deg f,$$

associated to f. We think of Γ as adding to Γ_0 the points on the line $(X_0 = 0)$ at infinity.

(ii) The *normalization*

$$\phi: C \to \Gamma \subset \mathbb{P}^2$$

of Γ.

Sometimes we refer to C itself as the normalization of Γ. It is a compact Riemann surface that classically was thought of as "the Riemann surface associated to the algebraic function $y(x)$ where $f(x, y(x)) \equiv 0$." What this means is that, via the projection $\Gamma_0 \to \mathbb{C}$ on the x-axis, we may picture Γ_0 as a branched covering of the complex x-plane. Then C is realized as a branched covering of the x-sphere $\mathbb{P}^1 = \mathbb{C} \cup \{\infty\}$ that agrees with $\Gamma_0 \to \mathbb{C}$ outside the singularities of Γ_0 and the points lying over $x = \infty$. The construction of $C \to \mathbb{P}^1$ that is pertinent to the following exercises is given, e.g., in Weyl [1], we shall refer to it as *completing* Γ_0 *to a compact Riemann surface.*

We consider the following affine algebraic curves:

(a) $y^2 = x^3 + 1$.
(b) $y^3 + x^3 = 1$.
(c) $x^2y^2 + x^2 + y^2 = 0$.
(d) $y^2 = x^6 - 1$.
(e) $y^2 = x^5 - 1$.
(f) $y^3 = x^5 - 1$.
(g) $y^3 = (x^2 + 1)^2(x^3 - 1)$.
(h) $y^7 = x^2(x - 1)$.
(i) $y + x^3 + xy^3 = 0$.

A-1. Show that the affine curves (a), (b), (d), (e), (f), and (i) are smooth. Show that (a) may be completed to a compact Riemann surface by adding one point to the affine curve, (b) by adding three points, (d) by adding two points, (e) and (f) by adding one point, and (i) by adding two.

A-2. Show that the curves (c) and (h) are singular exactly at the origin, while (g) is singular at $(\pm i, 0)$ ($i = \sqrt{-1}$). Show that in the normalizations of (g) and (h) there is exactly one point lying over each singular point, while for (c) there are two over each singular point.

A-3. Let $\Gamma = C$ be the compact Riemann surface associated to the curve in (a) above. The curve C can be realized as a branched covering $\pi: C \to \mathbb{P}^1$ of the x-sphere. Set, respectively

$$p_1, p_2, p_3 = (-1, 0), (-\omega, 0), (-\omega^2, 0), \qquad \omega = e^{2\pi i/3},$$

$$q_1, q_2 = (0, 1), (0, -1),$$

$$r = \pi^{-1}(\infty),$$

$$s_1, s_2 = (2, 3), (2, -3).$$

(i) Find the branch points of $\pi: C \to \mathbb{P}^1$ and show that the genus of C is equal to one.
(ii) Establish the linear equivalences

$$2p_1 \sim 2p_2 \sim 2p_3 \sim q_1 + q_2 \sim 2r \sim s_1 + s_2,$$

$$p_1 + p_2 + p_3 \sim 3r,$$

$$q_1 + s_1 \sim q_2 + s_2.$$

(iii) Determine $\mathscr{L}(p_1 + q_1)$ and describe the complete linear system $|p_1 + q_1|$.
(iv) Find the points $t \in C$ such that, respectively,

$$t \sim p_1 + q_2 - r,$$

$$t \sim 2s_1 - r.$$

A-4. Let $\Gamma = C$ be the compact Riemann surface associated to the curve in (b) above. The curve C can be realized as a branched covering $\pi: C \to \mathbb{P}^1$ of the x-sphere.

(i) Find the branch points and ramification points of $\pi: C \to \mathbb{P}^1$ and show that the genus of C is equal to one.

(ii) For any two points $p, q \in C$ find the complete linear series $|p + q|$. (*Hint*: Draw lines.)

(iii) Find a map $\eta: C \to \mathbb{P}^1$ of degree two such that $\eta((1, 0)) = \eta((0, 1))$, and determine the branch points of η.

(iv) Using A-3(i) and (ii) show that the curves in (a) and (b) are isomorphic, and find an explicit isomorphism.

A-5. Let C be the compact Riemann surface associated to the curve in (d) above, realized as a branched covering $\pi: C \to \mathbb{P}^1$ of the x-sphere. We shall denote also by (x, y) the point on C corresponding to a smooth point $(x, y) \in \Gamma_0$. Set

$$r_\alpha = (\xi^\alpha, 0), \quad \xi = e^{\pi i/3},$$
$$p + q = \pi^{-1}(\infty),$$
$$s_1, s_2 = (0, i), (0, -i), \quad \text{respectively.}$$

(i) Show that

$$p + q \sim 2r_\alpha,$$
$$\sum_{\alpha=0}^{5} r_\alpha \sim 3p + 3q.$$

(ii) Determine the complete linear system $|r_0 + r_2 + r_4|$. Find the divisor D on C such that $D + r_1 \sim r_0 + r_2 + r_4$.

(iii) For $D = p + q + r_0 + r_3$ find $\mathscr{L}(D)$. Describe the map

$$\phi_D: C \to \mathbb{P}^2$$

given by $|D|$, find the equation of $\phi_D(C)$, and determine the singularities of $\phi_D(C)$.

(iv) Determine the necessary and sufficient conditions that the Laurent tails

$$f_\alpha = \frac{a_\alpha}{y} \quad \text{at } r_\alpha, \qquad \alpha = 0, \ldots, 5,$$
$$f_6 = \frac{a_6}{x} \quad \text{at } s_1,$$
$$f_7 = \frac{a_7}{x} \quad \text{at } s_2,$$

should be the principal parts of a global meromorphic function on C.

A-6. Let C be the compact Riemann surface associated to the curve in (f) above, realized as a branched covering $\pi: C \to \mathbb{P}^1$ over the x-sphere. Set

$$p = \pi^{-1}(\infty),$$
$$r_\alpha = (\eta^\alpha, 0), \qquad \eta = e^{2\pi i/5}.$$

(i) Find the ramification divisor of π and compute the genus of C.
(ii) Establish the linear equivalences

$$3p \sim 3r_\alpha,$$
$$\sum_{\alpha=0}^{4} r_\alpha \sim 5p.$$

(iii) Determine the space $H^0(C, K)$ of holomorphic differentials on C. (*Hint*: Find $\mathscr{L}(D)$ where $D = (dx)$.)
(iv) Describe the canonical map

$$\phi_K: C \to \mathbb{P}^3$$

and determine the equations of its image.
(v) Using part (iii) show that, for $D = \sum_{\alpha=0}^{4} r_\alpha$

$$h^0(C, K(-D)) = 1.$$

From the Riemann–Roch theorem conclude that $r(D) = 2$, and by using residuation determine the complete linear system $|D|$.
(vi) For $E = 3p$ show that $|E|$ is the unique g_3^1 on C, and that $|2E| = |K|$.

A-7. Let C be the compact Riemann surface associated to the curve in (g) above with map $\pi: C \to \mathbb{P}^1$ onto the x-sphere. Set

$$p = \pi^{-1}(\infty),$$
$$q_1, q_2 = \text{points on } C \text{ lying over } (i, 0), (-i, 0) \text{ on } \Gamma,$$
$$r_\alpha = (\omega^\alpha, 0), \qquad \omega = e^{2\pi i/3}.$$

(i) Find the ramification divisor of π and show that $g(C) = 4$.
(ii) Show that

$$3p \sim 3q_i \sim 3r_2,$$
$$\sum_{0}^{2} r_\alpha + \sum_{1}^{2} q_\alpha \not\sim 5p.$$

(iii) Determine $H^0(C, K)$, describe the canonical map

$$\phi_K: C \to \mathbb{P}^3,$$

and find the equations of $\phi_K(C)$.

(iv) For

$$\begin{cases} E = 3p, \\ D = K - E, \end{cases}$$

show that

$$E \not\sim D.$$

From the Riemann–Roch theorem, conclude that $r(D) = 1$.

(v) Using residuation, find the pencil $|D|$ and describe the corresponding map ϕ_D.

(vi) Using part (vi) of A-6 and part (iv) above, show that the normalizations of the curves in (f) and (g) are *not* isomorphic.

A-8. Let $\Gamma = C$ be the compact Riemann surface associated to the curve in (i) above, and denote by $\pi_x, \pi_y: C \to \mathbb{P}^1$ the branched coverings obtained by projecting Γ_0 on the x- and y-axes, respectively. Compute the genus $g(C)$ by:

(i) the Riemann–Hurwitz formula applied to π_x;
(ii) the Riemann–Hurwitz formula applied to π_y;
(iii) the genus formula for $C \subset \mathbb{P}^2$ (you must show that C is smooth).

Note. The formula for the genus of a smooth plane curve of degree d is $g = (d - 1)(d - 2)/2$; this is discussed in Appendix A below.

A-9. Let C be the compact Riemann surface associated to the curve in (h) above.

(i) Compute the genus $g(C)$.
(ii) Determine $H^0(C, K)$, and describe the canonical map ϕ_K and its image.
(iii) Show that the normalizations of the curves in (h) and (i) above are *not* isomorphic.

A-10. Show that the curve Γ_0 in (c) above is rational by

(i) the Riemann–Hurwitz formula applied to $C \to \mathbb{P}^1$;
(ii) giving an explicit rational parametrization.

(*Hint*: Look at the conics through the three nodes of Γ.)

B. Projections

The following sequence of exercises concerns projections, and for this we will use the notations:

$L \to C$ is a line bundle on a smooth curve;
$V \subset H^0(C, L)$ is a linear subspace and $\mathscr{D} = \mathbb{P}V$ the corresponding linear system with associated map

$$\phi = \phi_{\mathscr{D}}: C \to \mathbb{P}V^*;$$

for any point $x \in \mathbb{P}V^*$ we set

$$\mathscr{D}_x = \mathbb{P}(\operatorname{Ann} x) \subset \mathscr{D},$$

and for $p \in C$ we set

$$\mathscr{D}(-p) = \{D - p\colon D \in \mathscr{D} \text{ and } D \geq p\}.$$

We assume that $\mathscr{D}$ has no base points.

B-1. Show that p is a base point of $\mathscr{D}_{\phi(p)}$ and

$$\mathscr{D}(-p) = \mathscr{D}_{\phi(p)}(-p).$$

(This is just to get familiar with the notation.)

B-2. Recall that the projection

$$\pi_x\colon \mathbb{P}V^* - \{x\} \to H$$

from $x \in \mathbb{P}V^*$ to a hyperplane H not containing x is defined by

$$\pi_x(y) = \overline{xy} \cap H,$$

where $\overline{xy}$ is the line joining x to y. Show that

$$\pi_x \phi_{\mathscr{D}} = \phi_{\mathscr{D}_x}$$

(this is true even when $x \in \phi_{\mathscr{D}}(C)$).

B-3. Let $\Gamma = \phi_{\mathscr{D}}(C)$, $x \in \mathbb{P}V^*$, and define the *multiplicity* $\operatorname{mult}_x(\Gamma)$ of Γ at x to be the degree of the base locus of $\mathscr{D}_x$. Assuming that $\pi_x\colon \Gamma \to \pi_x(\Gamma)$ is generically one-to-one, show that

$$\deg \pi_x(\Gamma) = \deg(\Gamma) - \operatorname{mult}_x(\Gamma).$$

B-4. Show that $\mathscr{D} = |K(p + q)|$ has the property that $\phi_{\mathscr{D}}$ is not an embedding but $\phi_{\mathscr{D}_x}$ is if C is non-hyperelliptic and x is the singular point of $\phi_{\mathscr{D}}(C)$.

B-5. Let $C \subset \mathbb{P}^r$ $(r \geq 3)$ be a smooth curve. Show that a general projection

$$\pi\colon \mathbb{P}^r - \mathbb{P}^{r-4} \to \mathbb{P}^3$$

maps C biregularly to a smooth curve.

B-6. Let $C \subset \mathbb{P}^3$ be a non-degenerate smooth curve. Show that a general projection with center $p \in \mathbb{P}^3$

$$\pi\colon \mathbb{P}^3 - \{p\} \to \mathbb{P}^2$$

maps C birationally to a plane curve having ordinary double points as its only singularities.

(*Hint*: Show that the set of trisecant lines to C and the set of tangent lines to C describe two surfaces S and S'. Then take $p \notin S \cup S'$.)

Note. It may be reassuring to the reader to know that the image of a complex analytic variety under a proper holomorphic map is again a complex analytic variety of the appropriate dimension. (Remmert's proper mapping theorem, cf. Gunning–Rossi [1].)

C. Ramification and Plücker Formulas

In the following sequences of exercises, we will describe the notion of ramification of a map $C \to \mathbb{P}^r$, or, more generally, of a linear series on C. For the following, then, C will be a smooth curve of genus g, L a line bundle of degree d, $V \subset H^0(C, L)$ an $(r + 1)$-dimensional vector space of sections, and $p \in C$ a point.

C-1. Show that the set

$$\Sigma_{p,V} = \{\operatorname{ord}_p \sigma\}_{\sigma \in V - \{0\}}$$

consists of exactly $(r + 1)$ non-negative integers.

The sequence $\alpha_0 \leq \alpha_1 \leq \cdots \leq \alpha_r$ defined by

$$\Sigma_{p,V} = \{\alpha_0, 1 + \alpha_1, 2 + \alpha_2, \ldots, r + \alpha_r\}$$

is called the *ramification sequence* of V at p; we say that p is an *inflectionary point* for V if $(\alpha_0, \ldots, \alpha_r) \neq (0, \ldots, 0)$. We shall sometimes write $\alpha_i(p, V)$ for α_i defined above.

C-2. Show that only finitely many points of C are inflectionary points for a given series V.

C-3. Show that a point $p \in C$ is a Weierstrass point (cf. Exercises E-1 to E-13) if and only if it is an inflectionary point for the canonical series $V = H^0(C, K)$, and that the weight of p is

$$w(p) = \sum \alpha_i.$$

C-4. If $\phi: C \to \mathbb{P}^r$ is any map, given locally by a vector-valued function

$$\phi(z) = [v(z)],$$

we may define a map

$$\phi^{(k)}: C \to G(k + 1, r + 1) \hookrightarrow \mathbb{P}\Lambda^{k+1}\mathbb{C}^{r+1}$$

by setting

$$\phi^{(k)}(z) = [v(z) \wedge v'(z) \wedge \cdots \wedge v^{(k)}(z)].$$

Show that:

(a) the vector $v(z) \wedge \cdots \wedge v^{(k)}(z)$ does not vanish identically, and hence $\phi^{(k)}(z)$ is defined for all z;
(b) $\phi^{(k)}(z)$ is independent of the choice of function $v(z)$; and
(c) $\phi^{(k)}(z)$ is independent of the choice of local coordinate z on C.

The map $\phi^{(k)}$ is called the kth *associated map* of ϕ; the k-plane $\Lambda \subset \mathbb{P}^r$ given by the image $\phi^{(k)}(p)$ is called the *osculating k-plane* to $\phi(C)$ at $\phi(p)$.

C-5. If $\phi = \phi_V$ for V a linear system as above, show that the osculating k-plane at $\phi(p)$ is the unique k-plane having maximal order of contact with $\phi(C)$ at $\phi(p)$, and that this order of contact is $k + \alpha_k$.

C-6. Show that the kth associated map $\phi^{(k)}$ is an immersion at p if and only if $\alpha_{k+1} = \alpha_k$; and that, in general, the differential $d\phi^{(k)}$ vanishes to order exactly $\alpha_{k+1} - \alpha_k$ at p.

C-7. In terms of the identifications

$$G(k + 1, V) = G(r - k, V^*),$$

prove the duality

$$(\phi^{(r-1)})^{(r-1)} = \phi,$$

and hence

$$(\phi^{(r-1)})^{(k)} = \phi^{(r-1-k)}.$$

C-8. In terms of the identification of the tangent space to the Grassmannian $G(k, V)$ at Λ with $\mathrm{Hom}(\Lambda, V/\Lambda)$, show that the image of the differential $d\phi^{(k)}$ at each uninflected point has rank 1, with kernel $\phi^{(k-1)}(p)$ and image $\phi^{(k+1)}(p)$. Using this, show that ϕ *is determined by* $\phi^{(k)}$ *for any k*.

C-9. Using the preceding exercise, show that in general a map $\psi: C \to G(k + 1, r + 1)$, such that the planes $\psi(p)$ do not contain a fixed vector, is the kth associated map of some map $\phi: C \to \mathbb{P}^r$ if and only if rank $d\psi \leq 1$ at each point (cf. Griffiths–Harris [3]).

(*Hint*: Do the case $k = 2$ first.)

There are a series of formulas relating the ramification indices of a linear system and the degrees of the associated maps, called the Plücker formulas; the next exercises give a derivation of these formulas.

C-10. Let E_L^{k+1} be the vector bundle of rank $k + 1$ on C whose fiber at a point p is the space of sections of L near p modulo those vanishing to order $k + 1$—that is,

$$(E_L^{k+1})_p = \Gamma(L/L(-(k + 1)p)).$$

(For a formal definition, let $\Delta \subset C \times C$ be the diagonal, π_1 and $\pi_2: C \times C \to C$ the projection maps, and take

$$E_L^{k+1} = \pi_{2,*}(\pi_1^* L \otimes \mathcal{O}_{C \times C}/\mathcal{I}_\Delta^{k+1}).)$$

Show that the first Chern class of E_L^{k+1} is

$$c_1(E_L^{k+1}) = (k + 1)\, d + \binom{k+1}{2}(2g - 2).$$

C-11. Let $V \otimes \mathcal{O}_C$ be the trivial vector bundle on C with fiber V, and $\tau_k: V \otimes \mathcal{O}_C \to E_L^{k+1}$ the natural evaluation map. If $k \leq r$, show that Coker τ_k is a sheaf supported on the inflectionary points of V, and that, in general, the dimension of the stalk of Coker τ_k at p is the sum of the ramification indices $\sum_{i=0}^{k} \alpha_i$.

C-12. Show that the kernel of τ_k is locally free; and that as a subbundle of $V \otimes \mathcal{O}_C$ it has fiber at each p given by

$$(\text{Ker } \tau_k)_p = \phi^{(k)}(p)^{\perp}.$$

Conclude that if d_k is the degree of

$$\phi^{(k)}: C \to G(k+1, r+1) \subset \mathbb{P}\Lambda^{k+1}\mathbb{C}^{r+1},$$

then

$$d_k = -c_1(\text{Ker } \tau_k).$$

C-13. From the preceding three exercises, and the exact sequence of sheaves

$$0 \to \text{Ker } \tau_k \to V \otimes \mathcal{O} \to E_L^{k+1} \to \text{Coker } \tau_k \to 0,$$

deduce the basic relation

$$(*) \qquad \sum_{p \in C} \sum_{i=0}^{k} \alpha_i(p, V) = -d_k + (k+1)d + \binom{k+1}{2}(2g-2),$$

and, in case $k = r$, the relation

$$(**) \qquad \sum_{p, i} \alpha_i(p, V) = (r+1)d + \binom{r+1}{2}(2g-2).$$

Taking second differences in the relation $(*)$, we have what are usually referred to as the Plücker formulas:

$$\sum_{p \in C} (\alpha_{k+1} - \alpha_k) = -d_{k+1} + 2d_k - d_{k-1} + 2g - 2.$$

C-14. From the relation $(**)$, deduce that *the only nondegenerate curve $C \subset \mathbb{P}^r$ with no inflectionary points is the rational normal curve.*

C-15. Combining the relation $(**)$ with exercise C-3 above, prove again (cf. Exercise E-8 of this chapter) that the sum of the weights of the Weierstrass points on C is $g^3 - g$.

In case $r = 1$, the Plücker formula $(**)$ is just the Riemann–Hurwitz formula. In the following exercise we look at the case $r = 2$.

C-16. Let $C \subset \mathbb{P}^2$ be a plane curve with only ordinary singularities—that is, nodes and cusps—and assume that the same is true for the dual curve $C^* = \phi^{(1)}(C) \subset \mathbb{P}^{2*}$. Let d, δ, κ, b, and f be the degree and number of nodes, cusps, bitangents and flexes of C, respectively, and d^*, δ^*, κ^*, b^*, and f^* the corresponding quantities for C^*. Show that

$$\delta^* = b, \qquad \kappa^* = f, \quad b^* = \delta \quad \text{and} \quad f^* = \kappa,$$

and, using (**) above and the *genus formula* $g = \binom{d-1}{2} - (\delta + \kappa)$, derive the classical Plücker formulas for plane curves

$$d^* = d(d-1) - 2\delta - 3\kappa,$$

$$d = d^*(d^*-1) - 2b - 3f.$$

D. Miscellaneous Exercises on Linear Systems

In the following series of exercises, C is a smooth curve of genus g and D is a divisor of degree d. Assume that $|D|$ is base-point-free and let ϕ_D be the map associated to $|D|$.

D-1. Show that the map ϕ_D is an embedding if, and only if,

$$r(D - p - q) = r(D) - 2$$

for all $p, q \in C$ (including the case $p = q$).

D-2. Show that ϕ_D is an embedding if $d \geq 2g + 1$. If $d = 2g$, then show that ϕ_D will fail to be an embedding if, and only if, $D \sim K + p + q$ for some $p, q \in C$.

D-3. Show that if D is a general divisor of degree $d \geq g + 3$, then ϕ_D is an embedding.
Hint: Choose D so that $u(D) \notin (u(K) - W_{g-3}(C)) + W_2(C)$. (Here, for $A, B \subset J(C)$, we set $A + B = \{a + b \colon a \in A \text{ and } b \in B\}$ and $-A = \{-a \colon a \in A\}$.)

D-4. Suppose that $g = 2$. Show that C cannot be embedded in $\mathbb{P}^r$ as a curve of degree $d < 5$. If $d = 5$, then describe the image of C under the map

$$\phi_D \colon C \to \mathbb{P}^3.$$

How many quadric and cubic surfaces contain $\phi_D(C)$? Among the cubics, how many are "new" (i.e., not in the ideal generated by the quadric)?

D-5. Show that the quadric containing the curve $\phi_D(C) \subset \mathbb{P}^3$ of the preceding exercise is singular if $D \sim 2K + p$ for some $p \in C$, and smooth otherwise.

D-6. Suppose that $g = 3$. Show that the smallest degree of an embedding of C in a projective space is 4 if C is non-hyperelliptic and 6 if C is hyperelliptic.

D-7. Suppose that $g = 4$. Show that the smallest degree of an embedding of C in a projective space is 6 if C is non-hyperelliptic and 7 if C is hyperelliptic.

D-8. Show that if $m > 1$ and $g > 1$

$$h^0(C, K_C^{\otimes m}) = (g - 1)(2m - 1).$$

Use this to conclude that a canonical curve of genus 4 lies on one quadric surface $Q \subset \mathbb{P}^3$ and on one cubic surface S not containing Q. Conclude that the canonical curve is equal to $Q \cap S$.

D-9. Let C be a hyperelliptic curve of genus $g \geq 2$. Show that every special linear series g_d^r is a sum of r copies of the hyperelliptic g_2^1 plus $d - 2r$ fixed points. Conclude that no curve of genus $g \geq d$ can have a g_2^1 and a base-point-free g_d^1 other than the g_2^1.

(*Hint*: Using the canonical model of C show that any collection $p_1, \ldots, p_d$ of distinct points that fails to impose independent conditions on $|K|$ must contain a pair of conjugates under the hyperelliptic involution. Bear in mind that any set of points on a rational normal curve is in linear general position.)

E. Weierstrass Points

The following series of exercises deals with the Weierstrass points on a smooth curve C of genus g. We let $\mathcal{M}(C)$ denote the field of meromorphic functions on C.

Definition. We set $\mathbb{N} = \{n \in \mathbb{Z}: n \geq 0\}$ and, for any point $p \in C$, define the *Weierstrass semigroup* $H_p \subset \mathbb{Z}$ by

$$H_p = \{n: \text{there exists } f \in \mathcal{M}(C) \text{ with } (f)_\infty = np\},$$

and the *Weierstrass gap sequence* $G_p \subset \mathbb{N}$ by

$$\begin{aligned} G_p &= \mathbb{N} - H_p \\ &= \{n: \text{there is } \textit{no } f \in \mathcal{M}(C) \text{ with } (f)_\infty = np\}. \end{aligned}$$

E-1. Show that, for any point $p \in C$, either:

$$r(np) = r((n-1)p) \quad \text{and} \quad n \in G_p,$$

or

$$r(np) = r((n-1)p) + 1 \quad \text{and} \quad n \in H_p.$$

Using the Riemann–Roch theorem conclude that

$$\text{cardinality of } G_p = g.$$

E-2.

(i) Show that

$$G_p = \{n \in \mathbb{N}: \text{there exists } \omega \in H^0(C, K) \text{ with } \mu_p(\omega) = n - 1\}.$$

Conclude again that G_p has cardinality g.

(ii) Show that for any point $p \in C$, $2g - 1 \in G_p$ if and only if $\mathcal{O}((2g - 2)p) \cong K$.

E-3. If C is hyperelliptic with $\pi: C \to \mathbb{P}^1$ the two-sheeted map, show that:

(i) p is a branch point of π if and only if $G_p = \{1, 3, 5, \ldots, 2g - 1\}$; and
(ii) p is not a branch point of π if and only if $G_p = \{1, 2, 3, \ldots, g\}$.

Definitions.

(i) A point $p \in C$ is a *Weierstrass* point if its gap sequence is not $\{1, 2, \ldots, g\}$.
(ii) $p \in C$ is a *normal* Weierstrass point if its gap sequence is $\{1, 2, \ldots, g - 1, g + 1\}$.
(iii) In general, if $H \subset \mathbb{N}$ is a semigroup such that $G = \mathbb{N} - H$ has cardinality g define its *weight* $w(H)$ by setting

$$w(H) = \sum_{n \in G} n - g(g + 1)/2.$$

Finally, for $p \in C$, we set

$$w(p) = w(H_p).$$

E-4. Show that the following are equivalent:

$$\begin{cases} p \text{ is a Weierstrass point,} \\ w(p) \neq 0, \\ r(gp) \neq 0. \end{cases}$$

Show that p is a normal Weierstrass point if and only if $w(p) = 1$.

E-5. If $H = \{a_1, a_2, \ldots\} \subset \mathbb{N}$ is any semigroup such that $G = \mathbb{N} - H$ has g elements, show that

$$w(H) = \sum_{n=1}^{\infty} (g - a_n + n)$$

(here we assume that $a_1 < a_2 < \cdots$).

E-6. Show that

$$w(p) = \sum_{n=1}^{\infty} (r(np) - \max\{0, n - g\})$$

(thus the weight $w(p)$ measures the overall failure of the sequence of divisors $\{np\}$ to adhere to the generic behavior $r(np) = \max\{0, n - g\}$).

Note. This problem may be notationally messy. It may be helpful to make an array of dots

```
. . . . . .
. . . .
. .
.
```

in which the lengths of the columns are the numbers $r(np) - \max(0, n - g)$; the lengths of the rows will then be the gap values.

E-7.

(i) Show that, if $H \subset \mathbb{N}$ is any semigroup with $G = \mathbb{N} - H$ having g elements, then

$$w(H) \leq g(g-1)/2,$$

with equality holding if and only if $2 \in H$.

(ii) If $g \geq 4$ and $w(H) = 2$, then show that either

$$H = \{g-1, g+2, g+3, \ldots\},$$

or

$$H = \{g, g+1, g+3, g+4, \ldots\}.$$

E-8. Choose a basis $\omega_1, \ldots, \omega_g$ for $H^0(C, K)$ and, in terms of a local coordinate z on C, set

$$\omega_\alpha = f_\alpha(z)dz, \qquad \alpha = 1, \ldots, g.$$

Consider the Wronskian

$$W(z) = \det \begin{pmatrix} f_1(z) & \cdots & f_g(z) \\ f'_1(z) & \cdots & f'_g(z) \\ \vdots & & \vdots \\ f_1^{(g-1)}(z) & \cdots & f_g^{(g-1)}(z) \end{pmatrix}$$

(i) Show that the local expressions

$$W(z)(dz)^{g(g+1)/2}$$

give a global section $W \in H^0(C, K^{\otimes \binom{g+1}{2}})$.

(ii) Show that

$$\mu_p(W(z)) = w(p),$$

and conclude that

$$\sum_{p \in C} w(p) = (g-1)g(g+1).$$

E-9. Conclude from the preceding two exercises that every curve of genus g possesses at least $2g+2$ distinct Weierstrass points, and that C is hyperelliptic if and only if it has exactly $2g+2$ distinct Weierstrass points.

E-10. For each of the curves (c)–(i) in exercise batch A, locate the Weierstrass points and determine their weights.

E-11. Let $C \subset \mathbb{P}^2$ be the *Fermat curve* given by

$$X_0^d + X_1^d + X_2^d = 0$$

in homogeneous coordinates. Let ξ be a dth root of -1 and

$$p = [1, \xi, 0].$$

Show that p is a Weierstrass point with gap sequence

$$\begin{array}{lll} 1, & 2, & 3, \ldots\ldots\ldots\ldots, d-2, \\ d+1, & d+2, & \ldots\ldots\ldots, 2d-3, \\ 2d+1, & 2d+2, & \cdots, 3d-4, \\ \vdots & \vdots & \\ \multicolumn{3}{l}{(d-4)d+1, (d-4)d+2,} \\ \multicolumn{3}{l}{(d-3)d+1.} \end{array}$$

(*Hint*: Observe that the line $L = \{X_1 - \xi X_0 = 0\}$ has contact of order d with C at p (i.e., $L \cdot C = dp$), and hence any curve of degree $k < d$ having contact of order $k + 1$ with p at d must contain L. Apply this to $|K_C| = |\mathcal{O}_C(d - 3)|$ using Exercise E-2.)

E-12. Continuing the preceding exercise, show that

$$w(p) = \binom{d+1}{4} - \binom{d-1}{2},$$

and that the total weight of all points $X_i = 1$, $X_j = \xi$, $X_k = 0$ is thus $3d\left(\binom{d+1}{4} - \binom{d-1}{2}\right)$.

E-13. Using Exercises E-8 and E-12, show that if $d = 4$ (in which case C is a canonical curve), then every Weierstrass point is of the form $X_i = 1$, $X_j = \xi$, $X_k = 0$, while if $d \geq 5$ there are other Weierstrass points of C. In case $d = 5$, locate these others.

F. Automorphisms

F-1. Show that every automorphism of $\mathbb{P}^1$ is of the form $\phi(z) = (az + b)/(cz + d)$, i.e., that

$$\operatorname{Aut}(\mathbb{P}^1) = \operatorname{PGL}(2, \mathbb{C}).$$

F-2. Let C be a curve of genus 1. Show that there is an exact sequence

$$0 \to T \to \operatorname{Aut}(C) \to F \to 0,$$

where T is the group of translations in the torus C, and F is a finite group of order 2, 4, or 6. (In fact, there is a unique curve of genus 1 such that $\# F = 4$ (resp., 6).)

Exercises F-3 through F-10 contain a proof of the following:

Theorem. *Let C be a smooth curve of genus $g \geq 2$ with automorphism group* $\mathrm{Aut}(C)$. *Then* $\mathrm{Aut}(C)$ *is a finite group of order at most* $84(g-1)$.

F-3. Let C be a hyperelliptic curve of genus $g \geq 2$ and denote by $\pi: C \to \mathbb{P}^1$ the two-sheeted covering map given by the hyperelliptic g_2^1. Denote also by τ the hyperelliptic involution (sheet interchange) and by $p_1, \ldots, p_{2g+2} \in \mathbb{P}^1$ the branch points of π. Show that any element of $\mathrm{Aut}(C)$ commutes with τ, and from this conclude that $\mathrm{Aut}(C)$ is a $\mathbb{Z}/2$-extension of the group of automorphisms of $\mathbb{P}^1$ preserving $\{p_1, \ldots, p_{2g+2}\}$. In particular, in this case Aut C is finite.

F-4. Let C be a curve of genus g, $\phi \in \mathrm{Aut}(C)$, and $p_1, \ldots p_{g+1}$ general points of C. Show that there exists a meromorphic function f on C with polar divisor $(f)_\infty = p_1 + \cdots + p_{g+1}$, and, counting zeros and poles of $f - \phi^* f$, that ϕ has at most $2g + 2$ fixed points. Using Weierstrass points, deduce that $\mathrm{Aut}(C)$ is finite.

F-5. For a simpler, albeit less elementary argument, show that if ϕ is any automorphism of C, the eigenvalues of

$$\phi^*: H^1(C, \mathbb{Z}) \to H^1(C, \mathbb{Z})$$

are all roots of unity, and use this and the Lefschetz fixed point theorem to prove that *any automorphism of C fixing all the Weierstrass points of C is the identity.* Again, conclude that $\mathrm{Aut}(C)$ is finite.

F-6. For a Weierstrass-point-free argument, show that:

(i) any automorphism of C acting trivially on $H^1(C, \mathbb{Z})$ is the identity (use the Lefschetz fixed point theorem); and
(ii) if $\phi \in \mathrm{Aut}\, C$, then the automorphism $\phi^*: H^1(C, \mathbb{Z}) \to H^1(C, \mathbb{Z})$ is both integral and unitary with respect to the Hermitian form $H(\omega, \omega') = \int_C \omega \wedge \overline{\omega}'$ on $H^{1,0}(C)$.

Conclude once more that $\mathrm{Aut}(C)$ is finite.

F-7. Let $\phi: C \to C$ be an automorphism of finite order n. Show that the orbit (or quotient) space $C/\{\phi\}$ has naturally the structure of a compact Riemann surface such that $C \to C/\{\phi\}$ is an n-sheeted covering totally ramified at the fixed points of ϕ. Show that if n is prime then these are all the ramification points.

F-8. Retaining the notations from the preceding exercise, and using the Riemann–Hurwitz formula show that the number α of fixed points of ϕ satisfies

$$\alpha \leq \frac{2g - 2 + 2n}{n - 1},$$

with inequality only if the quotient $C/\{\phi\} \cong \mathbb{P}^1$.

F-9. Show more generally that if G is any group of automorphisms of C of order n, $G_p \subset G$ the subgroup of elements fixing $p \in C$, then the quotient C/G has naturally the structure of a compact Riemann surface such that $C \to C/G$ is an n-sheeted cover with ramification indices

$$\nu_p = \operatorname{card}(G_p)$$

for each $p \in C$.

F-10. Continuing the preceding exercises, using again the Riemann–Hurwitz formula show that Aut(C) has order at most $84(g - 1)$, and that if equality holds then the quotient

$$C/\operatorname{Aut}(C)$$

is isomorphic to $\mathbb{P}^1$ and there are exactly three points of C fixed under non-trivial subgroups of Aut(C) (this proof requires non-trivial fiddling).

This completes the proof of the theorem stated above. In the remaining exercises on automorphisms we shall give applications and special cases of this theorem.

F-11. Show that a curve of genus $g = 2$ cannot have an automorphism of order 7. Conclude that the bound $84(g - 1)$ is not always sharp.

F-12. Show that a curve of genus $g = 3$ cannot have an automorphism of order 5.

F-13. Show that if a curve of genus $g = 3$ has an automorphism of order 7, then it must be the normalization of one of the two curves

$$\begin{cases} y^7 = x^2(x - 1), \\ y^7 = x(x - 1). \end{cases}$$

F-14. Let C be the normalization of the plane curve

$$y^3 = x^2(x^2 - 1).$$

Show that C is isomorphic to the normalization of the curve

$$y^2 = x^6 - 1,$$

but is *not* isomorphic to the normalizations of either $y^2 = x^6 - x$ or $y^2 = x(x^4 - 1)$. (*Hint*: use Exercise F-3.)

F-15. Show that if $g(C) = 2$, then Aut(C) has order at most 48, with equality if and only if C is the normalization of

$$y^2 = x(x^4 - 1).$$

F-16. Show that if $g(C) = 4$, then $\mathrm{Aut}(C)$ has order at most 120, with equality if and only if C is the curve given in homogeneous coordinates $X_0, \ldots, X_4$ in $\mathbb{P}^4$ by

$$\sum X_i = 0,$$
$$\sum X_i^2 = 0,$$
$$\sum X_i^3 = 0.$$

F-17. Show that if $g(C) = 3$, then the order of $\mathrm{Aut}(C)$ is 168 if and only if C is the curve

$$y^7 = x^2(x - 1)$$

(cf. Exercise F-13).

F-18. Show that, if an automorphism ϕ has order n and at least $(4n - 2)/(n - 1)$ fixed points, then these fixed points are Weierstrass points of C. In particular, this will be the case if C has five or more fixed points. (J. Lewittes.)

Note. Recall that an automorphism $\phi\colon C' \to C'$ is said to be a *deck transformation* for the covering $C' \xrightarrow{\alpha} C$ if ϕ commutes with α, i.e., $\alpha \circ \phi = \alpha$, and the covering $C' \xrightarrow{\alpha} C$ is called *Galois* (or *normal*) if the group of deck transformations acts transitively on the points of a general fiber.

F-19. Show that a covering $C' \xrightarrow{\alpha} C$ is Galois if and only if the corresponding extension of function fields $\mathcal{M}(C) \xrightarrow{\alpha^*} \mathcal{M}(C')$ is normal.

F-20. Suppose the covering $C' \xrightarrow{\alpha} C$ is unramified. Show that α is Galois if and only if for $p \in C'$ the image

$$\alpha_*(\pi_1(C', p)) \subset \pi_1(C, \alpha(p))$$

is a normal subgroup; and, conversely, every normal subgroup of finite index in $\pi_1(C)$ arises in this way.

F-21. Let $C' \xrightarrow{\alpha} C$ be an unramified Galois covering of Riemann surfaces and $\phi \in \mathrm{Aut}\, C$. Show that ϕ lifts to an automorphism of C' if and only if

$$\phi_*\colon \pi_1(C) \to \pi_1(C)$$

preserves the subgroup $\alpha_*\pi_1(C') \subset \pi_1(C)$ associated to the cover C'.

F-22. Using the preceding exercise, show that for any curve C of genus $g \geq 2$ and any n, there exists an n^{2g}-sheeted unramified Galois cover of C to which every automorphism of C lifts. Conclude, in particular, that if the estimate in the theorem preceding exercise F-3 is sharp for curves of genus g, it is sharp for curves of genus $h = n^{2g}(g - 1) + 1$; and hence, in particular, that it is sharp for infinitely many g.

Note. To see that the estimate is *not* sharp for infinitely many genera, in fact that the inequality $\mathrm{Aut}(C) \leq 8(g + 1)$ holds for infinitely many genera—see Accola [1].

G. Period Matrices

The periods of abelian integrals may always be given as ordinary calculus integrals. Sometimes, e.g., if the curve has a large symmetry group, they may even be evaluated.

G-1. Let C be the curve (a) of exercise batch A. Recalling the notations from Exercise A-3 let λ_1 be the (real) line segment joining -1 and $-\omega$ in the complex x-plane, and λ_2 the segment joining $-\omega$ and $-\omega^2$. Show that the inverse images $\pi^{-1}(\lambda_i) \subset C$ are closed curves that (with suitable orientations) give a symplectic basis for $H_1(C, \mathbb{Z})$.

G-2. Show that

$$\int_{-1}^{-\omega} \frac{dx}{\sqrt{x^3+1}} = \frac{1}{\omega} \int_{-\omega}^{-\omega^2} \frac{dx}{\sqrt{x^3+1}}$$

and conclude that C is isomorphic to $\mathbb{C}/(\mathbb{Z} + \mathbb{Z}\omega)$. ($\mathbb{Z} + \mathbb{Z}\omega$ is the lattice generated by 1 and ω.)

G-3.

(i) Let C now be the compact Riemann surface corresponding to the curve (d) of exercise batch A, and recall the notations used in Exercise A-5. Let λ_i be the line segment joining ξ^{i-1} to ξ^i in the complex x-plane and set $\Gamma_i = \pi^{-1}(\lambda_i)$. Orient the loops Γ_i and compute the intersection matrix $Q(\Gamma_i, \Gamma_j)$. Conclude that the Γ_i generate $H_1(C, \mathbb{Z})$.

(ii) In terms of the Γ_i find a normalized basis $\{\gamma_i\}$ for $H_1(C, \mathbb{Z})$.

G-4. Using the relations

$$\int_{\xi^{i-1}}^{\xi^i} \frac{dx}{\sqrt{x^6-1}} = \frac{1}{\xi} \int_{\xi^i}^{\xi^{i+1}} \frac{dx}{\sqrt{x^6-1}},$$

$$\int_{\xi^{i-1}}^{\xi^i} \frac{x\,dx}{\sqrt{x^6-1}} = \frac{1}{\xi^2} \int_{\xi^i}^{\xi^{i+1}} \frac{x\,dx}{\sqrt{x^6-1}},$$

find a basis for $H^0(C, K)$ normalized with respect to the basis $\{\gamma_i\}$ in the preceding exercise. From this write down the period matrix of C.

H. Elementary Properties of Abelian Varieties

A trivial but important fact is that any holomorphic mapping

$$f: X \to Y$$

between compact, complex manifolds induces a mapping

$$f^*: H^0(Y, \Omega_Y^q) \to H^0(X, \Omega_X^q)$$

on holomorphic q-forms.

H-1. Show that any holomorphic mapping between complex tori is a translation followed by a group homomorphism.

We recall that, if $\Lambda \subset \Lambda'$ are lattices of rank $2n$ in $\mathbb{C}^n$ and $A = \mathbb{C}^n/\Lambda$, $B = \mathbb{C}^n/\Lambda'$, then the induced map

$$A \to B$$

is an *isogeny*, and A and B are said to be *isogenous*.

H-2. Show that an equidimensional map between complex tori is an isogeny (equidimensional means that $\dim A = \dim B$ and the Jacobian of the map is generically of maximal rank).

An abelian variety A is *simple* if it has no non-trivial abelian subvarieties (non-trivial means of dimension not equal to 0 or $\dim A$).

H-3. Show that any non-constant holomorphic mapping between simple abelian varieties is an isogeny.

H-4. (Poincaré's Complete Reducibility Theorem). Show that a non-simple abelian variety is isogenous to a non-trivial product of abelian varieties. (*Note*: This requires essential use of the polarization.)

H-5. If $f: C_1 \to C_2$ is a non-constant holomorphic mapping between compact Riemann surfaces where C_2 has positive genus, show that $J(C_1)$ is isogenous to the product of $J(C_2)$ with another abelian variety.

H-6. Let E be the elliptic curve

$$y^2 = x^3 - 1$$

and C the hyperelliptic curve

$$y^2 = x^6 - 1.$$

Show that $J(C)$ is isogenous to $E \times E$.

(*Hint*: Consider the automorphisms $(x, y) \to (-x, y)$ and $(x, y) \to (-x, -y)$ of C.)

H-7. Let E be the elliptic curve

$$y^2 = x^4 - 1$$

and C the Fermat quartic given in affine coordinates by

$$x^4 + y^4 = 1.$$

Show that $J(C)$ is isogenous to $E \times E \times E$.

Remark. For C a smooth curve, the principally polarized abelian variety $J(C)$ is never a non-trivial product of principally polarized abelian varieties (here product means with polarizations). However, $J(C)$ may be a non-trivial product of complex tori.

Appendix A

The Riemann–Roch Theorem, Hodge Theorem, and Adjoint Linear Systems

The Riemann–Roch theorem is the basic result in the theory of algebraic curves. The usual proofs of this theorem are either potential theoretic, algebraic using repartitions, or a mixture of each using sheaf cohomology and Serre duality. In the first sequence of exercises below we will give an alternate proof, one that has proven useful pedagogically and that gives as a by-product the Hodge theorem and the completeness of adjoint series for plane curves with nodes.

The proof will be divided into two major steps, the first of which is the following existence result:

(*) **Theorem.** *Let C be a compact Riemann surface. Then there exists a holomorphic mapping*

$$f: C \to \mathbb{P}^2$$

whose image Γ is a plane curve with nodes and where $f: C \to \Gamma$ is birational.

The second step will be to prove the Riemann–Roch theorem for C by studying Γ. The reader may omit the first step and begin with Exercise 5 below.

To establish Theorem (*) we shall assume the following basic analytic fact, an elementary proof of which may be found in the appendix to Gunning–Rossi [1]:

(**) *Let C be a compact Riemann surface and $L \to C$ a holomorphic line bundle. Then*

$$h^1(C, L) < \infty.$$

1. Let $p_1, p_2, p_3, \ldots$ be a sequence of points (not necessarily distinct), and set $D_n = p_1 + \cdots + p_n$. Using (**) show that for n large enough,

$$h^1(C, L(D_n)) = 0,$$

and that

$$h^0(C, L(D_n)) = n - c$$

for some constant c.

(*Hint*: Consider the exact cohomology sequence of

$$0 \to L(D_{n-1}) \to L(D_n) \to L(D_n) \otimes \mathcal{O}_{p_n} \to 0$$

for $n = 1, 2, \ldots$. Here $L(D_n) = L \otimes \mathcal{O}(D_n)$ and $\mathcal{O}_{p_n}$ is the skyscraper sheaf $\mathcal{O}_C/\mathcal{O}_C(-p_n)$.)

2. Using the preceding exercise show that there exists a holomorphic embedding

$$\phi: C \to \mathbb{P}^r.$$

3. Let $C \subset \mathbb{P}^r$ be a one-dimensional complex submanifold (e.g., the image $\phi(C)$ in the preceding problem). Show that the arguments used in Exercises B-5 and B-6 of Chapter 1 may be adapted to show that:

(i) under a general projection $\mathbb{P}^r - \mathbb{P}^{r-4} \to \mathbb{P}^3$, C maps biholomorphically to its image; and

(ii) under a general projection $\mathbb{P}^r - \mathbb{P}^{r-3} \to \mathbb{P}^2$, C maps in a generically one-to-one fashion onto an *analytic curve* $\Gamma \subset \mathbb{P}^2$ (i.e., Γ is locally the zeros of an analytic function in open subsets of $\mathbb{P}^2$) with the following property: every point of Γ is either smooth or is the union of two smooth branches having distinct tangents (thus, $C \to \mathbb{P}^2$ is an *immersion*). We shall refer to the image of such an immersion as an *analytic plane curve with nodes*, and we shall call the original Riemann surface C (always connected) its *normalization*.

4. Let $\Gamma \subset \mathbb{P}^2$ be an analytic plane curve with nodes. By considering the projection

$$\pi: \Gamma \to \mathbb{P}^1$$

onto a general line, show that Γ is an irreducible algebraic curve $F(x_0, x_1, x_2) = 0$.

(*Hint*: In a suitable affine coordinate system the projection is $(x, y) \to x$. Both x and y are meromorphic functions on C, and we let $y_1(x), \ldots, y_d(x)$ be the values of y on the points in $\pi^{-1}(x)$. Now consider

$$f(x, y) = \sum_{\nu=1}^{d} (y - y_\nu(x)),$$

and show that $F(x_0, x_1, x_2) = x_0^d f(x_1/x_0, x_2/x_0)$ has the required properties.)

With this exercise we have completed the proof of Theorem (*).

Let $\Gamma \subset \mathbb{P}^2$ be an algebraic curve with nodes. We say that Γ *is in general position with respect to a coordinate system* $[x_0, x_1, x_2]$ in case the following conditions are satisfied:

(i) $[0, 0, 1] \notin \Gamma$;
(ii) the line $L_\infty = \{x_0 = 0\}$ is neither tangent to Γ nor passes through a node;
(iii) set $x = x_1/x_0$ and $y = x_2/x_0$ and consider the map

$$\pi_x \colon \Gamma \to \mathbb{P}^1$$

induced by the projection $(x, y) \to x$. Then over each point $x \in \mathbb{P}^1$ there is at most one simple branch point (two sheets interchanging) or one node (note that $\pi_x^{-1}(\infty) = L_\infty \cdot \Gamma$).

5. Show that Γ is in general position with respect to a general coordinate system.

(*Hint*: It is useful to interpret the branch points of $\pi \colon \Gamma \to \mathbb{P}^1$ as the vertical tangents, i.e., the points of Γ whose tangent line passes through the center $[0, 0, 1]$ of projection.)

From now on we will assume that $\Gamma \subset \mathbb{P}^2$ is an irreducible plane curve with nodes having homogeneous equation

$$F(x_0, x_1, x_2) = 0,$$

affine equation

$$f(x, y) = 0,$$

and in general position with respect to the coordinate system. We denote by $\phi \colon C \to \Gamma$ the normalization of Γ, and assume that Γ has degree d.

6. Let $D \subset \mathbb{P}^2$ be a plane curve of degree n given by a homogeneous polynomial $G(x_0, x_1, x_2)$ and set $g(x, y) = G(1, x, y)$. Assuming that Γ is not contained in D, define the divisor ϕ^*D to be the zero divisor of the meromorphic function $\phi^*(g)$ on C. Show that

$$\deg(\phi^*D) = nd$$

(Bezout's theorem).

(*Hint*: Note that

$$g(x, y) = G(x_0, x_1, x_2)/x_0^n$$

and that

$$\deg(nL_\infty \cdot \Gamma) = nd.$$

Now use the fact that the number of zeros equals the number of poles of a meromorphic function on C.)

7. Denote by

$$\pi: C \to \mathbb{P}^1$$

the composition of $\phi: C \to \Gamma$ with the projection of Γ on the x-axis, and by $\Delta \subset C$ the divisor lying over the nodes of Γ (thus $\deg \Delta = 2\delta$ where δ is the number of nodes). Let E be the plane curve with affine equation $(\partial f/\partial y)(x, y) = 0$. Show that the ramification divisor R of π is given by

$$R = \phi^*E - \Delta.$$

Using the Riemann–Hurwitz formula conclude that

$$g = (d-1)(d-2)/2 - \delta$$

(*genus formula*).

8. Let D be the plane curve given by a homogeneous polynomial $G(x_0, x_1, x_2)$ of degree $d - 3$ and set $g(x, y) = G(1, x, y)$. Define the *Poincaré residue* to be

$$\text{(i)} \qquad \omega = \phi^*\left(\frac{g(x, y)\,dx}{(\partial f/\partial y)(x, y)}\right) = \phi^*\left(-\frac{g(x, y)\,dy}{(\partial f/\partial x)(x, y)}\right)$$

(why is the equality true?). Show that ω is a meromorphic differential with divisor

$$\text{(ii)} \qquad (\omega) = \phi^*D - \Delta.$$

Using this prove the *adjunction formula*

$$\text{(iii)} \qquad K_C \cong \mathcal{O}_C(d-3)(-\Delta),$$

and the inequality

$$\begin{aligned} h^0(C, K_C) &\geq h^0(\mathbb{P}^2, \mathcal{O}(d-3)) - \delta \\ &= (d-1)(d-2)/2 - \delta \\ &= g. \end{aligned}$$

9. Show that, with ω defined as in (i), statement (ii) is true even if Γ is *not* in general position with respect to the coordinate system; conclude (iii) directly.

10. Denote by $H^1_{\mathrm{DR}}(C)$ the first deRham cohomology (with complex coefficients) of C and by $H^{1,0}(C) \subset H^1_{\mathrm{DR}}(C)$ the image of the natural map

$$j\colon H^0(C, K_C) \to H^1_{\mathrm{DR}}(C).$$

Prove the *Hodge theorem*

$$H^1_{\mathrm{DR}}(C) = H^{1,0}(C) \oplus \overline{H^{1,0}(C)}.$$

(*Hint*: For $\omega, \psi \in H^0(C, K_C)$ it is clear that

$$\begin{cases} \dfrac{\sqrt{-1}}{2} \displaystyle\int_C \omega \wedge \overline{\omega} > 0, \qquad \omega \neq 0, \\ \displaystyle\int_C \omega \wedge \psi = 0. \end{cases}$$

From this conclude that

$$j \oplus \bar{j}\colon H^0(C, K_C) \oplus \overline{H^0(C, K_C)} \to H^1_{\mathrm{DR}}(C)$$

is injective, and then by the inequality in Exercise 8 that $j \oplus \bar{j}$ is an isomorphism.)

11. Let $p_1, \ldots, p_\delta \in \mathbb{P}^2$ be the nodes of Γ:

(i) Show that $p_1, \ldots, p_\delta$ impose independent conditions on the linear system $|\mathcal{O}_{\mathbb{P}^2}(d-3)|$ of plane curves of degree $d-3$.
(ii) Show that the system $\Sigma \subset |\mathcal{O}_{\mathbb{P}^2}(d-3)|$ of curves passing through $p_1, \ldots, p_\delta$ cuts out the complete canonical series $|K_C|$; i.e.

$$|K_C| = \{\phi^* D - \Delta\colon D \in \Sigma\}.$$

(*Hint*: By the Hodge theorem equality must hold in the inequality in Exercise 8.)

Remark. Assertions (i) and (ii) are the cornerstones of the classical theory of plane curves; we have established them by a transcendental argument.

Let D be any divisor of degree m on C. In the remaining exercises of this group we will construct the complete linear system $|D|$ and establish the Riemann–Roch theorem and completeness of the adjoint system.

12. Write $D = D' - D''$ where D', D'' are effective divisors of respective degrees m', m''. Let G be a plane curve of degree n containing the nodes

$p_1, \ldots, p_\delta$ of Γ and the image of D'; i.e.

$$\phi^*G = D' + \Delta + E$$

for some effective divisor E with

$$\deg E = nd - 2\delta - m'.$$

Show that the plane curves H of degree n satisfying

$$\phi^*H \geq \Delta + E + D''$$

form a linear system $\mathscr{D} \subset |\mathcal{O}_{\mathbb{P}^2}(n)|$ with

$$\dim \mathscr{D} \geq (n + 1)(n + 2)/2 - 1 - nd + \delta + m.$$

13. Given $H \in \mathscr{D}$ write

$$\phi^*H = \Delta + E + D'' + B,$$

and show that

$$B \sim D.$$

From this conclude that

$$\begin{aligned} r(D) &\geq \dim \mathscr{D} - h^0(\mathbb{P}^2, \mathcal{O}(n - d)) \\ &\geq (n + 1)(n + 2)/2 - 1 - nd + \delta + m - (n - d + 1)(n - d + 2)/2 \\ &= -(d - 1)(d - 2)/2 + \delta + m \\ &= m - g. \end{aligned}$$

In particular, if $m \geq g$ then $r(D) \geq 0$. The assertion

$$r(D) \geq \deg D - g$$

is generally known as *Riemann's inequality*.

14. Assuming that $D = p_1 + \cdots + p_m$ is effective use the residue theorem in the form

$$\sum_i \operatorname{Res}_{p_i}(f\omega) = 0, \qquad f \in \mathscr{L}(D) \quad \text{and} \quad \omega \in H^0(C, K)$$

to show that

$$r(D) \leq m - g + h^0(C, K(-D)).$$

15. Using the preceding two exercises, establish the *Riemann–Roch theorem*

$$r(D) = \deg D - g + h^0(C, K(-D))$$

for any divisor D.

(*Hint*: Treat the cases $h^0(C, \mathcal{O}(D)) \neq 0$, $h^0(C, K(-D)) \neq 0$, and $h^0(C, \mathcal{O}(D)) = h^0(C, K(-D)) = 0$ separately.)

16. Show that the system $\Sigma_n \subset |\mathcal{O}_{\mathbb{P}^3}(n)|$ of curves passing through $p_1, \ldots, p_\delta$ cuts out the complete linear system $|K_C(n - d + 3)|$ for all n (this is called the *completeness of linear system of adjoint curves*).

§1. Applications of the Discussion About Plane Curves with Nodes

This series of exercises will give some properties of plane curves $\Gamma \subset \mathbb{P}^2$, possibly with nodes. We recall that a set $S = \{p_1, \ldots, p_d\}$ of distinct points in $\mathbb{P}^2$ is said to impose independent conditions on curves of degree n if

$$h^0(\mathbb{P}^2, \mathcal{I}_S(n)) = h^0(\mathbb{P}^2, \mathcal{O}_{\mathbb{P}^2}(n)) - d,$$

where $\mathcal{I}_S \subset \mathcal{O}_{\mathbb{P}^2}$ is the ideal sheaf of the zero-dimensional variety S.

17. Show that any $n + 1$ distinct points impose independent conditions on curves of degree n. Show that $n + 2$ distinct points fail to impose independent conditions on curves of degree n if, and only if, they lie on a line.

18. Let $C \subset \mathbb{P}^2$ be a smooth plane curve of degree $d \geq 4$. Using the preceding exercise and the geometric form of the Riemann–Roch theorem, show that:

(i) C does not have a g^1_m for $m \leq d - 2$;
(ii) if $|E|$ is a g^1_{d-1}, then $E = D - p$ where $p \in C$ and $D \in |\mathcal{O}_C(1)|$;
(iii) $h^0(C, \mathcal{O}_C(1)) = 3$ (i.e., the g^2_d on C cut out by the lines is complete), and $|\mathcal{O}_C(1)|$ is the unique g^2_d on C.

19. Improve Exercise 17 by showing that if a set S of $2n + 1$ distinct points in $\mathbb{P}^2$ fail to impose independent conditions on curves of degree n, then S must include $n + 2$ collinear points.

20. Now suppose that $\Gamma \subset \mathbb{P}^2$ is an irreducible curve of degree d with δ nodes and no other singularities and C its normalization. Show that:

(i) if $\delta \leq d - 2$, then C does not have a g^1_m for $m \leq d - 3$;
(ii) if $\delta \leq d - 3$, then any g^1_{d-2} on C is cut out by lines through a node;
(iii) if $\delta \leq d - 3$, then the g^2_d cut out on Γ by lines is complete, and if $\delta < d - 3$ it is the unique g^2_d on the normalization C of Γ.

21. Let $\Gamma \subset \mathbb{P}^2$ be the plane quintic

$$\sum_{i \neq j} X_i^2 X_j^3 = 0.$$

Determine the genus $g(C)$ and the space $H^0(C, K)$ of holomorphic differentials for the normalization C of Γ. Then give the canonical map and the equations of its image. (Upon doing this exercise you will know the standard Cremona transformation of $\mathbb{P}^2$, if you don't already.)

22. Let $\Gamma \subset \mathbb{P}^2$ be an irreducible quintic curve with δ nodes, C its normalization. Show that the g_5^2 cut out on C by the lines is complete if and only if $\delta \leq 3$.

23. Let $\Gamma \subset \mathbb{P}^2$ be an irreducible sextic curve with δ nodes, C its normalization. Show that the g_6^2 cut out on C by lines is complete if $\delta \leq 5$, incomplete if $\delta \geq 7$, and if $\delta = 6$ then it is incomplete if and only if all six nodes lie on a conic.

24. More generally, let $\Gamma \subset \mathbb{P}^2$ be an irreducible plane curve of degree d with nodes $r_1, \ldots, r_\delta$, C its normalization. Show that the g_d^2 cut out on C by lines is complete if and only if $r_1, \ldots, r_\delta$ impose independent conditions on curves of degree $d - 4$.

§2. Adjoint Conditions in General

The analysis of plane curves carried out in the above exercises under the assumption that Γ has only nodes can be carried out in more general circumstances.

Maintaining the notations of those exercises, if $r \in \Gamma$ is any singularity in the finite plane, $p_1, \ldots, p_k$ the points of C lying over r, we define the *adjoint divisor* Δ_r of r to be the divisor on C

$$\Delta_r = \sum \alpha_i p_i,$$

where

$$-\alpha_i = \operatorname{mult}_{p_i}\left(\frac{dx}{(\partial f/\partial y)(x, y)}\right).$$

We define the *adjoint ideal* $I_r \subset \mathcal{O}_{\mathbb{P}^2, r}$ to be the ideal of functions $g(x, y)$ such that the divisor

$$(\phi^* g) \geq \Delta_r;$$

that is, the ideal of functions g such that the differential

$$\omega_g = \phi^*\left(\frac{g(x, y)\,dx}{(\partial f/\partial y)(x, y)}\right)$$

is regular near $\phi^{-1}(r)$, and we say that g *satisfies the adjoint conditions at* r if $g \in I_r$. Globally, the definitions are analogous, e.g., the adjoint divisor Δ of $\phi: C \to \Gamma \subset \mathbb{P}^2$ is the sum

$$\Delta = \sum_{r \in \Gamma_{\text{sing}}} \Delta_r.$$

Finally, the *number* δ_r *of adjoint conditions at* r is the index in $\mathcal{O}_{\mathbb{P}^2, r}$ of the ideal I_r—that is, the dimension of $\mathcal{O}_{\mathbb{P}^2, r}/I_r$ as a complex vector space—and the *number of adjoint conditions* overall is the sum of these numbers; i.e.,

$$\delta = \sum_{r \in \Gamma_{\text{sing}}} \delta_r.$$

25. Show that the ramification divisor of the projection map $\pi: C \to \mathbb{P}^1$ is $R - \Delta$; similarly, show that the divisor of the meromorphic 1-form $\pi^*\,dx$ is

$$(\pi^*\,dx) = R - 2D_\infty - \Delta,$$

i.e.

$$K_C = \phi^*\mathcal{O}_\Gamma(d-3)(-\Delta).$$

Either way, deduce the formula

$$g(C) = \frac{(d-1)(d-2)}{2} - \tfrac{1}{2}\deg \Delta.$$

26. Show that for $g \in \mathbb{C}[x, y]$ the 1-form

$$\omega_g = \phi^*\left(\frac{g(x, y)\,dx}{(\partial f/\partial y)(x, y)}\right)$$

is holomorphic on C if and only if $\deg g \le d - 3$ *and* g satisfies the adjoint conditions at every singularity of Γ. Conclude that

$$h^0(C, K) \ge \frac{(d-1)(d-2)}{2} - \delta$$

and hence that

$$\delta \ge \tfrac{1}{2}\deg \Delta.$$

27. Suppose now that $r \in \Gamma$ lies in the finite plane, with $p_1, \ldots, p_k$ lying over r; let

$$\omega = \phi^*\left(\frac{dx}{\partial f/\partial y}\right)$$

and let V be the $(\deg \Delta_r)$-dimensional vector space

$$V = H^0(C, \mathcal{O}_C/\mathcal{O}_C(-\Delta_r)).$$

Define a symmetric bilinear pairing

$$\psi : V \times V \to \mathbb{C}$$

by

$$\psi(g, h) = \sum_i \operatorname{Res}_{p_i}(gh\omega).$$

Show that ψ is well defined and non-degenerate (i.e., the induced map $\tilde{\psi}: V \to V^*$ is an isomorphism).

28. Let $g(x, y)$ be any polynomial. Show by the residue theorem that

$$\sum_i \operatorname{Res}_{p_i}(\phi^* g\omega) = 0.$$

29. Let I_r be as above the adjoint ideal at r, W the δ_r-dimensional vector space

$$W = H^0(\mathcal{O}_\Gamma/I_r),$$

and $\tilde{W} = \phi^* W \subset V$ the image of W under the injection

$$\phi^*: H^0(\mathcal{O}_\Gamma/I_r) \to H^0(\mathcal{O}_C/\mathcal{O}_C(-\Delta_r)).$$

Show by Exercises 27 and 28 that $\tilde{W}$ *is isotropic for* ψ, i.e., $\psi(\tilde{W}, \tilde{W}) \equiv 0$ or, equivalently

$$\psi^* \tilde{W} \subset \operatorname{Ann}(\tilde{W}).$$

30. Deduce from the preceding exercise and Exercise 26 above the *Gorenstein relation*

$$\deg \Delta_r = 2\delta_r$$

and likewise the global relation

$$g(C) = \frac{(d-1)(d-2)}{2} - \delta.$$

31. Using Exercises 26 and 30, deduce that:

(a) the adjoint ideals of Γ impose independent conditions on curves of degree $d - 3$;
(b) the linear system of plane curves of degree $d - 3$ satisfying the adjoint conditions cuts out the complete canonical series on C; and
(c) more generally, the linear system of plane curves of degree n satisfying the adjoint conditions cuts out on C the complete series $|K_C(n - d + 3)|$.

32. Examples. Find the adjoint divisor and conditions for the following singularities; verify in each of these cases that $\deg \Delta_r = 2\delta_r$:

(a) a node: two smooth arcs meeting transversely (e.g., $y^2 - x^2 = 0$);
(b) an nth-order node: two smooth arcs having contact of order n (e.g., $y^2 - x^{2n} = 0$);
(c) an nth-order cusp (e.g., $y^2 - x^{2n+1} = 0$);
(d) an ordinary triple point: three smooth arcs meeting transversely at r (e.g., $xy(x - y) = 0$);
(e) a triple point with an infinitely near double point: three smooth arcs, two of which are simply tangent (e.g., $(y + x^2)(y - x^2)x = 0$);
(f) $y^3 - x^5 = 0$;
(g) $y^p - x^q = 0$ for p, q any pair of relatively prime integers.

Chapter II

Determinantal Varieties

In this chapter we have collected a few foundational results about determinantal varieties that we will need in later chapters. The main application of the results in this chapter will be to the varieties of special divisors on curves. These varieties have in fact a natural determinantal structure defined in terms of the Brill–Noether matrix, and their study is the central theme of this book. Another ubiquitous example of determinantal variety is the one of rational normal scrolls.

§1. Tangent Cones to Analytic Spaces

We first need to review a number of ancillary results dealing with the notion of tangent cone to an analytic space at a point.

If Y is a complex manifold and $X \subset Y$ an effective divisor (i.e., an analytic subspace given locally by one equation), then it is easy to define the tangent cone to X at a point $p \in X$. In holomorphic coordinates $z_1, \ldots, z_n$ on Y centered at p we expand a local defining equation $f(z)$ of X as a series of homogeneous forms

$$f(z) = \sum_{k=0}^{\infty} f_k(z).$$

Regarding $z_1, \ldots, z_n$ as coordinates on the tangent space $T_p(Y)$, the tangent cone is the algebraic subvariety

$$\mathcal{T}_p(X) \subset T_p(Y)$$

defined by the equation

$$f_h(z) = 0,$$

where h is the minimum integer k such that $f_k \neq 0$. For an arbitrary analytic space X, denoting by m the maximal ideal in the local ring $\mathcal{O}_{X,p}$, the *tangent cone to X at p* is defined to be

$$\mathcal{T}_p(X) = \operatorname{Spec}\left(\bigoplus_{i=0}^{\infty} m^i/m^{i+1}\right).$$

The natural surjective ring homomorphism

$$\bigoplus_i \operatorname{Sym}^i(m/m^2) \to \bigoplus_i m^i/m^{i+1}$$

defines an inclusion

$$\mathcal{T}_p(X) \subset T_p(X)$$

of the tangent cone in the Zariski tangent space to X at p. The *projectivized tangent cone to X at p* is defined by

$$\mathbb{P}\mathcal{T}_p(X) = \operatorname{Proj}\left(\bigoplus_{i=0}^{\infty} m^i/m^{i+1}\right).$$

In other words, the projectivized tangent cone is the exceptional divisor in the blow-up of X at p, and it is naturally an algebraic subvariety of the projectivized Zariski tangent space $\mathbb{P}T_p(X)$. The *multiplicity* $\operatorname{mult}_p(X)$ of X at p is then defined to be the degree of $\mathbb{P}\mathcal{T}_p(X)$ in $\mathbb{P}T_p(X)$. It is important to notice that $\mathbb{P}\mathcal{T}_p(X)$ is a Cartier divisor on the blow-up $\tilde{X}$ of X at p. In other terms, $\mathbb{P}\mathcal{T}_p(X)$ is an analytic subspace of $\tilde{X}$ defined, locally, by the vanishing of a single holomorphic function.

In case X is given as an analytic subspace of a complex manifold Y there is another convenient description of the tangent cones to X which parallels the one we have given for divisors. Let p be a point of X. Denote by I the ideal of X in $\mathcal{O}_{Y,p}$, and by M and $m = M/I$ the maximal ideals in $\mathcal{O}_{Y,p}$ and $\mathcal{O}_{X,p}$, respectively. We set

$$J_k = (M^k \cap I)/(M^{k+1} \cap I).$$

It is straightforward to check that

$$J = \bigoplus J_k \subset \bigoplus M^k/M^{k+1} = \bigoplus \operatorname{Sym}^k(M/M^2)$$

is a graded ideal. This is usually called the *ideal of initial forms* of I. We claim that J is, indeed, the ideal of $\mathcal{T}_p(X)$ in $T_p(Y)$. In fact, for any k,

$$\begin{aligned} m^k/m^{k+1} &\cong (M^k + I)/(M^{k+1} + I) \\ &\cong M^k/(M^{k+1} + M^k \cap I) \\ &\cong (M^k/M^{k+1})/J_k, \end{aligned}$$

and therefore,

$$\bigoplus m^k/m^{k+1} \cong \left(\bigoplus \operatorname{Sym}^k(M/M^2)\right)/J,$$

which is what we had to prove.

Another useful remark, in the above situation, is the following. Let $\tilde{Y}$ be the blow-up of Y at p, and denote by $E = \mathbb{P}T_p(Y)$ the exceptional divisor on $\tilde{Y}$. Then $\mathbb{P}\mathcal{T}_p(X)$ is the intersection of E with the blow-up $\tilde{X}$ of X at p. In particular, the line bundle associated to $\mathbb{P}\mathcal{T}_p(X)$ is the restriction of $\mathcal{O}(E)$ to $\tilde{X}$.

There is a standard situation where projectivized tangent cones can be easily computed; this will occur time and again in the following chapters. Specifically, suppose we are given a proper holomorphic map

$$f\colon X \to Y$$

between complex manifolds. We shall assume that f maps X bimeromorphically onto its image. Let p be a point of $f(X)$; we wish to compute the projectivized tangent cone to $f(X)$ at p under the additional assumption that the (scheme-theoretic) fiber $f^{-1}(p)$ is smooth. This assumption implies that the vector bundle homomorphism

$$df\colon N \to T_p(Y) \times f^{-1}(p)$$

is injective, where N is the normal bundle to $f^{-1}(p)$ in X. On the other hand df, being injective, induces a well-defined morphism

$$g\colon \mathbb{P}N \to \mathbb{P}T_p(Y).$$

The lemma we want to prove is the following:

(1.1) Lemma. *Let $f\colon X \to Y$ be a proper holomorphic map between complex manifolds which maps X bimeromorphically onto $Z = f(X)$. Let p be a point of Z such that the fiber $f^{-1}(p)$ is smooth. Denote by N the normal bundle to $f^{-1}(p)$ in X and let g be as above. Denote by γ and η the analytic cycles associated to $\mathbb{P}N$ and $\mathbb{P}\mathcal{T}_p(Z)$, respectively. Then $g_*(\gamma) = \eta$.*

To prove this we first blow up Z and Y at p, and X along $f^{-1}(p)$. We then get a commutative diagram

$$\begin{array}{ccccc}
G & \longrightarrow & E & \subset & F \\
\cap & & \cap & & \cap \\
\tilde{X} & \xrightarrow{\tilde{f}} & \tilde{Z} & \subset & \tilde{Y} \\
\downarrow & & \downarrow & & \downarrow \\
X & \xrightarrow{f} & Z & \subset & Y
\end{array}$$

where G, E, and F are the exceptional divisors. As we shall presently show the existence of $\tilde{f}$ is ensured by the smoothness of $f^{-1}(p)$. Choose coordinates $z_1, \ldots, z_n$ on X in such a way that G is locally defined by the equations $z_1 = \cdots = z_k = 0$; likewise, choose coordinates $w_1, \ldots, w_m$ on Y, centered at p. In these coordinates the map f can be written

$$w_\alpha = f_\alpha(z_1, \ldots, z_n), \qquad \alpha = 1, \ldots, m.$$

Let

$$\xi_i z_j = \xi_j z_i, \qquad i, j = 1, \ldots, k,$$

be the local equations of $\tilde{X}$ in $\mathbb{P}^{k-1} \times X$. Similarly, let

$$\eta_\alpha w_\beta = \eta_\beta w_\alpha, \qquad \alpha, \beta = 1, \ldots, m, \tag{1.2}$$

be the local equations of $\tilde{Y}$ in $\mathbb{P}^{m-1} \times Y$. In these coordinates the lifting $\tilde{f}$ is given by

$$\begin{aligned} \eta_\alpha &= \sum_{i=1}^{k} \frac{\partial f_\alpha}{\partial z_i}(0, \ldots, 0, z_{k+1}, \ldots, z_n)\xi_i \\ &\quad + \frac{1}{2} \sum_{i,j=1}^{k} \frac{\partial^2 f_\alpha}{\partial z_i \, \partial z_j}(0, \ldots, 0, z_{k+1}, \ldots, z_n)\xi_i z_j + \cdots, \\ w_\alpha &= f_\alpha(z_1, \ldots, z_n). \end{aligned}$$

A trivial computation shows that (1.2) is satisfied. On the other hand, to say that $f^{-1}(p)$ is smooth means that the Jacobian matrix

$$\left(\frac{\partial f_\alpha}{\partial z_i}(0, \ldots, 0, z_{k+1}, \ldots, z_n)\right)_{\substack{\alpha = 1, \ldots, m \\ i = 1, \ldots, k}}$$

has rank k, and therefore the η_α's cannot simultaneously vanish.

By what we already remarked we have

$$\begin{aligned} E &= \mathbb{P}\mathcal{T}_p(Z), \\ F &= \mathbb{P}T_p(Y); \end{aligned}$$

furthermore,

$$G = \mathbb{P}N.$$

With these identifications it is apparent from the local expression of $\tilde{f}$ that the restriction

$$\tilde{f}|_G \colon G \to F$$

is nothing but g. Notice, moreover, that Z is reduced and irreducible and $\tilde{f}$, being proper, maps $\tilde{X}$ bimeromorphically *onto* $\tilde{Z}$. It follows, in particular, that $g = \tilde{f}|_G$ maps G onto E. Assuming, for the time being, that $f^{-1}(p)$, and hence G, is connected, this implies that

$$g_*(\gamma) = \lambda\eta,$$

where λ is a rational number. We are thus reduced to showing that the degrees of $g_*(\gamma)$ and η are the same. We recall that, if $\mathcal{O}_G(1)$ denotes the tautological line bundle on $G = \mathbb{P}N$,

$$\mathcal{O}_F(1) \cong \mathcal{O}_{\tilde{Y}}(-F) \otimes \mathcal{O}_F,$$
$$\mathcal{O}_G(1) \cong \mathcal{O}_{\tilde{X}}(-G) \otimes \mathcal{O}_G,$$

while the local expressions for $\tilde{f}$ show that

$$\tilde{f}^*\mathcal{O}_{\tilde{Y}}(F) \cong \mathcal{O}_{\tilde{X}}(G).$$

Therefore, we have

$$\begin{aligned}
\deg(\eta) &= \int_\eta c_1(\mathcal{O}_F(1))^{n-1} \\
&= \int_\eta c_1(\mathcal{O}_{\tilde{Y}}(-F))^{n-1} \\
&= \int_{\tilde{Z}} c_1(\mathcal{O}_{\tilde{Y}}(-F))^{n-1} \cdot c_1(\mathcal{O}_{\tilde{Z}}(E)) \\
&= -\int_{\tilde{Z}} c_1(\mathcal{O}_{\tilde{Y}}(-F))^n \\
&= -\int_{\tilde{X}} c_1(\mathcal{O}_{\tilde{X}}(-G))^n \\
&= \int_G c_1(\mathcal{O}_G(1))^{n-1}.
\end{aligned}$$

The next to last equality follows from the fact that $\tilde{f}$ maps $\tilde{X}$ bimeromorphically onto $\tilde{Z}$. On the other hand, we have

$$\begin{aligned}
\deg(g_*(\gamma)) &= \int_G g^*c_1(\mathcal{O}_F(1))^{n-1} \\
&= \int_G c_1(\mathcal{O}_G(1))^{n-1}.
\end{aligned}$$

This disposes of the case when $f^{-1}(p)$ is connected. In general, notice that, upon shrinking X and Y if necessary, we may suppose that each connected component of X contains exactly one component of G. The general case follows at once by applying to each connected component of X the special case we have just proved. The proof of Lemma (1.1) is thus completed.

Corollary. *With the hypotheses and the notation of Lemma* (1.1)

$$g(\mathbb{P}N) = \mathbb{P}\mathcal{T}_p(Z)$$

as sets.

To prove the corollary it suffices to observe that $\mathbb{P}\mathcal{T}_p(Z)$, being a Cartier divisor in the irreducible n-dimensional variety $\tilde{Z}$, has pure dimension $n - 1$, so that the set-theoretical equality follows from the equality of cycles in the statement of the lemma.

In special cases the conclusion of the above corollary can be considerably sharpened.

(1.3) Lemma. *Suppose the hypotheses of Lemma* (1.1) *are met. Assume moreover that*

$$h = df\colon N \to T_p(Y)$$

is a birational map of N onto its image $h(N)$, and that $h(N)$ is normal. Then

$$h(N) = \mathcal{T}_p(Z)$$

as schemes.

Proof. By the corollary we just proved $h(N)$ and $\mathcal{T}_p(Z)$ have the same support; in terms of their affine coordinate rings this means that

$$\bigoplus_{i\geq 0} m^i/m^{i+1} \xrightarrow{\alpha} \Gamma(h(N), \mathcal{O}_{h(N)}) = \Gamma(N, \mathcal{O}_N)$$

is onto. Here m stands for the maximal ideal in $\mathcal{O}_{Z,p}$ and the equality on the right-hand side follows from the normality of $h(N)$ and the fact that N and $h(N)$ are birational. We let $I \subset \mathcal{O}_X$ be the ideal sheaf of $f^{-1}(p)$. Thus,

$$\Gamma(N, \mathcal{O}_N) = \bigoplus_{i\geq 0} \Gamma(X, I^i/I^{i+1}),$$

and all we have to prove is that

$$\bigoplus_{i\geq 0} m^i/m^{i+1} \to \bigoplus_{i\geq 0} \Gamma(X, I^i/I^{i+1})$$

is injective. Suppose not: this means that there is a function f_1 on Z vanishing at p to order exactly equal to i, whose pull-back to X vanishes to order at least $i + 1$ along $f^{-1}(p)$. By the surjectivity of α we can inductively construct, for any $h \geq 1$, functions $g_h \in m^{i+h}$ such that the pull-back to X of $f_1 + g_1 + \cdots + g_h$ vanishes to order at least $i + h + 1$ along $f^{-1}(p)$. Passing to completions, this shows that

$$\hat{\mathcal{O}}_{Z,p} \to \varprojlim \Gamma(X, \mathcal{O}_X/I^i)$$

is not injective. By the theorem on formal functions (cf. Hartshorne [2, page 277]; here the result is proved in an algebraic setting, but the argument works also in an analytic one) the right-hand side is isomorphic to $\widehat{(f_* \mathcal{O}_X)_p}$. Thus the homomorphism

$$\hat{\mathcal{O}}_{Z,p} \to \widehat{(f_* \mathcal{O}_X)_p}$$

is not injective. This is absurd since, by the birationality assumption on f, $\mathcal{O}_{Z,p}$ injects into $(f_* \mathcal{O}_X)_p$. Q.E.D.

§2. Generic Determinantal Varieties: Geometric Description

Let $M(m, n) = M$ be the variety of $m \times n$ complex matrices, and, for $0 \leq k \leq \min(m, n)$, denote by $M_k(m, n) = M_k$ the locus of matrices of rank at most k. This is called a *generic determinantal variety*. We also set

$$\tilde{M}_k = \tilde{M}_k(m, n) = \{(A, W) \in M \times G(n - k, n) : A \cdot W = 0\}.$$

Projection onto $G(n - k, n)$ exhibits $\tilde{M}_k$ as an algebraic vector bundle over $G(n - k, n)$ of rank mk. Therefore $\tilde{M}_k$ is smooth, connected, and has dimension $k(m + n - k)$. Now let

$$\pi : M \times G(n - k, n) \to M$$

be the projection. Clearly π maps $\tilde{M}_k$ properly onto M_k, showing that M_k is an irreducible algebraic subvariety of M. Moreover, when $A \in M_k - M_{k-1}$, there is exactly one point of $\tilde{M}_k$ lying above A, namely the pair $(A, \ker A)$. This shows

Proposition. *M_k is an irreducible algebraic subvariety of M of codimension $(m - k)(n - k)$.*

The next step will be to compute the tangent space to $\tilde{M}_k$ at any point (A, W). We shall do this by "differentiating" the relation $A \cdot W = 0$. We begin by recalling the standard identification between the tangent space to

$G(n-k, n)$ at W and $\mathrm{Hom}(W, \mathbb{C}^n/W)$, which can be given as follows. A tangent vector to $G(n-k, n)$ at W is a morphism from Spec ($\mathbb{C}[\varepsilon]$) to $G(n-k, n)$ centered at W. Upon choosing a basis $w_1, \ldots, w_{n-k}$ for W, this can be lifted to a morphism

$$(w_1 + \varepsilon v_1, \ldots, w_{n-k} + \varepsilon v_{n-k})$$

into the variety of $(n-k)$-frames in $\mathbb{C}^n$. The v_i's are uniquely determined modulo W, and the homomorphism

$$\phi: W \to \mathbb{C}^n/W$$

corresponding to the given tangent vector is defined by

$$\phi(w_i) = \text{class of } v_i \text{ in } \mathbb{C}^n/W.$$

Now, a tangent vector to $\tilde{M}_k$ at (A, W) is a couple (B, ϕ), where B is an $m \times n$ matrix and ϕ is as above, such that

$$(A + \varepsilon B)(w_i + \varepsilon v_i) = 0, \qquad i = 1, \ldots, n-k.$$

This amounts to saying that $A \cdot W = 0$ and, moreover,

$$Av_i + Bw_i = 0, \qquad i = 1, \ldots, n-k,$$

that is,

$$(A\phi + B) \cdot W = 0.$$

Thus

$$\pi_*(T_{(A,W)}(\tilde{M})) = \{B \in M : B \cdot W \subset A \cdot \mathbb{C}^n\}, \tag{2.1}$$

which has dimension equal to

$$(n-k)(m - \mathrm{rank}(A)).$$

In particular, when $A \in M_k - M_{k-1}$, the differential of π at $(A, \ker A)$ is injective, showing that A is a smooth point of M_k and that

$$T_A(M_k) = \{B \in M : B \cdot \ker A \subset A \cdot \mathbb{C}^n\}.$$

In general, the set-theoretic fiber of

$$\pi: \tilde{M}_k \to M$$

over any $A \in M_k$ can be identified with $G(n-k, \ker A)$. The above dimension count shows that, if W is an $(n-k)$-plane in $\ker A$, then the differential of π maps the normal space at (A, W) to the set-theoretic fiber over A injectively into $T_A(M)$. Thus the scheme-theoretic fiber $\pi^{-1}(A)$ is smooth, and therefore the hypotheses of Lemma (1.1) are met by

$$\pi: \tilde{M}_k \to M$$

at any $A \in M_k$. Hence, by the corollary to Lemma (1.1) and formula (2.1), as a set, the tangent cone to M_k at A is

$$\mathscr{T}_A(M_k) = \{B \in M: \text{for some } W \in G(n-k, \ker A), B \cdot W \subset A \cdot \mathbb{C}^n\} \tag{2.2}$$

(actually we will see later that this is an equality of schemes). A first consequence is that, when $A \in M_{k-1}$, $\mathscr{T}_A(M_k)$ spans M as a vector space. In fact, given any non-zero vector $v \in \mathbb{C}^n$, and any vector $w \in \mathbb{C}^m$, there is a $B \in \mathscr{T}_A(M_k)$ such that $Bv = w$. That this is true follows from the remark that there is certainly an element $W \in G(n-k, \ker A)$ such that $v \notin W$; for such a W, the conditions

$$B \cdot W \subset A \cdot \mathbb{C}^n,$$
$$B \cdot v = w$$

can be simultaneously met. Recalling that, as we proved above, $M_k - M_{k-1}$ is smooth, we conclude

Proposition. *The singular locus of M_k is exactly equal to M_{k-1}.*

A priori, formula (2.2) only describes the tangent cones to M_k as sets. What we would like to remark here is that, as we announced above, these tangent cones are indeed reduced. We shall prove this by showing that the hypotheses of Lemma (1.3) are satisfied by

$$\pi: \tilde{M}_k \to M$$

at any $A \in M_k$. The point is that the support of $\mathscr{T}_A(M_k)$, namely

$$\{B \in M: \text{for some } W \in G(n-k, \ker A), B \cdot W \subset A \cdot \mathbb{C}^n\}$$

is nothing but the product of a linear space times a generic determinantal variety. To see this we choose complementary subspaces H_2 to $H_1 = \ker A$ and K_2 to $K_1 = A \cdot \mathbb{C}^n$, and notice that to give an $m \times n$ matrix B such that, for some $W \in G(n-k, \ker A)$, $B \cdot W \subset A \cdot \mathbb{C}^n$, is the same as giving arbitrary linear mappings

$$\alpha_1 : H_1 \to K_1,$$
$$\alpha_2 : H_2 \to K_2,$$
$$\alpha_3 : H_2 \to K_1,$$

plus a linear mapping

$$\alpha_4 : H_1 \to K_2$$

of rank at most $k - h$, where $h = \operatorname{rank}(A)$. Thus the support of $\mathscr{T}_A(M_k)$ can be identified with the product

$$V \times M_{k-h}(m-h, n-h),$$

where V is a linear space of dimension $h(m+n-h)$. What is more, the normal bundle to the fiber $\pi^{-1}(A) = G(n-k, \ker A)$, that is,

$$\{(B, W) : B \in M, W \in G(n-k, \ker A), B \cdot W \subset A \cdot \mathbb{C}^n\},$$

is nothing but the product

$$V \times \tilde{M}_{k-h}(m-h, n-h),$$

and hence maps birationally to the support of $\mathscr{T}_A(M_k)$. At this point, to finish with the proof that $\mathscr{T}_A(M_k)$ is reduced, it would be enough to show that *generic determinantal varieties are normal.* However, the elementary linear algebra that we have been using so far is not sufficient to prove this. What is needed is a rather more sophisticated analysis of the ideal of M_k; this will be carried out in the next section, where a proof of the normality of generic determinantal varieties will also be found.

§3. The Ideal of a Generic Determinantal Variety

The main purpose of the present section is to study the ideal of a generic determinantal variety. The first results to be proved are quite classical and go under the names of first and second fundamental theorem of invariant theory for the general linear group (cf. Weyl [2]). These will be proved by making a systematic use of standard bases as introduced by Hodge and others. In the same vein, we shall next prove the more recent result, due to Eagon–Hochster, asserting that generic determinantal varieties are Cohen–Macaulay.

Departing from tradition we shall prove the second fundamental theorem of invariant theory first. Using the notation of the previous section this can be stated as follows.

Second Fundamental Theorem of Invariant Theory. *The ideal of M_k in the affine coordinate ring of M is generated by the $(k+1) \times (k+1)$ minors.*

We shall denote the ideal generated by the $(k+1) \times (k+1)$ minors by the symbol I_{k+1}. From the very definition of M_k it is clear that the support of the variety defined by I_{k+1} is M_k. Therefore it suffices to prove the following.

Proposition. *For any k the ideal I_k is equal to its radical.*

The proof of this fact relies on a detailed analysis of the homogeneous coordinate ring of the Grassmannian $G(k, N)$ under the Plücker embedding. This ring will be denoted by the symbol R.

To set up, to any $N \times k$, rank k matrix A we associate the point of $G(k, N)$ corresponding to the vector subspace of $\mathbb{C}^N$ spanned by the columns of A. Given a matrix

$$\begin{pmatrix} x_{11} & \cdots & x_{1k} \\ \vdots & & \vdots \\ x_{N1} & \cdots & x_{Nk} \end{pmatrix}$$

of indeterminates, the Plücker coordinates for $G(k, N)$ are the $k \times k$ minors of this matrix. Given indices $i_1, \ldots, i_k$, with $1 \leq i_j \leq N$, for any j, we shall denote by

$$[i_1 \cdots i_k]$$

the determinant of the $k \times k$ matrix whose jth row is the i_jth row of $(X_{\lambda\mu})$. We also introduce the following notation:

$$S(a_1, \ldots, a_t) = \text{Symmetric group on the letters } a_1, \ldots, a_t.$$

The first basic results we need are the famous

Plücker Relations. *Let $s \leq k$, and set*

$$S = S(i_s, \ldots, i_k, j_1, \ldots, j_s),$$
$$S' = S(i_s, \ldots, i_k) \times S(j_1, \ldots, j_s) \subset S.$$

Then

$$\sum_{\bar{\sigma} \in S/S'} \varepsilon(\sigma)[i_1 \cdots i_{s-1}\sigma(i_s) \cdots \sigma(i_k)] \cdot [\sigma(j_1) \cdots \sigma(j_s) j_{s+1} \cdots j_k] = 0.$$

Proof. First observe that the sum is well defined since each summand depends only on the class $\bar{\sigma}$ of σ modulo S'. Writing

$$\begin{pmatrix} x_{11} & \cdots & x_{1k} \\ \vdots & & \vdots \\ x_{N1} & \cdots & x_{Nk} \end{pmatrix} = \begin{pmatrix} v_1 \\ \vdots \\ v_N \end{pmatrix},$$

the above sum may be viewed as an alternating $(k+1)$-form in $v_{i_s}, \ldots, v_{i_k}$, $v_{j_1}, \ldots, v_{j_s}$; since $\bigwedge^{k+1} \mathbb{C}^k = 0$, it must vanish identically. Q.E.D.

Using the Plücker relations we are now going to exhibit an explicit basis for R, the homogeneous coordinate ring of $G(k, N)$, as a complex vector space (actually the same computation works over the integers, too).

Given a monomial

$$[i_{11} \cdots i_{1k}] \cdots [i_{h1} \cdots i_{hk}] \in R,$$

we shall find it convenient to write it in matrix notation as

$$\begin{bmatrix} i_{11} & \cdots & i_{1k} \\ \vdots & & \vdots \\ i_{h1} & \cdots & i_{hk} \end{bmatrix}.$$

Such a monomial will be said to be *standard*[1] if the rows are increasing, i.e.,

$$i_{t1} < i_{t2} < \cdots < i_{tk} \quad \text{for any } t,$$

and the columns are non-decreasing, i.e.,

$$\begin{matrix} i_{1s} \\ \wedge\!\!\backslash \\ i_{2s} \\ \wedge\!\!\backslash \\ \vdots \\ \wedge\!\!\backslash \\ i_{hs} \end{matrix} \quad \text{for any } s.$$

Lemma. *The standard monomials generate R as a complex vector space.*

[1] The reader will notice the slight ambiguity in this terminology. To be precise we should have defined standard tableaux and called standard the monomials associated to such tableaux. For our modest purposes, however, such a subtle distinction would prove more of a hindrance than an advantage.

To prove this we begin by ordering monomials lexicographically. First of all we order Plücker coordinates as follows. We say that

$$[i_1 \cdots i_k] < [j_1 \cdots j_k]$$

if there exists s such that $1 \leq s \leq k$ and

$$i_1 = j_1, \ldots, i_{s-1} = j_{s-1}, \qquad i_s < j_s.$$

This is a total order on the Plücker coordinates, with minimal element $[1 \cdots k]$. We then order monomials lexicographically, i.e.,

$$\begin{bmatrix} i_{11} & \cdots & i_{1k} \\ \vdots & & \vdots \\ i_{h1} & \cdots & i_{hk} \end{bmatrix} < \begin{bmatrix} j_{11} & \cdots & j_{1k} \\ \vdots & & \vdots \\ j_{h1} & \cdots & j_{hk} \end{bmatrix}$$

if there is t such that $1 \leq t \leq h$ and

$$[i_{v1} \cdots i_{vk}] = [j_{v1} \cdots j_{vk}] \quad \text{if} \quad v < t,$$
$$[i_{t1} \cdots i_{tk}] < [j_{t1} \cdots j_{tk}].$$

To prove our lemma is clearly suffices to show that any non-standard monomial can be expressed as a linear combination of monomials which precede it in the lexicographical order. Indeed it suffices to check this for a non-standard monomial of degree two

$$\begin{bmatrix} i_1 & \cdots & i_k \\ j_1 & \cdots & j_k \end{bmatrix}.$$

Clearly, up to a change of sign, we may assume that the rows are increasing. Now, to say that this monomial is non-standard means that there is an s such that

$$j_1 < \cdots < j_s < i_s < \cdots < i_k.$$

But now the Plücker relation

$$\sum \varepsilon(\sigma) \begin{bmatrix} i_1 \cdots i_{s-1}\sigma(i_s) \cdots \sigma(i_k) \\ \sigma(j_1) \cdots \sigma(j_s) j_{s+1} \cdots j_k \end{bmatrix} = 0$$

expresses our monomial as a linear combination of "smaller" ones. Our second basic step is

Lemma. *The standard monomials give a basis of R as a complex vector space.*

We shall in fact prove a more general result. Consider the flag

$$F_1 \subset F_2 \subset \cdots \subset F_N = \mathbb{C}^N,$$

where F_i is spanned by the first i vectors in the canonical basis of $\mathbb{C}^N$. Consider also the Schubert varieties

$$V_{[i_1 \cdots i_k]} = \{W \in G(k, N) \colon \dim W \cap F_{i_s} \geq s, 1 \leq s \leq k\}.$$

Clearly, $V_{[j_1 \cdots j_k]}$ is contained in $V_{[i_1 \cdots i_k]}$ if and only if $i_1 \geq j_1, \ldots, i_k \geq j_k$. Moreover,

$$V_{[1 \cdots k]} = \{F_k\},$$

$$V_{[N-k+1 \cdots N]} = G(k, N).$$

We shall denote by $R_{[i_1 \cdots i_k]}$ the homogeneous coordinate ring of $V_{[i_1 \cdots i_k]}$. We will prove our lemma by proving more generally the following:

Lemma. *Those standard monomials T such that $T \cdot [i_1 \cdots i_k]$ is not standard form a basis for the ideal of $V_{[i_1 \cdots i_k]}$. Thus the restrictions to $V_{[i_1 \cdots i_k]}$ of those standard monomials T such that $T \cdot [i_1 \cdots i_k]$ is standard form a basis for $R_{[i_1 \cdots i_k]}$.*

Proof. A simple computation shows that if

$$\begin{bmatrix} s_1 & \cdots & s_k \\ i_1 & \cdots & i_k \end{bmatrix}$$

is non-standard then $[s_1 \cdots s_k]$ vanishes identically on $V_{[i_1 \cdots i_k]}$. Therefore our lemma will be proved if we can show that the restrictions to $V_{[i_1 \cdots i_k]}$ of those standard monomials T such that $T \cdot [i_1 \cdots i_k]$ is standard are linearly independent. We proceed by induction. The statement is trivial for $V_{[1 \cdots k]}$. Assume it is true for all Schubert varieties contained in $V_{[i_1 \cdots i_k]}$ and suppose we have a relation

$$\sum a_h T_h = 0 \quad \text{on} \quad V_{[i_1 \cdots i_k]}, \qquad a_h \neq 0.$$

A simple computation shows that $[i_1 \cdots i_k]$ does not vanish on the irreducible variety $V_{[i_1 \cdots i_k]}$. We may therefore assume that one of the monomials T_h, say T_1, is of the form

$$T_1 = \begin{bmatrix} i_{11} & \cdots & i_{1k} \\ \vdots & & \vdots \\ i_{r1} & \cdots & i_{rk} \end{bmatrix},$$

with

$$[i_{r1} \cdots i_{rk}] \neq [i_1 \cdots i_k].$$

By the induction hypothesis T_1 does not vanish on $V_{[i_{r1} \cdots i_{rk}]}$. On the other hand, we know that T_h vanishes on $V_{[i_{r1} \cdots i_{rk}]}$ if $T_h \cdot [i_{r1} \cdots i_{rk}]$ is not standard. Thus

$$\sum a_{h_i} T_{h_i} = 0 \quad \text{on} \quad V_{[i_{r1} \cdots i_{rk}]},$$

where the T_{h_i} are those among the T_h such that $T_h \cdot [i_{r1} \cdots i_{rk}]$ is standard. Since T_1 is one of these monomials, the above is a non-empty relation. This contradicts the induction hypothesis. Q.E.D.

We are now in a position to prove the proposition and therefore our theorem. To each matrix $(a_{ij}) \in M$ we may associate the element of $G(n, n + m)$ spanned by the columns of

$$\begin{pmatrix} a_{11} & \cdots & a_{1n} \\ \vdots & & \vdots \\ a_{m1} & \cdots & a_{mn} \\ 0 & \cdots & 1 \\ \vdots & & \vdots \\ 1 & \cdots & 0 \end{pmatrix}.$$

This yields on isomorphism between M and the open subset of $G(n, n + m)$ where $[m + 1 \cdots n + m]$ does not vanish. The affine coordinate ring $\mathbb{C}[x_{ij}]$ of M is therefore isomorphic to a quotient of R, namely

$$\mathbb{C}[x_{ij}] = R/([m + 1 \cdots m + n] \pm 1).$$

Thus a basis for $\mathbb{C}[x_{ij}]$ is given by all the standard monomials

$$\begin{bmatrix} i_{11} & \cdots & i_{1n} \\ \vdots & & \vdots \\ i_{h1} & \cdots & i_{hn} \end{bmatrix}.$$

such that $[i_{h1} \cdots i_{hn}] \neq [m + 1 \cdots m + n]$. On the other hand, any $s \times s$ minor of (x_{ij}) is a Plücker coordinate. More precisely, let us consider the minor constructed out of rows $i_1, \ldots, i_s$ and columns $r_1 \cdots r_s$. It is straightforward to see that this minor is nothing but the Plücker coordinate

$$[i_1 \cdots i_s i_{s+1} \cdots i_n],$$

where $\{i_{s+1}, \ldots, i_n\}$ is the complement in $\{m + 1, \ldots, m + n\}$ of

$$\{n + m - r_s + 1, \ldots, n + m - r_1 + 1\}.$$

Therefore a basis of the determinantal ideal I_k is given by all standard monomials

$$\begin{bmatrix} i_{11} & \cdots & i_{1n} \\ \vdots & & \vdots \\ i_{t1} & \cdots & i_{tn} \end{bmatrix},$$

such that $i_{1k} \leq m$ and $[i_{t1} \cdots i_{tm}] \neq [m + 1 \cdots m + n]$. Reciprocally, a basis for the quotient ring $D_k = \mathbb{C}[x_{ij}]/I_k$ consists of all those standard monomials

$$\begin{bmatrix} i_{11} & \cdots & i_{1n} \\ \vdots & & \vdots \\ i_{t1} & \cdots & i_{tn} \end{bmatrix},$$

such that $i_{1k} > m$ and $[i_{t1} \cdots i_{tn}] \neq [m + 1 \cdots m + n]$.

To conclude the proof of the proposition suppose

$$(\sum a_h T_h)^r \in I_k,$$

where the T_h are standard monomials not belonging to I_k. We may and will assume that T_1 is the maximum of the T_h in the lexicographical order. But then

$$(\sum a_h T_h)^r = a_1^r T_1^r + \Sigma,$$

where Σ stands for a linear combination of monomials which precede T_1^r in the lexicographical order. Using the Plücker relations we can rewrite Σ as a linear combination of standard monomials which also precede T_1^r. Since T_1^r is a standard monomial which does not belong to I_k, the only way out is that $a_1 = 0$. Repeating this argument, one shows that all the a_h must be zero. This concludes the proof of the proposition and hence of the theorem.

We now turn to the first fundamental theorem of invariant theory. The general linear group $G = \mathrm{GL}(k, \mathbb{C})$ acts on

$$M(m, k) \times M(k, n)$$

by

$$g(A, B) = (Ag^{-1}, gB).$$

Also, matrix multiplication yields a map

$$M(m, k) \times M(k, n) \to M(m, n).$$

Clearly, the image of this map is $M_k(m, n)$. Let S be the coordinate ring of $M(m, k) \times M(k, n)$ and $D = D_{k+1}$ the coordinate ring of $M_k(m, n)$. We want to prove the

First Fundamental Theorem of Invariant Theory. *The ring D is the ring of invariants of S under the action of G. In symbols*

$$D = S^G.$$

Corollary. *The variety M_k is normal.*

To prove the corollary observe that since $M(m, k) \times M(k, n)$ is smooth, it is enough to show that D is integrally closed in S. Suppose $a \in S$ satisfies a relation of integral dependence over D; then all of its translates by the action of G satisfy the same relation. Hence there is a finite number of these. But G is connected so that a must be G-invariant.

We now proceed to prove the first fundamental theorem. First of all notice that, if the theorem holds for a given choice of m and n, it also holds for any smaller values of m and n. We may therefore assume that m and n are large, in particular, larger than k. We shall often write pairs of matrices (A, B), of size $m \times k$ and $k \times n$, respectively, in block form as follows:

$$(A, B) = \left(\begin{pmatrix} A' \\ A'' \end{pmatrix}, (B', B'')\right),$$

where A' and B' are $k \times k$ matrices, and we set

$$d(A, B) = \det(A'B').$$

We need the following:

Lemma. $$D[1/d] = S[1/d]^G.$$

Proof. One inclusion is trivial. We denote by W the open subset of $M(m, k) \times M(k, n)$ where d does not vanish. We also denote by V the linear space of those pairs (A, B) such that A' is the identity matrix. The group G acts on $G \times V$ by left multiplication on the first factor. We then have a G-invariant isomorphism

$$G \times V \to W,$$

$$(g, u) \to gu,$$

an explicit inverse of which is given by

$$\left(\begin{pmatrix} A' \\ A'' \end{pmatrix}, (B', B'')\right) \to \left(\begin{pmatrix} I \\ A''A'^{-1} \end{pmatrix}, (A'B', A'B'')\right).$$

Clearly the ring of invariants of $G \times V$ is just the coordinate ring of V. Hence any G-invariant regular function f on W, i.e., any element of $S[1/d]^G$, is a regular function of $A''A'^{-1}$, $A'B'$, $A'B''$. But

$$A''A'^{-1} = \frac{1}{d} A''B',$$

proving that $f \in D[1/d]$. Q.E.D.

To conclude the proof of the first fundamental theorem we consider a G-invariant element $a \in S$. By the lemma we just proved that a belongs to $D[1/d]$, and therefore there is an integer k such that $d^k a \in D$. Thus it suffices to show that, whenever $f \in S$ and $df \in D$, then $f \in D$. We will do this by using the explicit basis for D constructed in the proof of the second fundamental theorem. Recall that we have identified $M(m, n)$ with an open subset of $G(n, m + n)$ by associating to a matrix (a_{ij}) the n-plane spanned by the columns of

$$\begin{pmatrix} a_{11} & \cdots & a_{1n} \\ \vdots & & \vdots \\ a_{m1} & & a_{mn} \\ 0 & \cdots & 1 \\ \vdots & & \vdots \\ 1 & \cdots & 0 \end{pmatrix}.$$

We then have

$$d = [1 \cdots k \quad m + 1 \cdots m + n - k].$$

Recalling the standard basis we constructed for D in the course of the proof of the second fundamental theorem, let us now write

$$df = \sum \alpha_i T_i,$$

where each T_i is a standard monomial of type

$$\begin{bmatrix} i_{11} & \cdots & i_{1n} \\ \vdots & & \vdots \\ i_{t1} & \cdots & i_{tn} \end{bmatrix},$$

such that $i_{1,k+1} > m$ and $[i_{t1} \cdots i_{tn}] \neq [m+1 \cdots m+n]$. In particular, the regular function $\sum \alpha_i T_i$ vanishes on the hypersurface $d = 0$ and therefore there is a positive integer h such that

$$(\sum \alpha_i T_i)^h = d\gamma,$$

where $\gamma \in D$. Now write γ as a linear combination of standard monomials

$$\gamma = \sum \beta_j S_j.$$

We may assume that T_1 is the maximum among the T_i's in the lexicographical order. We may then write, as in final step of the proof of the second fundamental theorem,

$$\begin{aligned}(\sum \alpha_i T_i)^h &= \alpha_1^h T_1^h + \Sigma \\ &= \sum \beta_j \, dS_j,\end{aligned}$$

where Σ stands for a linear combination of standard monomials which precede the standard monomial T^h in the lexicographical order. Since dS_i is a standard monomial for any i, we conclude that T_1^h, and hence T_1, is divisible by d. Repeating this argument shows that all the T_i's are actually divisible by d. Therefore

$$f = \sum \alpha_i (T_i/d)$$

belongs to D. This concludes the proof of the first fundamental theorem of invariant theory.

The last results of this section are due to Eagon–Hochster and to Hochster, Laksov, and Musili. Our proof will closely follow Musili's.

(3.1) Theorem. *M_k is Cohen–Macaulay.*

As we have already done, we identify a matrix $(a_{ij}) \in M$ with the point of $G(n, n+m)$ spanned by the columns of the matrix

$$\begin{pmatrix} a_{11} & & \cdots & a_{1n} \\ \vdots & & & \vdots \\ a_{m1} & & \cdots & a_{mn} \\ 0 & & \cdots & 1 \\ \vdots & & \cdot^{\cdot^{\cdot}} & \vdots \\ 0 & 1 & & \\ 1 & 0 & \cdots & 0 \end{pmatrix}.$$

In this fashion M is identified with one of the standard coordinate patches of $G(n, n+m)$. Let now F be the span of the last n vectors in the canonical basis of $\mathbb{C}^{n+m}$. The determinantal variety M_k is described inside M by the Schubert condition

$$M_k = \{\Lambda \in M : \dim(\Lambda \cap F) \geq n - k\},$$

so that M_k can be thought of as an open subset of a Schubert variety. Thus our theorem follows from the following more general one.

(3.2) Theorem. *The cone over a Schubert variety is Cohen–Macaulay. In fact, if V is a union of Schubert varieties of codimension one in a fixed Schubert variety, the cone over V is Cohen–Macaulay.*

This theorem will be proved by a double induction on the dimension and number of components of V. We fix the flag

$$F_1 \subset F_2 \subset \cdots \subset F_N = \mathbb{C}^N,$$

where F_i is spanned by the first i vectors in the canonical basis of $\mathbb{C}^N$. We denote by $V_{[i_1 \cdots i_k]}$ the Schubert variety of those points W in $G(k, N)$ such that

$$\dim(W \cap F_{i_s}) \geq s, \qquad s = 1, \ldots, k,$$

and by $R_{[i_1 \cdots i_k]}$ the homogeneous coordinate ring of $V_{[i_1 \cdots i_k]}$, as we have already done in the proof of the second fundamental theorem of invariant theory. As we observed there, the ideal of $V_{[i_1 \cdots i_k]}$ is generated (as an ideal) by those Plücker coordinates $[j_1 \cdots j_k]$ such that

$$\begin{bmatrix} j_1 & \cdots & j_k \\ i_1 & \cdots & i_k \end{bmatrix}$$

is not standard, that is, such that $i_l < j_l$ for some l.

We begin our inductive proof of the theorem by noticing that its conclusion is trivially true for $V_{[1 \cdots k]}$, which consists of a single point. We next fix a Plücker coordinate

$$\tau = [i_1 \cdots i_k],$$

and set

$$\tau_s = [i_1 \cdots i_{s-1} \quad i_s - 1 \quad i_{s+1} \cdots i_k],$$

$$\tau_{st} = [i_1 \cdots i_{s-1} \quad i_s - 1 \quad i_{s+1} \cdots i_{t-1} \quad i_t - 1 \quad i_{t+1} \cdots i_k].$$

Clearly the Schubert varieties of codimension one in V_τ are exactly the varieties V_{τ_s}. Since the ideal of each V_{τ_s} is generated by standard monomials, the ideal of $\bigcup_{s=1}^k V_{\tau_s}$ is also generated by standard monomials, and hence is generated (as an ideal) by Plücker coordinates. The coordinate $[j_1 \cdots j_k]$ belongs to the ideal of V_{τ_s} if and only if it belongs to the ideal of V_τ or else $j_s = i_s$. Thus $[j_1 \cdots j_k]$ belongs to the ideal of $\bigcup_{s=1}^k V_{\tau_s}$ if and only if it belongs to the ideal of V_τ, or else it equals τ; in particular, $R_\tau/(\tau)$ is the homogeneous coordinate ring of $\bigcup_{s=1}^k V_{\tau_s}$. We also notice that

$$V_{\tau_s} \cap V_{\tau_t} = V_{\tau_{st}}.$$

We would like to prove that the cone over any union of Schubert varieties of codimension one in V_τ is Cohen–Macaulay, assuming the analogous statement to be already proved for all Schubert varieties of dimension lower than that of V_τ. We begin with a single Schubert variety V_{τ_s}. By one of our previous remarks, the ring $R_{\tau_s}/(\tau_s)$ is the homogeneous coordinate ring of the union of all Schubert varieties contained in V_{τ_s}, hence it is Cohen–Macaulay by induction hypothesis. Since τ_s is not a zero divisor, R_{τ_s} is also Cohen–Macaulay. We next consider

$$V = V_{\tau_{s1}} \cup \cdots \cup V_{\tau_{sh}}.$$

We may assume, inductively, that the cone over

$$V' = V_{\tau_{s1}} \cup \cdots \cup V_{\tau_{sh-1}}$$

is Cohen–Macaulay. Notice that

$$V' \cap V_{\tau_{sh}} = \bigcup_{i=1}^{h-1} V_{\tau_{sish}}$$

is a union of Schubert varieties of codimension one in $V_{\tau_{sh}}$, and hence the cone over it is Cohen–Macaulay by induction hypothesis. Denoting by R', R'', R''' the homogeneous coordinate rings of V, V', $V \cap V_{\tau_{sh}}$, respectively, we have an exact sequence of R-modules

$$0 \to R' \to R'' \oplus R_{\tau_{sh}} \to R''' \to 0.$$

Since the image of τ_{s_h} in $R'' \oplus R_{\tau_{sh}}$ is not a zero divisor, part (c) of the following elementary result implies that R' is also Cohen–Macaulay, finishing the proof of our theorem.

Lemma. *Let*

$$0 \to L \xrightarrow{\beta} M \xrightarrow{\alpha} N \to 0$$

be an exact sequence of modules over a Noetherian ring A. Let $f_0, \ldots, f_d$ be elements of A. Then:

(a) *If $f_0, \ldots, f_d$ is a regular sequence for M and N, it is a regular sequence for L.*
(b) *If $f_0, \ldots, f_d$ is a regular sequence for L and N, it is a regular sequence for M.*
(c) *If $f_0, \ldots, f_d$ is a regular sequence for M, $f_1, \ldots, f_d$ is a regular sequence for N and $f_0 N = \{0\}$, then $f_0, \ldots, f_d$ is a regular sequence for L.*

Proof. When $d = 0$, (a) is obvious; to prove (b), suppose $f_0 x = 0$, for some x in M. Then $\alpha(x) = 0$ by hypothesis, therefore $x \in L$ and, again by our assumption, $x = 0$. Now, to prove (a) and (b) by induction on d, it suffices to show that

$$0 \to L/f_0 L \xrightarrow{\beta'} M/f_0 M \xrightarrow{\alpha'} N/f_0 N \to 0$$

is exact, that is, that β' is injective. To do this, suppose x is an element of L such that $\beta(x)$ belongs to $f_0 M$. Thus there is a y in M such that $\beta(x) = f_0 y$. Therefore $f_0 \alpha(y) = 0$, $\alpha(y) = 0$, y belongs to L, and x belongs to $f_0 L$.

Part (c) of the lemma now follows from parts (a) and (b). Clearly

$$\mathrm{Tor}_A^1(M, A/f_0 A) = 0; \qquad N/f_0 N \cong N.$$

Moreover, from the long exact sequence of Tor's for N and

$$0 \to A/J \xrightarrow{\times f_0} A \to A/f_0 A \to 0,$$

where J is the annihilator of f_0, we deduce that

$$\mathrm{Tor}_A^1(N, A/f_0 A) \cong N/JN = N.$$

In fact, since f_0 is not a zero divisor in M, $JM = 0$, and hence $JN = 0$. Thus we deduce from our original exact sequence another exact sequence

$$0 \to N \to L/f_0 L \to M/f_0 M \xrightarrow{\gamma} N \to 0.$$

Denoting by K the kernel of γ, we conclude, by (a), that $f_1, \ldots, f_d$ is a regular sequence for K and, by (b), that it is regular for $L/f_0 L$ as well. Since f_0 is not a zero divisor in L, this concludes the proof of the lemma.

§4. Determinantal Varieties and Porteous' Formula

Generic determinantal varieties are the prototypes of what one calls a determinantal variety. We proceed with a formal definition. Let

$$\phi: E \to F$$

be a homomorphism between holomorphic vector bundle of ranks n, m over an analytic space X. Upon choosing local trivializations of E and F over an open set U, the homomorphism ϕ is represented by an $m \times n$ matrix A of holomorphic functions. This corresponds to a morphism

$$f: U \to M = M(m, n).$$

We denote by U_k the preimage of $M_k = M_k(m, n)$. In other terms, the ideal of U_k is generated by the $(k + 1) \times (k + 1)$ minors of A. It is immediate to see that U_k does not depend on the choice of trivialization, and therefore there is a well-defined analytic subspace

$$X_k(\phi) \subset X$$

such that $X_k(\phi) \cap U = U_k$ for any U. The variety $X_k(\phi)$ is what we call the kth *determinantal variety* or kth *degeneracy locus* associated to ϕ: it is supported on the set

$$\{p \in X: \operatorname{rank}(\phi_p) \leq k\}.$$

It is clear from the definition that $X_k(\phi)$, when non-empty, has codimension at most $(m - k)(n - k)$.

The generic determinantal variety M_k is nothing but the kth determinantal variety associated to the morphism

$$\phi: \mathcal{O}_M^n \to \mathcal{O}_M^m,$$

where

$$\phi_A(v) = A \cdot v.$$

It is convenient to consider, beside the variety $X_k(\phi)$, the analogue $\tilde{X}_k(\phi)$ of the desingularization $\tilde{M}_k$ of M_k defined in the previous section. This can be intrinsically defined as follows. We let

$$\pi: G(n - k, E) \to X$$

be the Grassmann bundle of $(n - k)$-planes in the fibers of E. We also denote by S and Q, respectively, the universal subbundle and quotient bundle on

$G(n-k, E)$. Thus if x is a point of X and W is an $(n-k)$-plane in E_x, i.e., a point of $\pi^{-1}(x)$, the fibers of S and Q over W are W itself and E_x/W, respectively. The bundles S and Q sit in an exact sequence

$$0 \to S \to \pi^*E \to Q \to 0.$$

Composing the lifted homomorphism

$$\pi^*(\phi): \pi^*E \to \pi^*F,$$

with the inclusion of S in π^*E gives a homomorphism

$$\tilde{\phi}: S \to \pi^*F.$$

We shall denote by $\tilde{X}_k(\phi)$ the subvariety of $G(n-k, E)$ defined by the vanishing of $\tilde{\phi}$. Thus the support of $\tilde{X}_k(\phi)$ is simply the set of all couples (x, W), where x is a point of X_k and W is an $(n-k)$-plane contained in the kernel of ϕ_x. It is a simple exercise in words to check that, if U and f are as in the definition of $X_k(\phi)$, then $\tilde{X}_k(\phi) \cap \pi^{-1}(U)$ is nothing but the fiber product $\tilde{M}_k \times_M U$. In what follows we shall often write $X_k, \tilde{X}_k$ instead of $X_k(\phi), \tilde{X}_k(\phi)$, when no confusion is likely.

The theorem stating that generic determinantal varieties are Cohen–Macaulay (Theorem (3.1)) implies that the varieties X_k are also Cohen–Macaulay, provided they have the "correct" codimension. More precisely, we have:

(4.1) Proposition. *Let X be a complex manifold and let*

$$\phi: E \to F$$

be a homomorphism of holomorphic vector bundles of ranks n and m, respectively. If $X_k(\phi)$ has codimension $(m-k)(n-k)$, it is Cohen–Macaulay.

It follows, in particular, that if the hypotheses of the above proposition are satisfied, then $X_k(\phi)$ has no embedded components. The proposition is an immediate consequence of Theorem (3.1) plus the following simple algebraic statement.

Lemma. *Let $f: X \to Y$ be a morphism between complex manifolds. Suppose Z is a Cohen–Macaulay analytic subspace of Y of pure codimension k. Then, if $f^{-1}(Z)$ has pure codimension k, it is Cohen–Macaulay.*

Proof. Let $\Gamma \subset X \times Y$ be the graph of f. Projection onto X induces an isomorphism between Γ and X, and $f^{-1}(Z)$ is just the projection of the subspace Z' of Γ cut out by $X \times Z$. On the other hand, since X is smooth, $X \times Z$

is Cohen–Macaulay, while Γ is locally a complete intersection, by the smoothness of Y. In fact, if $x_1, \ldots, x_n$ are local coordinates for X and $y_1, \ldots, y_m$ local coordinates for Y, the equations of Γ are, locally,

$$y_i = f_i(x_1, \ldots, x_n), \qquad i = 1, \ldots, m,$$

where $(f_1, \ldots, f_m)$ is a local coordinate expression of f. Now our hypothesis on the codimension of $f^{-1}(Z)$ implies that

$$\begin{aligned}\operatorname{codim}(Z', X \times Z) &= \dim X + \dim Y - k - \dim(Z') \\ &= n + m - k - (n - k) \\ &= m.\end{aligned}$$

Thus Z' is of pure codimension m and locally defined by m equations in the Cohen–Macaulay space $X \times Z$. This implies that Z', and hence $f^{-1}(Z)$, are Cohen–Macaulay as well (cf. Matsumura [1, pp. 107–108]). Q.E.D.

The main purpose of this section is to prove a general formula, due to Porteous [1] and Kempf–Laksov [1] that expresses the fundamental classes of the determinantal loci of any holomorphic bundle mapping $E \to F$ as a polynomial in the Chern classes of the bundles E and F. To explain this we first recall a few notations from the theory of Chern classes. In our discussion we shall use integral cohomology throughout; however, it is important to notice that, in an algebraic setting, Porteous' formula and its proof go through verbatim if one uses, instead, Chern classes with values in the Chow ring of algebraic cycles modulo rational equivalence.

For any complex vector bundle F over a space Y,

$$c_t(F) = 1 + c_1(F)t + c_2(F)t^2 + \cdots$$

is the Chern polynomial of F. By the Whitney product formula, c_t extends to a homomorphism from the Grothendieck group $K(Y)$ to the multiplicative group of the invertible elements in the power series ring $H^*(Y)[[t]]$. We shall sometimes improperly write $-F$ to denote the negative of the *class* of F in $K(Y)$. Then we have

$$c_t(-F) = \frac{1}{c_t(F)} = 1 - c_1(F)t + (c_1^2(F) - c_2(F))t^2 + \cdots.$$

Similarly, we write $E - F$ for the difference of the classes of E and F in $K(Y)$.

We are now in the position to explain Porteous' formula. Let E and F be holomorphic vector bundles of respective ranks n and m over a complex manifold X. If

$$\phi\colon E \to F$$

is any holomorphic bundle mapping, we denote by x_k the fundamental class of $X_k(\phi)$. Porteous' formula expresses x_k in terms of the Chern classes of E and F. To state it, we introduce a piece of notation. For any formal series

$$a(t) = \sum_{k=-\infty}^{+\infty} a_k t^k,$$

we set

$$\Delta_{p,q}(a) = \det\begin{pmatrix} a_p & \cdots & a_{p+q-1} \\ \vdots & & \vdots \\ a_{p-q+1} & \cdots & a_p \end{pmatrix}.$$

We then have

(4.2) Porteous' Formula. *Under the assumption that $X_k(\phi)$ is empty or has the expected dimension* $\dim X - (n-k)(m-k)$, *we have*

$$\begin{aligned} x_k &= \Delta_{m-k,n-k}(c_t(F-E)) \\ &= (-1)^{(m-k)(n-k)}\Delta_{n-k,m-k}(c_t(E-F)). \end{aligned}$$

As we already announced, when X is a smooth quasi-projective variety, and ϕ is a homomorphism of algebraic vector bundles, Porteous' formula remains valid *in the Chow ring of X*. We also remark that, by an argument using universal bundles, it is not difficult to establish the existence of a formula expressing x_k as a polynomial $P(c(E), c(F))$ in the Chern classes of E and F; what requires effort is determining this polynomial. A hint is provided by Giambelli's formula from Schubert calculus (cf. page 205 of Griffiths–Harris [1]) which gives Porteous' formula in case F is trivial. We will concern ourselves first with finding the polynomial P and then, in the final discussion, consider what exactly has been proved.

From this point of view, the difficulty in establishing Porteous' formula is primarily technical, and readily overcome. The Chern classes of a vector bundle $E \to X$ describe the loci where one or more sections of X will be linearly dependent, and in some cases this is sufficient to solve our problem: i.e., the locus X_0 is the zero locus of ϕ, considered as a section of $\mathrm{Hom}(E, F)$, so that

$$x_0 = c_{mn}(E^* \otimes F);$$

and, in case $m = n$, we have that X_{m-1} is the zero locus of $\bigwedge^m \phi$, so that

$$\begin{aligned} x_{m-1} &= c_1\left(\textstyle\bigwedge^m E^* \otimes \bigwedge^m F\right) \\ &= c_1(F) - c_1(E). \end{aligned}$$

The problem is that, in general, X_k is not so readily expressed as such a locus—or, rather, since we can always write

$$X_k = \{\bigwedge^{k+1}\phi = 0\},$$

that, as a section of $\bigwedge^{k+1} E^* \otimes \bigwedge^{k+1} F$, $\bigwedge^{k+1} \phi$ vanishes in the wrong codimension.

The solution to this problem is straightforward and elegant; we must ask not only for the fundamental class of X_k, but also for the class $\tilde{x}$ of $\tilde{X}_k$ in $G(n-k, E)$. Keeping the notations introduced at the beginning of this section, recall that $\tilde{X}_k$ is just the zero locus of the bundle homomorphism $\tilde{\phi}$, which we may view as a section of $\operatorname{Hom}(S, \pi^*F)$. Accordingly

$$\tilde{x} = c_{m(n-k)}(S^* \otimes \pi^*F),$$

and we may compute the class of X_k as

$$x_k = \pi_*(\tilde{x}),$$

where

$$\pi_*\colon H^*(G(n-k, E), \mathbb{Z}) \to H^*(X, \mathbb{Z})$$

is the Gysin homomorphism.

Our work is now clearly cut out for us: in the following two discussions, we derive a formula for the top Chern class of the tensor product of two vector bundles; and in the third we apply this formula to the bundles S^* and π^*F on $G(n-k, E)$, and evaluate the image of the class we get under the Gysin map π_* to arrive at Porteous' formula.

(i) Sylvester's Determinant

We recall here a classical formula for the resultant of two polynomials. If k is a field, $p(t) = \sum_{i=0}^{n} p_i t^i$ and $q(t) = \sum_{i=0}^{m} q_i t^i$ are two polynomials with coefficients in k, then p and q have a common root if and only if the $(m+n) \times (m+n)$ determinant

$$R(p, q) = \begin{vmatrix} p_0 & p_1 & \cdots & p_n & 0 & \cdots & 0 \\ 0 & p_0 & p_1 & \cdots & p_n & 0 \cdots & 0 \\ \vdots & & & & & & \vdots \\ 0 & \cdots & 0 & p_0 & & \cdots & p_n \\ q_0 & \cdots & q_m & 0 & & \cdots & 0 \\ 0 & q_0 & \cdots & q_m & 0 & \cdots & 0 \\ \vdots & & & & & & \vdots \\ 0 & & \cdots & & 0 & q_0 \cdots & q_m \end{vmatrix}$$

vanishes. The proof is clear: recognizing the rows of this determinant as the coefficients of the polynomials $p, tp, \ldots, t^{m-1}p, q, tq, \ldots, t^{n-1}q$, we see that if p and q have a common root, then these polynomials do not span the $(m+n)$-dimensional vector space of polynomials of degree not exceeding $m+n-1$, and hence must be linearly dependent; while a linear relation among the rows says that there exist polynomials $a(t)$, $b(t)$ of degrees $m-1$ and $n-1$ such that

$$a(t)p(t) + b(t)q(t) = 0;$$

since $a(t)$ can vanish at at most $m-1$ of the roots of q, $p(t)$ must vanish at the remaining one(s). Alternatively, we can argue that the first m and last n rows of this matrix generate the subspace of $\mathbb{P}^{m+n-1}$ spanned by the roots of p and q, respectively, on a rational normal curve C; since any $m+n$ points on such as curve are independent, two such secant planes can fail to span $\mathbb{P}^{m+n-1}$ only if they have a point of C in common—that is, if p and q have a common root.

We note that, if $p_0 = 1$, the above determinant can be reduced somewhat by multiplying on the right by the unipotent matrix

$$\begin{pmatrix} s_0 & s_1 & s_2 & \cdots & s_{m+n-1} \\ 0 & s_0 & s_1 & \cdots & s_{m+n-2} \\ 0 & 0 & s_0 & \cdots & \\ \vdots & & & \ddots & \vdots \\ & & & & s_1 \\ 0 & 0 & & \cdots & s_0 \end{pmatrix},$$

where $\sum s_i t^i = 1/p(t)$; we obtain that $R(p, q)$ equals

$$\begin{vmatrix} 1 & 0 & & & & \cdots & 0 \\ 0 & 1 & & & & & \vdots \\ \vdots & & & & & & \vdots \\ 0 & \cdots & 0 & 1 & 0 & \cdots & 0 \\ r_0 & r_1 & & & & & r_{m+n-1} \\ 0 & r_0 & & & & & r_{m+n-2} \\ \vdots & & & & & & \vdots \\ 0 & \cdots & & r_0 & & \cdots & r_m \end{vmatrix}$$

$$= \begin{vmatrix} r_m & \cdots & r_{m+n-1} \\ \vdots & & \vdots \\ r_{m-n+1} & \cdots & r_m \end{vmatrix}$$

$$= \Delta_{m,n}(q(t)/p(t)),$$

where $\sum r_i t^i = q(t)/p(t)$.

We can also exchange the first m and the last n rows of our determinant to find

$$\Delta_{m,n}(q(t)/p(t)) = (-1)^{mn}\Delta_{n,m}(p(t)/q(t));$$

the analogous operations in general yield the formula

$$\Delta_{m,n}(f(t)) = (-1)^{mn}\Delta_{n,m}\left(\frac{1}{f(t)}\right)$$

for any monic power series $f(t)$.

We note also that the polynomials $\Delta_{m,n}$ provide a criterion for the rationality of a power series: $f(t) = \sum_{i=0}^{\infty} a_i t^i$ is the power series of a rational function if and only if

$$\Delta_{m,n}(f) = 0$$

for all sufficiently large m and n.

(ii) The Top Chern Class of a Tensor Product

We can rephrase the above discussion in terms of symmetric functions, as follows. If $\alpha_1, \ldots, \alpha_n$ and $\beta_1, \ldots, \beta_m$ are indeterminates, $p_1, \ldots, p_n$, and $q_1, \ldots, q_m$ their elementary symmetric functions, so that

$$p(t) = 1 + \sum_{i=1}^{n} p_i t^i = \prod (1 + \alpha_i t),$$

$$q(t) = 1 + \sum_{i=1}^{m} q_i t^i = \prod (1 + \beta_i t),$$

then the identity

$$\prod_{i,j} (\beta_j - \alpha_i) = \Delta_{m,n}(q(t)/p(t))$$
$$= (-1)^{mn}\Delta_{n,m}(p(t)/q(t))$$

holds. In fact, we already have

$$\prod_{i,j} (\beta_j - \alpha_i) = A(p_1, \ldots, p_n, q_1, \ldots, q_m)$$

for some polynomial A of weighted degree mn; since $\Delta_{m,n}(q(t)/p(t))$ is such a polynomial vanishing whenever A does, it follows that

$$\Delta_{m,n}(q(t)/p(t)) = \lambda A$$

for some scalar λ, which may be seen to be 1 by setting $\alpha_1 = \cdots = \alpha_n = 0$.

This statement in turn gives us directly the

Lemma. *If E and F are complex vector bundles on a space X, of ranks n and m respectively, then*

$$\begin{aligned} c_{mn}(E^* \otimes F) &= \Delta_{m,n}(c_t(F)/c_t(E)) \\ &= (-1)^{mn}\Delta_{n,m}(c_t(E)/c_t(F)). \end{aligned}$$

Proof. We may assume that E and F split into direct sums of line bundles:

$$E = \oplus L_i, \qquad c_1(L_i) = \alpha_i,$$

$$F = \oplus M_i, \qquad c_1(M_i) = \beta_i,$$

so that

$$c_t(E) = \prod (1 + \alpha_i t); \qquad c_t(F) = \prod (1 + \beta_i t).$$

We have

$$E^* \otimes F = \bigoplus_{i,j} L_i^* \otimes M_j$$

and correspondingly

$$\begin{aligned} c_{mn}(E^* \otimes F) &= \prod_{i,j} c_1(L_i^* \otimes M_j) \\ &= \prod_{i,j} (\beta_j - \alpha_i) \\ &= \Delta_{m,n}(c_t(F)/c_t(E)). \quad \text{Q.E.D.} \end{aligned}$$

(iii) Porteous' Formula

We now have all the tools necessary to derive Porteous' formula; all we require to make the computation is a little luck.

As in the definition of the loci $\tilde{X}_k(\phi)$, we let

$$\pi\colon G(n-k, E) \to X$$

be the Grassmann bundle of $(n-k)$-planes in the fibers of E, and denote by S and Q the universal subbundle and quotient bundle on $G(n-k, E)$, respectively. As we already observed we have

$$\begin{aligned} x_k &= \pi_*(c_{m(n-k)}(S^* \otimes \pi^* F)) \\ &= \pi_* \Delta_{m, n-k}(\pi^* c_t(F)/c_t(S)), \end{aligned}$$

by our formula for the top Chern class of a tensor product. To express this quantity in terms of the Chern classes of E and F, we need to notice that, for any monomial $\prod c_i(Q)^{\alpha_i}$ we have

$$\pi_*\left(\prod_{i=0}^{k} c_i(Q)^{\alpha_i}\right) = 0 \quad \text{if} \quad \sum i\alpha_i < k(n-k).$$

The above direct image is a multiple of the fundamental class of X when $\sum i\alpha_i = k(n-k)$, while the Gysin images of monomials in $c_i(Q)$ of (weighted) degree greater than $k(n-k)$ are, of course, more complicated; and here is where the luck comes in: since

$$c(S)c(Q) = \pi^*c(E),$$

we may write

$$\begin{aligned} x_k &= \pi_*\Delta_{m,n-k}(\pi^*c_t(F)/c_t(S)) \\ &= \pi_*\Delta_{m,n-k}\left(\pi^*\left(\frac{c_t(F)}{c_t(E)}\right)\cdot c_t(Q)\right), \end{aligned}$$

and now we notice that, inasmuch as the entries of the $(n-k)\times(n-k)$ determinant Δ are all linear combinations, with coefficients in $H^*(X, \mathbb{Z})$, of the Chern classes $c_0(Q), \ldots, c_k(Q)$, by our previous remark, *only those terms in the entries of the determinant* Δ *which contain the factor* $c_k(Q)$ *will contribute to its Gysin image* $\pi_*\Delta$. Thus, by the push-pull[1] formula we have

$$x_k = \Delta_{m-k,n-k}(c_t(F)/c_t(E))\cdot(\pi_*c_k(Q)^{n-k}),$$

and, to conclude, we only have to show that

$$\pi_*(c_k(Q)^{n-k}) = 1_X.$$

Since Q restricts to the universal quotient bundle on each fiber of π, this will follow from the

Lemma. *If Q is the universal quotient bundle on the Grassmannian $G(n-k, n)$, then*

$$\int_{G(n-k,n)} c_k(Q)^{n-k} = 1.$$

[1] We recall the content of the push-pull formula. This states that, for any map $f: X \to Y$ such that f_* is defined, and for any integral cohomology classes α on X and β on Y, one has: $f_*(\alpha\cdot f^*\beta) = f_*(\alpha)\cdot\beta$.

Proof. This is a fairly standard result in the Schubert calculus of the Grassmannian. In fact the integral on the left is just the number of zeros common to $n - k$ general sections of Q. Any vector $v \in \mathbb{C}^n$ determines a section of Q, which vanishes precisely at those $(n - k)$-planes W which contain v. If we choose $n - k$ linearly independent vectors $v_1, \ldots, v_{n-k} \in \mathbb{C}^n$, the corresponding sections of Q vanish simultaneously only at the span of $v_1, \ldots, v_{n-k}$.

Q.E.D.

(iv) What Has Been Proved

We want to make one observation here about what has been established by the preceding computation. The content of Porteous' formula is that the class

$$\Delta_{m-k,n-k}(c_t(F - E))$$

is supported on X_k, and agrees with the fundamental class of X_k when this locus is empty or has the right codimension. In fact, what we *have* established are the analogous statements for the locus $\tilde{X}_k$, i.e.,

In case the locus $\tilde{X}_k$ is of the "right" dimension or empty, its fundamental class is:

$$\Delta_{m,n-k}(\pi^* c_t(F - E) \cdot c_t(Q)),$$

and since if $\tilde{X}_k$ is of the right dimension, X_k must also be, and the map $\tilde{X}_k \to X_k$ is generically one-to-one, it follows that:

In case $\tilde{X}_k$ is of the right dimension or empty, the fundamental class of X_k is:

$$\Delta_{m-k,n-k}(c_t(F - E)).$$

Since X_k is empty if and only if $\tilde{X}_k$ is, this firmly establishes Porteous' formula when X_k is empty, or, which is the same, shows that $\Delta_{m-k,n-k}(c_t(F - E))$ is supported on X_k. In general the problem is that, if $\tilde{X}_k$ is of the wrong dimension, our formula for its fundamental class is completely without meaning, as is then the remainder of the calculation. But *it is possible that $\tilde{X}_k$ have the wrong dimension, and X_k the right dimension, even so.* This will occur, in fact, exactly when the general fiber of $\tilde{X}_k \to X_k$ is positive-dimensional, i.e., when a component of X_k is contained in X_{k-1}.

To actually prove Porteous' formula, in the sense given above, by means of the calculation made here, one needs to use the powerful refined intersection theory developed by Fulton [2] and MacPherson which is admirably suited to problems of "excess intersection." Alternatively, one can employ a different approach altogether, working directly on X, as in Kempf–Laksov [1]. As it stands, however, *our calculation establishes Porteous' formula only under the additional assumption that, when X_k is non-empty, none of its components be entirely contained in X_{k-1}.* Incidentally, we may notice that this extra hypothesis will always be satisfied in the cases in which we will apply Porteous' formula.

§5. A Few Applications and Examples

As a first illustration of the usefulness of Porteous' formula we shall evaluate the degree of the generic determinantal variety $M_k(m, n)$. Since this is a cone with vertex at the zero matrix we may, equivalently, evaluate the degree of the projectivization V of $M_k(m, n)$ in $\mathbb{P}^N$, where $N = mn - 1$. Denoting by x_{ij}, $i = 1, \dots, m$, $j = 1, \dots, n$, the homogeneous coordinates in $\mathbb{P}^N$, the matrix (x_{ij}) can be viewed as a matrix with entries in $H^0(\mathbb{P}^N, \mathcal{O}(1))$, and hence gives a vector bundle homomorphism

$$\phi\colon \mathcal{O}^n_{\mathbb{P}^N} \to \mathcal{O}^m_{\mathbb{P}^N}(1).$$

The variety V is just the kth determinantal variety associated with ϕ. Since it has the "correct" dimension $k(n + m - k) - 1$, its fundamental class can be computed by means of Porteous' formula. Replacing ϕ with its transpose, if necessary, we may and will assume that $m \geq n$. Denoting by ξ the class of a hyperplane we have

$$\begin{aligned} c_t(\mathcal{O}^m(1)) &= c_t(\mathcal{O}(1))^m \\ &= (1 + \xi t)^m \\ &= \sum_{i=0}^{m} \binom{m}{i} \xi^i t^i. \end{aligned}$$

Therefore the fundamental class of V is

$$\Delta_{m-k,n-k}(c_t(\mathcal{O}^m(1)) =$$

$$\det \begin{pmatrix} \binom{m}{m-k}\xi^{m-k} & \cdots & \binom{m}{m+n-2k-1}\xi^{m+n-2k-1} \\ \vdots & & \vdots \\ \binom{m}{m-n+1}\xi^{m-n+1} & \cdots & \binom{m}{m-k}\xi^{m-k} \end{pmatrix},$$

or, which is the same, the degree of V is

$$\det\begin{pmatrix} \binom{m}{m-k} & \binom{m}{m-k+1} & \cdots & \binom{m}{m+n-2k-1} \\ \binom{m}{m-k-1} & & & \\ \vdots & & & \vdots \\ \binom{m}{m-n+1} & & \cdots & \binom{m}{m-k} \end{pmatrix}.$$

To evaluate this determinant, we first replace the second column with the sum of the first two columns, the third column with the sum of the second and third columns, etc., then in the resulting matrix we replace the third column with the sum of the second and third columns, the fourth column with the sum of the third and fourth columns, and so on, until we get

$$\deg(M_k(m, n))$$

$$= \det\begin{pmatrix} \binom{m}{m-k} & \binom{m+1}{m-k+1} & \cdots & \binom{m+n-k-1}{m+n-2k-1} \\ \binom{m}{m-k-1} & & & \\ \vdots & & & \vdots \\ \binom{m}{m-n+1} & & \cdots & \binom{m+n-k-1}{m-k} \end{pmatrix}$$

$$= \prod_{i=0}^{n-k-1} \frac{(m+i)!}{(k+i)!} \det\begin{pmatrix} \frac{1}{(m-k)!} & \cdots & \frac{1}{(m+n-2k-1)!} \\ \vdots & & \vdots \\ \frac{1}{(m-n-1)!} & \cdots & \frac{1}{(m-k)!} \end{pmatrix}$$

$$= \prod_{i=0}^{n-k-1} \frac{(m+i)!}{(k+i)!(m-k+i)!}$$

$$\times \det\begin{pmatrix} 1 & 1 & \cdots \\ m-k & m-k+1 & \cdots \\ (m-k)(m-k-1) & (m-k+1)(m-k) & \cdots \\ \vdots & \vdots & \end{pmatrix}.$$

The determinant in the last expression is nothing but the Vandermonde determinant

$$\det\begin{pmatrix} 1 & 1 & \cdots \\ m-k & m-k+1 & \cdots \\ (m-k)^2 & (m-k+1)^2 & \cdots \\ \vdots & \vdots & \end{pmatrix}$$

in disguise, hence its value is

$$\prod_{n-k-1\geq i>j\geq 0} [(m-k+i)-(m-k+j)] = \prod_{i=0}^{n-k-1} i!.$$

In conclusion we find the following formula for the degree of $M_k(m, n)$:

$$\deg(M_k(m, n)) = \prod_{i=0}^{n-k-1} \frac{(m+i)!\,i!}{(k+i)!(m-k+i)!}. \tag{5.1}$$

Recalling the determinantal description of the tangent cones to $M_k(m, n)$ we gave in Section 2, the above formula also makes it possible to compute the multiplicity of $M_k(m, n)$ at any one of its points. The result is

(5.2) Proposition. *If A is an $m \times n$ matrix of rank $h \leq k$, the multiplicity of $M_k(m, n)$ at A is*

$$\begin{aligned}\operatorname{mult}_A(M_k(m, n)) &= \deg M_{k-h}(m-h, n-h)\\ &= \prod_{i=0}^{n-k-1} \frac{(m-h+i)!\,i!}{(k-h+i)!\,(m-k+i)!}.\end{aligned}$$

In this book a recurrent example of determinantal variety is provided by the so-called *rational normal scrolls*. These can be introduced in three equivalent ways. The first one has a geometrical flavor and generalizes the classical construction of a quadric in three-space.

Take k complementary linear subspaces

$$L_i \subset \mathbb{P}^n, \qquad i = 1, \ldots, k,$$

with $\dim L_i = a_i$ and such that not all the a_i's are equal to zero. If $a_i \neq 0$ choose a rational normal curve $C_i \subset L_i$ and an isomorphism

$$\phi_i\colon \mathbb{P}^1 \to C_i.$$

If $a_i = 0$, we set $C_i = L_i$ and let ϕ_i be the constant map. The variety

$$X_{a_1,\ldots,a_k} = \bigcup_{t \in \mathbb{P}^1} \overline{\phi_1(t), \ldots, \phi_k(t)}$$

swept out by the $(k-1)$-planes spanned by the corresponding points of the C_i's is then called a rational normal scroll.

Alternatively the variety $X_{a_1,\ldots,a_k}$ may be described as the image of the projective bundle

$$\mathbb{P}(E) = \mathbb{P}(\mathcal{O}_{\mathbb{P}^1}(-a_1) \oplus \cdots \oplus \mathcal{O}_{\mathbb{P}^1}(-a_k))$$

under the map given by $\mathcal{O}_{\mathbb{P}(E)}(1)$. Then, for each i, the image of the direct summand $\mathcal{O}_{\mathbb{P}^1}(-a_i)$ of E maps to the rational curve C_i. It is not hard to see that the degree of the scroll $X_{a_1,\ldots,a_k}$ is given by

$$\deg X_{a_1,\ldots,a_k} = \sum_{i=1}^{k} a_i = n - k + 1.$$

This is the smallest possible degree of an irreducible non-degenerate k-fold $\mathbb{P}^n$. Conversely, we have the following classical result

(5.3) Theorem. *Any irreducible non-degenerate k-fold of degree $n-k+1$ in $\mathbb{P}^n$ is either a rational normal scroll, a cone over the Veronese surface in $\mathbb{P}^5$, or a quadric of rank greater than* 4.

The reader may find a proof of this theorem in Harris [2] or, for the case $k = 2$, which is the relevant one, on p.525 of Griffiths–Harris [1].

For us the interest of scrolls, at this junction, stems from the fact that they admit a very nice and classical determinantal description which has the advantage of clearly exhibiting their defining ideal. Assume that $a_1 = \cdots = a_{h-1} = 0$, $a_i \neq 0$, $i = h, \ldots, k$. Choose in $\mathbb{P}^n$ homogeneous coordinates

$$X_0^{(1)}, \ldots, X_{a_1}^{(1)}, X_0^{(2)}, \ldots, X_{a_2}^{(2)}, \ldots, X_0^{(k)}, \ldots, X_{a_k}^{(k)}$$

in such a way that $X_0^{(i)}, \ldots, X_{a_i}^{(i)}$ are homogeneous coordinates in L_i, $i = 1, \ldots, k$. Consider the matrix

$$M_{a_1,\ldots,a_k} = \begin{pmatrix} X_0^{(h)} & \cdots & X_{a_h-1}^{(h)} & \cdots & X_0^{(k)} & \cdots & X_{a_k-1}^{(k)} \\ X_1^{(h)} & \cdots & X_{a_h}^{(h)} & \cdots & X_1^{(k)} & \cdots & X_{a_k}^{(k)} \end{pmatrix}.$$

We then have the following

Proposition. *The scroll $X_{a_1,\ldots,a_k}$ is the determinantal variety whose ideal is generated by the 2-by-2 minors of the matrix $M_{a_1,\ldots,a_k}$.*

Let Y be the determinantal variety defined by the vanishing of the 2-by-2 minors of $M_{a_1,\dots,a_k}$. First of all observe that the set-theoretic equality between $X_{a_1,\dots,a_k}$ and Y follows immediately from the geometrical description of $X_{a_1,\dots,a_k}$ and from the very well-known fact that, up to a change of coordinates, the rational normal curve

$$C_i \subset L_i \cong \mathbb{P}^{a_i}, \qquad i = h, \dots, k,$$

is given by

$$C_i = \left\{ \operatorname{rank}\begin{pmatrix} X_0^{(i)} & \cdots & X_{a_i-1}^{(i)} \\ X_1^{(i)} & \cdots & X_{a_i}^{(i)} \end{pmatrix} \le 1 \right\}.$$

Now the set-theoretical equality we just proved implies, in particular, that Y has the "correct" codimension. By (4.1) this implies that Y (and also the cone over Y) are Cohen–Macaulay, so that Y has no embedded components. Therefore to prove the proposition it suffices to show that the degree of Y is equal to the degree of $X_{a_1}, \dots, {}_{a_k}$, i.e., equal to $n - k + 1$. On the other hand, Y can be thought of as a linear section of the generic determinantal variety $M_1(n - k + 1, 2)$, and formula (5.1) exactly tells us that

$$\deg(M_1(n - k + 1, 2)) = n - k + 1,$$

proving the proposition.

It may be instructive to introduce the matrix defining a scroll $X = X_{a_1,\dots,a_k}$ in a more intrinsic way. Let L be the restriction to X of the hyperplane bundle on $\mathbb{P}^n$. The scroll X is ruled by a pencil of $(k - 1)$-planes which we denote by $|E|$. We then have

$$h^0(X, \mathcal{O}(E)) = 2, \qquad h^0(X, L(-E)) = n - k + 1.$$

The second equality follows from the first and from the linear normality of X, which we shall prove in the next proposition. Let us consider the multiplication map

$$\mu\colon H^0(X, \mathcal{O}(E)) \oplus H^0(X, L(-E)) \to H^0(X, L).$$

It is then easy to show that, with a suitable choice of bases, the transpose of the matrix $M_{a_1,\dots,a_k}$ represents the dual map

$$\mu^*\colon H^0(X, L)^* \to \operatorname{Hom}(H^0(X, \mathcal{O}(E)), H^0(X, L(-E))^*).$$

As we just mentioned, one of the basic properties of rational normal scrolls is given by the following.

Proposition. *Rational normal scrolls are projectively normal.*

A direct proof is as follows. Let X be a rational normal scroll of degree k in $\mathbb{P}^n$. We proceed by induction on k. The case $k = 1$ is well known. Given an integer v the cohomology sequence of

$$0 \to \mathscr{I}_X(v) \to \mathcal{O}_{\mathbb{P}^n}(v) \to \mathcal{O}_X(v) \to 0$$

shows that the projective normality of X is equivalent to the vanishing statement

$$H^1(\mathbb{P}^n, \mathscr{I}_X(v)) = 0, \qquad v = 1, 2, \ldots$$

Let H be a general hyperplane in $\mathbb{P}^n$; then, by Theorem (5.3) $H \cap X$ is again a rational normal scroll, therefore our vanishing statement follows, from the induction hypothesis, by looking at the cohomology sequence of

$$0 \to \mathscr{I}_X(v - 1) \to \mathscr{I}_X(v) \to \mathscr{I}_{X \cap H}(v) \to 0.$$

Another approach to the projective normality of scrolls is by means of Serre's normality criterion, which reads as follows:

Serre's Criterion (cf. Serre [1, p. III-13]). *A Noetherian domain R is integrally closed if and only if the following two conditions hold:*

(i) *For every prime P of height one in R, R_P is a discrete valuation ring.*
(ii) *The associated primes of any principal ideal in R all have height one.*

We shall prove only the part of the criterion which we shall need here and in the following chapters, namely that (i) and (ii) imply that R is integrally closed. If we can show that

$$\bigcap_{ht(P)=1} R_P = R,$$

since the R_P are integrally closed, the same will be true of R. Suppose then that α and β are elements of R such that α/β belongs to R_P for every prime P of height one. In particular, if $\beta R = \bigcap Q_i$ is a primary decomposition of βR, and P_i stands for the radical of Q_i, condition (ii) implies that α belongs to βR_{P_i}, for every i. This means that there is a γ_i not belonging to P_i such that $\alpha\gamma_i$ belongs to βR; thus α belongs to βR, that is, α/β belongs to R. This finishes the proof.

In the context of analytic spaces, Serre's criterion can be formulated as follows.

(5.4) Proposition. *Let X be a reduced, irreducible analytic space. Then X is normal if and only if the following conditions hold:*

(i) *The singular locus of X has codimension two or more.*
(ii) *For every analytic function g on an open subset of X, the subspace defined by g has no embedded components.*

As we did for Serre's criterion, we shall only prove that (i) and (ii) imply normality. To use Serre's criterion we simply have to show that, for any point x on X, and for any height one prime ideal P in the local ring R of X at x, R_P is a discrete valuation ring. Let $\mathcal{F}$ be the conductor sheaf of the normalization of X. Then $\mathcal{O}_X/\mathcal{F}$ is supported on an analytic subset of X of codimension two or more. If F is the stalk of $\mathcal{F}$ at x, this implies that F is not contained in P, thus $FR_P = R_P$. But FR_P is the conductor of the integral closure of R_P. This means that R_P is integrally closed, hence a discrete valuation ring, since P has height one. This finishes the proof of Proposition (5.4).

To show that rational normal scrolls are projectively normal, we may proceed as follows. First of all it is clear, either from the determinantal or geometric description of a rational normal scroll X, that its singularities occur in codimension two or more; clearly, this is also true of the cone over X. We also remarked that the cone over X is Cohen–Macaulay. In particular, complete intersections in the cone over X do not have embedded components. Normality of the cone over X, that is, projective normality of X, follows from Proposition (5.4).

Let us consider again the exact cohomology sequence of:

$$0 \to \mathcal{I}_X(1) \to \mathcal{I}_X(2) \to \mathcal{I}_{X \cap H}(2) \to 0,$$

where X is a rational normal scroll and H is a hyperplane. Since X is non-degenerate, we have that $h^0(X, \mathcal{I}_X(1)) = 0$. On the other hand $h^1(\mathbb{P}^n, \mathcal{I}_X(1)) = 0$, so that we get an isomorphism

$$I_X(2) \xrightarrow{\cong} I_{X \cap H}(2)$$

(here, and in the sequel, we denote by I_Y the homogeneous ideal of the projective variety Y and by $I_Y(h)$ its degree h summand). By taking the intersection of X with a general $(n-k)$-plane we then see that the dimension of $I_X(2)$ equals the number of linearly independent quadrics in $\mathbb{P}^{n-k}$ passing through $n-k+1$ points in general position. Hence

$$\dim I_X(2) = \binom{n-k+1}{2}.$$

Combining this with the determinantal description of X we obtain the following

Theorem. *Let X be a k-dimensional rational normal scroll contained in $\mathbb{P}^n$. Then there exists a matrix of linear forms in $\mathbb{P}^n$:*

$$M = \begin{pmatrix} l_1 & \cdots & l_{n-k+1} \\ l'_1 & \cdots & l'_{n-k+1} \end{pmatrix}$$

such that the ideal I_X is generated by the 2-by-2 minors of M. Moreover the $\binom{n-k+1}{2}$ quadrics

$$l_\alpha l'_\beta - l'_\alpha l_\beta = 0, \qquad \alpha < \beta, \quad \alpha, \beta = 1, \ldots, n - k + 1,$$

are linearly independent and form a basis of $I_X(2)$.

Bibliographical Notes

Lemma (1.3) appears in Kempf [2]. The approach to the first and second fundamental theorems of invariant theory has been suggested to us by C. De Concini. Both theorems appear in Weyl's book [2]. A modern treatment, valid also in characteristic p, can be found in De Concini–Procesi [1], or else in De Concini–Eisenbud–Procesi [1]. Fundamental papers in the theory are the one by Hodge [1] and the one by Doubilet–Rota–Stein [1]. The first proof of the Cohen–Macaulay property of determinantal varieties is to be found in Eagon–Hochster [1]. The analogous result for Schubert varieties is due to Musili [1], Laksov [1], and Hochster [1]. For Porteous' formula we refer to the original paper of Porteous [1] and to Kempf–Laksov [1], where it is proved under less restrictive hypotheses. Finally, a discussion of rational normal scrolls can be found in Harris [2].

Exercises

A. Symmetric Bilinear Maps

We shall use the following notations:

$\mathbb{P}^N = \mathbb{P}^{(n+1)(n+2)/2-1}$ is the projective space of symmetric quadratic forms on $\mathbb{C}^{n+1}$;

Q will denote either a point in $\mathbb{P}^N$ thought of as a quadric in $\mathbb{P}^n$, or a point in $\mathbb{C}^{(n+1)(n+2)/2} - \{0\}$ thought of as a symmetric linear map from $\mathbb{C}^{n+1}$ to its dual (the context will make clear which interpretation to use);

$V_k \subset \mathbb{P}^N$ is the determinantal variety of forms of rank $\leq k$;

G_k is the Grassmannian of $(n - k)$-planes in $\mathbb{P}^n$, or equivalently the Grassmannian of $(n - k + 1)$-planes in $\mathbb{C}^{n+1}$;

$\Lambda \in G_k$ will denote either a $\mathbb{P}^{n-k}$ in $\mathbb{P}^n$ of the corresponding $\mathbb{C}^{n-k+1}$ in $\mathbb{C}^{n+1}$ (again the context will make clear which interpretation to use);

for $Q \in P^N$, ker Q will be either the kernel of the symmetric map

$$Q: \mathbb{C}^{n+1} \to (\mathbb{C}^{n+1})^*,$$

or else the projection of this linear space to $\mathbb{P}^n$;

$$\begin{array}{ccc} & X_k = \{(\Lambda, Q) \in G_k \times \mathbb{P}^N : \Lambda \subset \ker Q\} & \\ \swarrow^{\pi_1} & & \searrow^{\pi_2} \\ G_k & & \mathbb{P}^N \end{array}$$

is the incidence variety with indicated projections.

A-1. Show that X_k is irreducible and smooth; and that

$$\dim X_k = k(n + 1 - k) + \binom{k+1}{2} - 1.$$

A-2. Show that π_2 is biholomorphic over $V_k - V_{k-1}$, and from this conclude that

$$\operatorname{codim} V_k = \binom{n-k+2}{2}.$$

A-3. In this exercise we will identify the tangent spaces to X_k, and for this we use the identifications (cf. Section 2 of Chapter II)

$$T_\Lambda(G_k) \cong \operatorname{Hom}(\Lambda, \mathbb{C}^{n+1}/\Lambda),$$
$$T_Q(\mathbb{P}^N) \cong \mathbb{C}^{N+1}/\mathbb{C}_Q,$$

where $\mathbb{C}_Q \subset \mathbb{C}^{N+1}$ is the line over $Q \in \mathbb{P}^N$. Points in $T_{(\Lambda, Q)}(G_k \times \mathbb{P}^N)$ will be denoted by (ϕ, R) where $\phi \in \operatorname{Hom}(\Lambda, \mathbb{C}^{n+1}/\Lambda)$ and $R \in \mathbb{C}^{N+1}/\mathbb{C}_Q$. Show that

$$T_{(\Lambda, Q)}(X_k) = \{(\phi, R) : R(v, w) + Q(\phi v, w) = 0 \text{ for all } v \in \Lambda,\ w \in \mathbb{C}^{n+1}\}.$$

A-4. Show that the projectivized tangent cone to V_k at $Q \in V_l - V_{l-1}$ has support given by

$$\operatorname{supp}\ \mathbb{P}\mathcal{T}_Q(V_k) = \{R : \operatorname{rank}(R|_{\ker Q}) \le l - k\}$$

and has multiplicity one.

A-5. A point $Q \in V_k - V_{k-1}$ may be pictured as a cone over a smooth quadric in $\mathbb{P}^{k-1}$ with vertex $\Lambda \cong \mathbb{P}^{n-k}$ where $\Lambda = \ker Q$. Show that

$$T_Q(V_K) \cong H^0(\mathbb{P}^N, \mathcal{I}_\Lambda(2)),$$

where the right-hand side is the vector space of quadrics in $\mathbb{P}^n$ that contain Λ.

B. Quadrics

In this sequence of exercises we will adopt the following notations:

V is an m-dimensional complex vector space and

$$Q: V \times V \to \mathbb{C}$$

a non-degenerate bilinear form;

we denote by

$$\tilde{Q}: V \to V^*$$

the isomorphism corresponding to Q;

a k-plane $\Lambda \subset V$ is *isotropic* for Q if

$$Q|_{\Lambda \times \Lambda} = 0;$$

$G(k, V) = G_k$ denotes the Grassmannian of k-planes in V and

$$\Sigma_k \subset G_k$$

the locus of isotropic subspaces (we write $\Sigma_k(Q)$ if there is a possibility of confusion).

In Exercises B-1 through B-7 we will assume Q to be symmetric.

B-1. Show that Σ_k is empty for $k > m/2$ and non-empty for $k = [m/2]$. Show also that for $k' < k \leq [m/2]$ every k'-dimensional isotropic subspace lies in a k-dimensional isotropic subspace.

B-2. Making the identification

$$T_\Lambda(G_k) \cong \mathrm{Hom}(\Lambda, V/\Lambda),$$

show that, for $k \leq [m/2]$, Σ_k is smooth of codimension $\binom{k+1}{2}$ in G_k with tangent space given by

$$T_\Lambda(\Sigma_k) = \{\phi \in \mathrm{Hom}(\Lambda, V/\Lambda): Q(\phi \cdot v, w) + Q(v, \phi \cdot w) = 0 \text{ for all } v, w \in \Lambda\}.$$

B-3. Show that, if $k < m/2$, the locus Σ_k is irreducible.

(*Hint*: If $v \in V$ is an isotropic vector for Q, set $W = \tilde{Q}(v)^\perp/\mathbb{C} \cdot v$ (thinking of $\tilde{Q}(v)$ as a vector in V^*) and note that Q induces a non-degenerate form Q_W on W such that

$$\Sigma_{k-1}(Q_W) \cong \{\Lambda \in \Sigma_k : v \in \Lambda\}.$$

Now use induction on k.)

B-4. Denote the automorphism group of Q by

$$\mathrm{Aut}(Q) = \{A \in \mathrm{GL}(V) : Q(Av, Aw) = Q(v, w) \text{ for all } v, w\}.$$

Note that $\mathrm{Aut}(Q)$ acts on Σ_k, and show that this action is transitive (i.e., for $\Lambda, \Lambda' \in \Sigma_k$ there exists $A \in \mathrm{Aut}(Q)$ such that

$$A\Lambda = \Lambda';$$

note that this proves that Σ_k is smooth).

B-5. Keeping the notations of the preceding exercises, assume that $m = 2n$ is even and $k = n$. Show that

$$(*) \qquad \dim(\Lambda \cap \Lambda') \equiv n \pmod 2$$

if and only if

$$\det A = +1.$$

Conclude that, if $m = 2n$, then Σ_n has two irreducible components Γ_1, Γ_2 such that the condition $(*)$ is equivalent to Λ, Λ' being in the same component.

B-6. From the above deduce the

Corollary. *For* $\Lambda_1, \Lambda_2, \Lambda_3 \in \Sigma_n$,

$$\dim \Lambda_1 \cap \Lambda_2 + \dim \Lambda_2 \cap \Lambda_3 + \dim \Lambda_1 \cap \Lambda_3 \equiv n \pmod 2.$$

B-7. Assuming still that $m = 2n$, let $\Lambda_0 \in \Sigma_n$. Show that

$$V_k(\Lambda_0) = \{\Lambda \in \Sigma_n : \dim \Lambda \cap \Lambda_0 \geq k \text{ and } \dim \Lambda \cap \Lambda_0 \equiv k \pmod 2\}$$

is a closed subvariety of codimension $k(k-1)/2$ in Σ_n.

(*Suggestion*: express $V_k(\Lambda_0)$ as the image of an incidence correspondence fibered over $G(k, \Lambda_0)$.)

Now let V be an m-dimensional complex vector space,

$$Q : V \times V \to \mathbb{C}$$

a non-degenerate alternating (skew-symmetric) bilinear form.

B-8. Show that m is even, and that $\Sigma_k \neq \varnothing$ if and only if $k \leq m/2$.

B-9. When $k = m/2$, show that Σ_k is smooth and irreducible of dimension $k(k+1)/2$.

B-10. Let $\mathrm{Aut}(Q)$ be the group of linear transformations $A : V \to V$ such that $Q(v, w) = Q(Av, Aw)$, for all $v, w \in V$. Show that $\det A = 1$ for all $A \in \mathrm{Aut}(Q)$.

Hint: Consider Q as an element of $\Lambda^2 V^*$, and look at $\underbrace{Q \wedge Q \wedge \cdots \wedge Q}_{m/2 \text{ times}}$.

C. Applications of Porteous' Formula

Most of the following exercises will involve the Grassmannian $G = G(2, 4)$ of lines in $\mathbb{P}^3$. We will denote by S the universal sub-bundle on G, and by c_1 and c_2 its Chern classes.

C-1. Let L be a linear functional on $\mathbb{C}^4$. Show that we may define a global section σ_L of S^* by setting

$$\sigma_L(\Lambda) = L|_\Lambda$$

for $\Lambda \in G(2, 4)$, and that these are all the global section of S^*.

C-2. Using the preceding exercise, show that the classes $-c_1$ and c_2 are represented by the cycles

$$\Sigma_1 = \{\Lambda : \Lambda \cap \Lambda_0 \neq (0)\},$$

and

$$\Sigma_2 = \{\Lambda : \Lambda \subset W_0\},$$

respectively, where Λ_0 is a fixed two-plane in $\mathbb{C}^4$, and W_0 a fixed three-plane. In particular, conclude that $c_2^2 = c_1^2 c_2 = 1$ in $H^8(G, \mathbb{Z}) \cong \mathbb{Z}$.

C-3. If Q is the universal quotient bundle on G, use the relation

$$c(S)c(Q) = 1$$

to express $c_i(Q)$ in terms of c_1 and c_2, and to obtain two relations on the classes c_1 and c_2. In particular, deduce that $c_1^4 = 2$.

C-4. Using the splitting principle, find the Chern classes of $\mathrm{Sym}^n S^*$ for $n = 2, 3, 4$, and 5.

C-5. Let $F \in \mathrm{Sym}^3(\mathbb{C}^4)^*$ be a homogeneous cubic polynomial on $\mathbb{C}^4$. Show that we may define a global section σ_F of $\mathrm{Sym}^3 S^*$ by

$$\sigma_F(\Lambda) = F|_\Lambda.$$

Use this and the preceding exercise to determine the number of lines on a cubic surface.

Remark. In case the cubic surface $\{F = 0\}$ is smooth—and only in this case—the zeros of σ_F are all simple.

C-6. In a similar vein, let $\{F_\lambda\}_{\lambda \in \mathbb{P}^1}$ be a general pencil of quartic surfaces. Representing the pencil $\{F_\lambda\}$ as a bundle map

$$\mathcal{O}^{\oplus 2} \to \mathrm{Sym}^4 S^*$$

on G, use Exercise C-4 above and Porteus' formula to determine the number of surfaces F_λ in the pencil that contain a line. (Answer: 320)

C-7. Similarly, find the number of quintic surfaces in a general net (two-dimensional linear system) that contain a line. (Answer: ?)

C-8. Using the same techniques as in the preceding problems, find the number of conic curves in a pencil that contain a line.

C-9. Find the number of reducible curves in a general net of plane cubic curves.

C-10. Find the number of reducible surfaces in a general web (three-dimensional linear system) of quadric surfaces in $\mathbb{P}^3$. Compare this to the answer obtained by representing the locus of reducible quadrics as the image of the multiplication map

$$\phi: H^0(\mathbb{P}^3, \mathcal{O}(1)) \times H^0(\mathbb{P}^3, \mathcal{O}(1)) \to H^0(\mathbb{P}^3, \mathcal{O}(2)).$$

C-11. Let $\Sigma \subset G(2, 4)$ be a curve. Show that the degree of the surface $X \subset \mathbb{P}^3$ swept out by the corresponding one-parameter family of lines is given by

$$\deg X = -\Sigma \cdot c_1.$$

C-12. Let $\{F_\lambda\}_{\lambda \in \mathbb{P}^1}$ be a general pencil of cubic surfaces. Using the preceding exercise, find the degree of the surface swept out by the lines of the surfaces F_λ.

D. Chern Numbers of Kernel Bundles

In the following, we will be dealing with a map of vector bundles $\phi: E \to F$ on a space X, and its kth degeneracy locus X_k. We will consider the case where the expected (and actual) dimension of X_k is 1, that is, rank $E = m$, rank $F = n$, and $\dim X = (m - k)(n - k) + 1$; in particular, $X_{k+1} = \varnothing$. In this case the kernel and cokernel of ϕ are bundles on X_k, denoted K and K'; we will be concerned with their first Chern classes.

D-1. Let $G = G(m - k, E) \xrightarrow{\pi} X$ be the Grassmannian bundle, $S \to G$ the universal sub-bundle, and

$$\tilde{\phi}: S \to \pi^* F$$

the induced map (cf. Section 4 of Chapter II). Assuming X_k is reduced, show that

$$(*) \qquad \begin{aligned} c_1(K) &= c_1(S) \cdot \tilde{X}_k \\ &= c_1(S) \cdot c_{n(m-k)}(\pi^* F - S), \end{aligned}$$

where $\tilde{X}_k$ is the zero locus of $\tilde{\phi}$.

D-2. Evaluate the product in $(*)$ above to obtain the formula

$$c_1(K) = -\det \begin{pmatrix} c_{n-r+1} & c_{n-r+2} & \cdots & c_{n+m-2r} \\ c_{n-r-1} & c_{n-r} & \cdots & c_{n+m-2r-2} \\ \vdots & & c_{n-r} & \vdots \\ c_{n-m+1} & \cdots & \cdots & c_{n-r} \end{pmatrix}$$

where $c_i = c_i(F - E)$.

D-3. Assuming X_k is smooth and using the tangent space computations of Section 2, show that the normal bundle to X_k in X is

$$N = K^* \otimes K'$$

and compute its first Chern class accordingly.

D-4. Let $A = (L_{ij})$ be a general $n \times (n+1)$ matrix of homogeneous linear forms on $\mathbb{P}^3$, and $C \subset \mathbb{P}^3$ the locus where rank $A \leq n - 1$. Find the genus of C.

D-5. Using the notations of Exercise Sequence C, show that the first Chern class of the tangent bundle to $G = G(2, 4)$ is

$$c_1(\Theta_G) = 4c_1.$$

Finally, if $\{F_\lambda\}$ is a pencil of cubic surfaces, use this and the set-up of Exercise C-12 to find the genus of the curve C formed by the lines on the surfaces F_λ. Compare this result with the one obtained by representing C as a 27-sheeted cover of $\mathbb{P}^1$ and using the Riemann–Hurwitz formula.

Note. For a further application of the formulas above, see Exercise C-7 of Chapter VIII.

Chapter III

Introduction to Special Divisors

In this chapter we shall begin our study of linear series on smooth curves. Our central question is: What are the limitations on the dimension $r(D)$ of a complete linear series $|D|$? The first result in this direction is the classical Clifford theorem. After discussing this, in a somewhat different vein, we shall prove Castelnuovo's bound on the genus of a curve in projective r-space. This will lead to Max Noether's theorem on the projective normality of canonical curves, to a detailed study of extremal curves in r-space and to a brief presentation of Petri's theory.

§1. Clifford's Theorem and the General Position Theorem

We begin by recalling some definitions and results from Chapter I. Let D be an effective degree d divisor on a smooth curve C of genus g. By the Riemann–Roch theorem we always have

$$d - g \leq r(D) \leq d,$$

and as we explained in Section 2 of Chapter I,

If $d > 2g - 2$ then $r(D) = d - g$.

If $d \leq 2g - 2$ and D is general then $\begin{cases} r(D) = 0 & \text{for } d \leq g, \\ r(D) = d - g & \text{for } d \geq g. \end{cases}$

In geometry one seldom encounters general divisors, and our main object of study will be the behavior of $r(D)$ for non-general D when $1 \leq d \leq 2g - 2$. The main general fact concerning special divisors is

Clifford's Theorem. (i) *If D is an effective divisor of degree d on C with $d \leq 2g - 1$, then*

$$r(D) \leq d/2;$$

(ii) *if equality holds then either D is zero, D is a canonical divisor, or C is hyperelliptic and D is linearly equivalent to a multiple of a hyperelliptic divisor.*

Proof. We note that, by the Riemann–Roch theorem and the assumption that $d \leq 2g - 1$, the Clifford inequality is trivially satisfied unless D is special; we then assume that $i(D) \neq 0$. We first observe that $r(D) \geq r$ if and only if there is a divisor in $|D|$ containing any r given points of the curve. An immediate consequence is that

$$r(D + D') \geq r(D) + r(D')$$

for any two effective divisors D and D' (indeed given $r(D) + r(D')$ general points of C we may find $E \in |D|$ containing the first $r(D)$ and $E' \in |D'|$ containing the last $r(D')$; then $E + E' \in |D + D'|$ contains the given set of points). Since D is special we may find an effective divisor D' such that $D + D' \in |K|$. Noting that $i(D') = r(D') + 1$ we may combine the above inequality with the Riemann–Roch theorem to obtain

$$r(D) + r(D') \leq g - 1,$$

$$r(D) - r(D') = d - g + 1.$$

The first part of Clifford's theorem follows by adding these relations.

The basic inequality above, applied to a pair of residual divisors D and D', may be interpreted as saying that the image of the bilinear map

$$\mu_0 \colon H^0(C, \mathcal{O}(D)) \otimes H^0(C, K(-D)) \to H^0(C, K)$$

has dimension at least equal to

$$l(D) + i(D) - 1,$$

that is, the sum of the dimensions of the two factors of the domain of μ_0, minus 1.

In fact, it is a remarkable theorem of H. Hopf that, given any setup of a linear map

$$v \colon A \otimes B \to C,$$

where A, B, C are *complex* vector spaces and v is injective on each factor separately, then

$$\dim v(A \otimes B) \geq \dim A + \dim B - 1.$$

This result, which is false over $\mathbb{R}$, was one of the earliest applications of topology to algebra.

The analysis of the map μ_0 in terms of the geometry of C is, perhaps, the central theme of this book. The first case of this is the second half of Clifford's theorem, in which we describe the circumstances under which equality may hold, that is, when μ_0 may have minimal rank $l(D) + i(D) - 1$. The main tool here will be the classical:

General Position Theorem. *Let $C \subset \mathbb{P}^r$, $r \geq 2$, be an irreducible non-degenerate, possibly singular, curve of degree d. Then a general hyperplane meets C in d points any r of which are linearly independent.*

Assuming the general position theorem we now complete the proof of Clifford's theorem.

Proof of (ii) *in Clifford's Theorem.* If equality holds in (i) then we must have

$$r(D) + r(D') = g - 1,$$

and then every canonical divisor is a sum $E + E'$, with $E \in |D|$ and $E' \in |D'|$. Since $\deg D + \deg D' = 2g - 2$, we may assume that $\deg D \leq g - 1$, and by the inequality of the first part either $D = 0$ or $r(D) > 0$. Assuming the latter, if C were non-hyperelliptic then, by the geometric Riemann–Roch theorem, every hyperplane section $E + E'$ of the canonical curve $C \subset \mathbb{P}^{g-1}$ would contain a set E of $d \leq g - 1$ dependent points. This contradicts the general position theorem. Suppose now that C is hyperelliptic. In Chapter I, after proving the geometric version of the Riemann–Roch theorem, we showed that any complete g^r_d on C is of the form

$$rg^1_2 + p_1 + \cdots + p_{d-2r},$$

provided that $d \leq g$. Since we are assuming that $2r(D) = d$, it follows immediately that $|D|$ is a multiple of the hyperelliptic g^1_2 on C. Q.E.D.

We now go back to the general position theorem. When $r = 2$ there is nothing to prove: therefore, we assume from now on that $r \geq 3$. Due to the importance of the theorem we shall give two arguments. The first is based on the following:

Lemma. *Let $C \subset \mathbb{P}^r$, $r \geq 3$, be as in the statement of the general position theorem. Then a general hyperplane meets C in d points, no three of which are collinear.*

We shall first show, in three steps, that the lemma implies the theorem.
(i) For a general point $p \in C$, the projection

$$\pi_p : C \to \mathbb{P}^{r-1}$$

from p is birational onto its image. Indeed, by the lemma we may find a pair of points $p, q \in C$ that do not lie on a trichord, and then π_p is one-to-one over $\pi_p(q)$.

(ii) Next we let $U \subset (\mathbb{P}^r)^*$ be the open set of hyperplanes transverse to C, and consider the incidence correspondence

$$I \subset C \times U$$

consisting of pairs (p, H) where $p \in H \cap C$. We claim that, if the general position theorem is false, then for a general pair $(p, H) \in I$ there is a dependent set of points

$$p = p_1, p_2, \ldots, p_r \in H \cap C.$$

Indeed, we first observe that I is irreducible of dimension r. Now we consider the subvariety $I_0 \subset I$ consisting of pairs (p, H) where $p \in H \cap C$ is part of a dependent set of points $p = p_1, \ldots, p_r$. It is clear that, if the general position theorem is false, then $\dim I_0 = r$. By irreducibility we must have $I_0 = I$ as desired.

(iii) Finally, for $p \in C$ a general point we consider the projection

$$\pi_p \colon C \to C' \subset \mathbb{P}^{r-1}.$$

By step (i), π_p is birational onto the image curve C'. By step (ii), if the general position theorem fails for $C \subset \mathbb{P}^r$, then it also fails for $C' \subset \mathbb{P}^{r-1}$. But the lemma is equivalent to the general position theorem in the case $r = 3$, and this contradiction shows that the lemma implies the general position theorem.

Proof of the Lemma. We first observe that the lemma is equivalent to the statement that *there are at most* ∞^1 *trichords to* C, which is in turn equivalent to the assertion that *not every pair of points of* C *lie on a trichord*; it is this last statement we shall prove. To see the equivalence of the last two italicized statements let $J \subset C^2 \times U$ be defined by

$$J = \{(p_1, p_2, H) \colon p_1, p_2 \in H \text{ and } \overline{p_1 + p_2} \text{ is a trichord}\}.$$

Since C^2 is irreducible and the fiber dimension of $J \to C^2$ is always $r - 2$, we see that the surjectivity of $J \to U$ is equivalent to $\dim J \geq r$, which is in turn equivalent to the surjectivity of $J \to C^2$. To prove that not every chord of C is a trichord, we argue by contradiction. If the statement is false, then given general points $p, q \in C$ the chord $\overline{pq}$ will meet C in a third point $v = u(p, q)$. We choose small arcs around each of the points p, q, and v, local parameters s on the first arc, t on the second, u on the third, and imagine the three arcs as given by the projections to $\mathbb{P}^r$ of $\mathbb{C}^{r+1}$-valued functions $p(s)$, $q(t)$, $v(u)$. Then we may determine $u(s, t)$ such that

$$p(s) \wedge q(t) \wedge v(u(s, t)) \equiv 0.$$

Differentiating this relation with respect to s and t gives

$$p' \wedge q \wedge v + p \wedge q \wedge v' \cdot \frac{\partial u}{\partial s} \equiv 0,$$

$$p \wedge q' \wedge v + p \wedge q \wedge v' \cdot \frac{\partial u}{\partial t} \equiv 0.$$

Since it is clearly the case that $\partial u/\partial s \not\equiv 0$ and $\partial u/\partial t \not\equiv 0$, we infer that

$$p' \wedge q \wedge v = \lambda(p \wedge q' \wedge v), \qquad \lambda \neq 0.$$

This in turn implies that the vectors p, p', q, q' lie in a $\mathbb{C}^3$; i.e., we have shown that if the lemma is false then any two tangent lines to C meet in a point. Now, since $r \geq 3$ and C is non-degenerate, it follows that the tangent lines do not lie in a fixed plane and hence this point must be the same for any pair of tangent lines. So that, if the lemma is false, all tangent lines to C pass through a fixed point p. But now projecting C from p would give a map with vanishing differential and therefore C would be a line through p. This contradiction establishes the lemma. Q.E.D.

Second Proof of the General Position Theorem. This approach introduces a new idea but yields a much broader statement. Briefly, the idea is that a general hyperplane section $\Gamma = H \cap C$ cannot possess both dependent and independent sets of r points, because *there is no way to distinguish, uniformly as H varies over all hyperplanes transverse to C, one subset of r points of Γ from another.* Since $C \subset \mathbb{P}^r$ is non-degenerate, Γ will contain some subset of r independent points. The general position theorem will follow once we make the italicized statement precise. As before we consider the incidence correspondence $I \subset C \times U$. The projection

$$\pi_2 : I \to U$$

is a d-sheeted topological covering. If we fix a base point $H_0 \in U$ and let $\Gamma_0 = C \cap H_0$ be the corresponding hyperplane section, then as is the case for any topological covering space there is the monodromy map

$$\pi_1(U, H_0) \to \operatorname{Aut}(\Gamma_0),$$

the image of which is called the monodromy group M of the covering $I \to U$. Our general principle can be expressed by the following:

Lemma. *The monodromy group of $I \to U$ is the full symmetric group.*

For any positive integer m, we set

$$I(m) = \{(p_1, \ldots, p_m, H) : H \in U, \text{ and all the } p_i\text{'s are distinct and belong to } H \cap C\}.$$

A statement equivalent to the above lemma is that the associated correspondence $I(d) \subset C^d \times U$ is connected. This is because an arc in $I(d)$ connecting two points in $I(d)$ lying over H_0 will project to a loop in $\pi_1(U, H_0)$ whose corresponding monodromy transformation takes one of the two orderings of Γ_0 into the other. Similarly, to say that $I(m)$ is connected is equivalent to saying that the monodromy group is m-times transitive; i.e., it acts transitively on the set of ordered subsets of Γ_0 consisting of m distinct points.

To show that M is the full symmetric group it will suffice to show that M is twice transitive and contains a simple transposition. We proceed in two steps. We first show that M is twice transitive. We set

$$\begin{aligned}\tilde{I}(2) &= \{(p_1, p_2, H): p_1, p_2 \in H \cap C \text{ and } p_1 \neq p_2\}\\ &\subset C \times C \times \mathbb{P}^{r*};\end{aligned}$$

this is a slightly enlarged version of $I(2)$. Then, since any two distinct points of C span a line, we see that $\tilde{I}(2)$ maps onto $(C \times C) - \Delta$, with all fibers being $\mathbb{P}^{r-2}$'s. Consequently $\tilde{I}(2)$ is irreducible, and the Zariski open subset $I(2) \subset \tilde{I}(2)$ is connected.

We now show that M contains a simple transposition. As is customary, we denote by C^* the dual of C, that is, the set of hyperplanes that are tangent to C. If $H_1 \in C^* - (C^*)_{\text{sing}}$ is a hyperplane that is simply tangent to C at one point, and if

$$\{H_t\}_{t \in \mathbb{C}, |t-1| < \varepsilon}$$

is a one-parameter family of hyperplanes with $H_t \in U$ for $t \neq 1$, and H_1 as as above, meeting C^* transversely at H_1, then it is easy to see that $H_t \cap C$ contains two points that come together to the point of tangency as $t \to 1$. These points interchange as t turns once around 1: Thus M contains a simple transposition. This proves the lemma.

Having established that M is the full symmetric group we know that $I(r)$ is connected. The subvariety

$$J = \{(p_1, \ldots, p_r; H): p_1, \ldots, p_r \text{ are dependent}\} \subset I(r)$$

is therefore a proper closed subvariety of $I(r)$, and, by irreducibility, $\dim J < r$. Thus J projects onto a proper subvariety of U and this proves the general position theorem.

As promised, this version yields a more general result.

Uniform Position Theorem. *Let $C \subset \mathbb{P}^r$, $r \geq 3$, be an irreducible, non-degenerate, possible singular, curve of degree d. If $\mathcal{D}$ is any linear system on C, and $\Gamma = H \cap C$ a hyperplane section general with respect to $\mathcal{D}$, then all subsets of* *m points of* Γ *impose the same number of conditions on* $\mathscr{D}$. *Equivalently, if* $\Gamma' \subset \Gamma$ *is any subset which fails to impose independent conditions on* $\mathscr{D}$, *then every divisor in* $\mathscr{D}$ *containing* Γ' *contains* Γ.

This property of Γ is called *uniform position.* Of course, applying the above theorem when $\mathscr{D}$ is the series cut out by hyperplanes in $\mathbb{P}^r$, we obtain the general position theorem.

§2. Castelnuovo's Bound, Noether's Theorem, and Extremal Curves

In this work we are primarily interested in exceptional special divisors. When C is non-hyperelliptic, these are the divisors whose points fail to be independent on the canonical curve $C \subset \mathbb{P}^{g-1}$. Before proceeding we make one remark. If D is a special divisor and D' is residual to D, then, in view of the Riemann–Roch theorem:

$$r(D) + g = r(D') + d + 1,$$

it will, at least in principle, suffice to consider divisors of degree $d \leq g - 1$. It is natural to pose the following question: What is the "most special" a divisor D can be? Because of the Riemann–Roch theorem, maximizing $i(D) = r(D') + 1$ is the same as maximizing $r(D)$. Taken literally, the above question has already been answered by Clifford's theorem: the bound in part (i) is, after all, sharp. But, as we saw in part (ii) of Clifford's theorem, the existence of g^r_d's on a hyperelliptic curve C that achieve this bound is simply a reflection of the fact that C possesses a g^1_2. Accordingly, we may refine our question by restricting our attention to linear series that are not composed with an involution[1]—i.e., those with the property that the mapping to projective space given by the linear series is birational onto the image curve—and ask for an upper bound $r(D) \leq R(g, d)$ under this additional hypothesis. For these reasons we consider an irreducible non-degenerate projective algebraic curve C_0 in $\mathbb{P}^r$ of degree d. If

$$\phi: C \to C_0 \subset \mathbb{P}^r$$

is the normalization map and $D \in |\phi^*\mathcal{O}_{C_0}(1)|$ is the pull-back of a hyperplane section, then the mapping ϕ is given by a linear subsystem of the complete system $|D|$; in particular, $r(D) \geq r$. Since we may expect the function $R(g, d)$ to be monotone increasing with d and decreasing with g, then our original question should be, and, as it will turn out, is equivalent to the problem: What

[1] Classically, the word "involution" means expressing C as a many-sheeted covering of another curve.

is the maximum genus of a curve C that admits a birational mapping ϕ onto a non-degenerate curve $C_0 \subset \mathbb{P}^r$ of degree d? When $r = 2$, there is a very classical answer provided by the genus formula

$$g = (d - 1)(d - 2)/2 - \sum_{p \in C_0} \delta_p,$$

where δ_p is a measure of the singularity of C_0 at p. In particular, since $\delta_p = 0$ if and only if p is a smooth point of C_0 we have that

$$g \leq (d - 1)(d - 2)/2,$$

with equality holding only for smooth plane curves. When $r \geq 3$, even for a smooth curve there is no formula for its genus in terms of its degree. The situation is therefore more complicated. However there is an inequality (known to Halphen [1] among others) for space curves, and this was extended to curves in $\mathbb{P}^r$ by G. Castelnuovo in 1889 [1]. We shall now derive his bound.

Castelnuovo's idea is to estimate the dimension

$$\alpha_l = r(lD)$$

of the complete linear systems associated to multiples of the hyperplane divisor D on C, and then to apply Clifford's theorem. To do this let

$$E_l \subset |lD|$$

be the linear subseries cut out on C by hypersurfaces of degree l and set

$$\beta_l = \dim E_l.$$

Let Γ be the pull-back to C of a general hyperplane section Γ_0 of C_0. Then, since E_l certainly contains the series $\Gamma + E_{l-1}$, we have

$$\beta_l - \beta_{l-1} \geq \dim E_l - \dim E_l(-\Gamma),$$

where $E_l(-\Gamma)$ is the subseries of divisors in E_l containing Γ. Hence, denoting by $\mathscr{I}_{C_0}$ the ideal sheaf of $C_0 \subset \mathbb{P}^r$, we have

$$\begin{aligned} \beta_l - \beta_{l-1} &\geq (h^0(\mathbb{P}^r, \mathcal{O}(l)) - h^0(\mathbb{P}^r, \mathscr{I}_{C_0}(l))) \\ &\quad - (h^0(\mathbb{P}^r, \mathscr{I}_{\Gamma_0}(l)) - h^0(\mathbb{P}^r, \mathscr{I}_{C_0}(l))) \\ &= h^0(\mathbb{P}^r, \mathcal{O}(l)) - h^0(\mathbb{P}^r, \mathscr{I}_{\Gamma_0}(l)), \end{aligned}$$

which is just the number ρ_l of conditions imposed by Γ_0 on hypersurfaces of degree l in $\mathbb{P}^r$ or, equivalently, on hypersurfaces of degree l contained in the hyperplane $H \cong \mathbb{P}^{r-1}$ which cuts out Γ_0. What is needed then is an estimate on ρ_l; and this is provided by

Lemma. *If $p_1, \ldots, p_d$ is a set of points in $\mathbb{P}^{r-1}$ such that any r among them are linearly independent, then $p_1, \ldots, p_d$ impose at least*

$$\min(d, k(r-1)+1)$$

independent conditions on the homogeneous polynomials of degree k.

Proof. We first assume that $k(r-1) < d$ and label the points as

$$\underbrace{p_1, \ldots, p_{r-1}}_{H_1}; \ldots; \underbrace{p_{(k-1)(r-1)+1}, \ldots, p_{k(r-1)}}_{H_k}; \ldots; p_d.$$

Let L_j be a non-zero linear function vanishing at the points in the group H_j. By the assumption of the lemma L_j will be non-zero at the remaining points, and in particular $L_j(p_d) \neq 0$. Then the homogeneous polynomial of degree k

$$F = L_1 \cdot \ldots \cdot L_k$$

will vanish at $p_1, \ldots, p_{k(r-1)}$ but not at p_d. Since our labeling of the points p_i is completely arbitrary, we may find a homogeneous form assuming preassigned values at any of $k(r-1)+1$ points selected from D, which implies the lemma when $k(r-1)+1 \leq d$. The remaining case is similar (and easier). Q.E.D.

According to the lemma, then, we have

$$\beta_l - \beta_{l-1} \geq \rho_l \geq \min(d, l(r-1)+1). \tag{2.1}$$

Thus, if we set

$$m = \left[\frac{d-1}{r-1}\right],$$

we have

$$\begin{cases} \alpha_0 \geq \beta_0 = 0, \\ \alpha_1 \geq \beta_1 \geq r, \\ \alpha_2 \geq \beta_2 \geq r + 2(r-1) + 1 = 2r - 1, \\ \vdots \\ \alpha_m \geq \beta_m \geq \sum_{l=1}^{m} l(r-1) + 1 = \binom{m+1}{2}(r-1) + m. \end{cases} \tag{2.2}$$

Applying Clifford's theorem to the series $|mD|$ we see that mD is non-special; hence by the Riemann–Roch theorem,

$$
\begin{aligned}
g &= \deg(mD) - r(mD) \\
&= md - \alpha_m \\
&\leq md - \binom{m+1}{2}(r-1) - m \\
&= \binom{m}{2}(r-1) + m\varepsilon,
\end{aligned}
\tag{2.3}
$$

where

$$d - 1 = m(r-1) + \varepsilon.$$

We have proved, then

Castelnuovo's Bound. *Let C be a smooth curve that admits a birational mapping onto a non-degenerate curve of degree d in $\mathbb{P}^r$. Then the genus of C satisfies the inequality*

$$g(C) \leq \pi(d, r),$$

where Castelnuovo's number $\pi(d, r)$ is defined by

$$\pi(d, r) = \frac{m(m-1)}{2}(r-1) + m\varepsilon,$$

where

$$m = \left[\frac{d-1}{r-1}\right]$$

$$d - 1 = m(r-1) + \varepsilon.$$

Observe that, for fixed r and large d, asymptotically

$$\pi(d, r) \sim \frac{d^2}{2(r-1)}.$$

A closer look at the argument for Castelnuovo's bound gives more than the bound itself. Since E_l is a subseries of $|\mathcal{O}_C(lD)|$, Γ imposes at least as many conditions on $|\mathcal{O}_C(lD)|$ as on E_l, so that we have

$$\alpha_l - \alpha_{l-1} \geq \rho_l \geq \min(d, l(r-1) + 1)$$

as well as (2.1). Thus if equality holds in (2.3), it follows that we must have

$$\alpha_l - \alpha_{l-1} = \beta_l - \beta_{l-1} = \min(d, l(r-1)+1).$$

In particular, we have $\alpha_k = \beta_k$ for all k; from this we infer that if equality holds in Castelnuovo's bound then the mappings

$$\mathrm{Sym}^k H^0(C, \mathcal{O}(D)) \to H^0(C, \mathcal{O}(kD))$$

are surjective for all $k \geq 1$. Equivalently, the image of

$$\phi: C \to \mathbb{P}^r = \mathbb{P}H^0(C, \mathcal{O}(D))^*$$

is a projectively normal curve that is biholomorphic to C under the mapping ϕ. (Recall that for curves, and only for curves, projective normality implies smoothness.) We shall refer to curves whose genus equals Castelnuovo's number as *extremal curves*. Since ϕ_L maps C biholomorphically to its image C_0, we may identify C with C_0. To obtain a clearer picture of extremal curves we shall divide them in three cases according to the possibilities

$$d < 2r, \qquad d = 2r, \qquad d > 2r.$$

(i) When $d < 2r$ it follows from Castelnuovo's bound—or better from Clifford's theorem—that D is non-special and consequently $r = d - g$. Then $d < 2r = 2d - 2g$ implies that $d > 2g$, and in this case it is well known that the map ϕ is an embedding.

(ii) When $d = 2r$, Castelnuovo's formula gives us $r = g - 1$, and therefore $C \subset \mathbb{P}^{g-1}$ is a canonical curve. A consequence of the projective normality of extremal curves is then the very famous

Max Noether's Theorem. *If C is a non-hyperelliptic curve, then the homomorphisms*

$$\mathrm{Sym}^l H^0(C, K) \to H^0(C, K^l)$$

are surjective for $l \geq 1$.

In particular, since

$$\dim \mathrm{Sym}^2 H^0(C, K) = g(g+1)/2,$$
$$\dim H^0(C, K^2) = 3g - 3,$$

we have that a canonical curve of genus g lies on exactly $(g-2)(g-3)/2$ linearly independent quadrics. These quadrics will be the subject of considerable discussion in the next section and in Section 4 of Chapter VI. Here we shall give two examples.

When $g = 4$ we find that a canonical curve lies on a *unique* quadric in $\mathbb{P}^3$. A similar count shows that there are five independent cubics containing the curve, hence at least one cubic containing the curve but not the quadric, so that a non-hyperelliptic curve of genus 4 is the complete interesection of a quadric and a cubic in $\mathbb{P}^3$.

When $g = 5$ we see that a canonical curve $C \subset \mathbb{P}^4$ lies on exactly three linearly independent quadrics and is generally the complete intersection of these quadrics. More precisely, we recall that a curve C is said to be *trigonal* in case it has a base-point-free g_3^1; i.e., there is a map $f: C \to \mathbb{P}^1$ of degree 3, and we will later show that a non-hyperelliptic, non-trigonal curve of genus 5 is the complete intersection of three quadrics in $\mathbb{P}^4$ and conversely.

(iii) When $d \geq 2r + 1$, we are in the range of exceptional special divisors; we shall first make a few remarks about these beautiful curves when $r \leq 3$, and then we shall discuss these in general.

First, when $r = 2$ Castelnuovo's bound gives the previously noted inequality

$$g \leq (d - 1)(d - 2)/2;$$

in this case extremal curves are smooth plane curves. The hyperplane section consists of d points on a line and there is not much more to be said about these here.

When $r = 3$ the estimate gives

$$g \leq \begin{cases} (k - 1)^2 & \text{when } d = 2k \text{ is even,} \\ k(k - 1) & \text{when } d = 2k + 1 \text{ is odd.} \end{cases} \tag{2.4}$$

Moreover, if equality holds, then a general hyperplane section gives a configuration D of $d \geq 5$ points in $\mathbb{P}^2$ that imposes only $\min(d, 2 \cdot 2 + 1) = 5$ conditions on the linear system of plane conics. In other words, any conic that passes through five points of D must contain all of D, and from this it follows that the points of D lie on a unique plane conic. Finally, since equality holds in (2.1) we have

$$\begin{aligned} h^0(C, \mathcal{O}_C(2)) &= h^0(C, \mathcal{O}_C(1)) + 5 \\ &= 9 \\ &= h^0(\mathbb{P}^3, \mathcal{O}_{\mathbb{P}^3}(2)) - 1, \end{aligned}$$

i.e., if C is an extremal curve of degree $d \geq 7$, then C lies on a unique quadric Q. In fact, we can say more: Assuming that the quadric is smooth we recall that it has two rulings $|L_1|, |L_2|$ by straight lines, and consequentely any curve C on Q has two numerical characters

$$m = C \cdot L_1,$$

$$n = C \cdot L_2,$$

whose sum $m + n$ is the degree of C. The canonical bundle K_Q is represented by the divisor $-2L_1 - 2L_2$, and by the adjunction formula the genus of C is

$$\begin{aligned} g &= \tfrac{1}{2}(K_Q \cdot C + C \cdot C) + 1 \\ &= mn - m - n + 1 \\ &= (m - 1)(n - 1). \end{aligned}$$

Assuming $m \geq n$, this is maximized when

$$m = n = k, \quad \text{in case} \quad d = 2k,$$

$$m - 1 = n = k, \quad \text{in case} \quad d = 2k + 1.$$

Moreover, in these cases equality holds in (2.4). In case the quadric Q is smooth we may draw the following conclusion.

Lemma. *An extremal space curve of degree $d = 2k$ is the complete intersection of a quadric and a surface of degree k. When $d = 2k + 1$ then, for some line L lying on the quadric, the sum $C + L$ is the complete intersection of a quadric and a surface of degree $k + 1$.*

If Q is not smooth, a similar argument can be used to prove the lemma.

We note that if C is any curve lying on a smooth quadric Q in $\mathbb{P}^3$, the points of a general hyperplane section lie on a conic and so impose the minimum number of conditions on the curves of degree k in that hyperplane, i.e., $\min\{d, 2k + 1\}$; the difference between extremal curves, i.e., those of "balanced type" (m, m) or $(m, m - 1)$, and the others, i.e., those of unbalanced type (m, n), where $|m - n| \geq 2$, is that the latter are not projectively normal.

Although the geometry of extremal curves in $\mathbb{P}^r$ for $r \geq 4$ is analogous to that of extremal curves in $\mathbb{P}^3$, it is certainly more delicate. In $\mathbb{P}^3$, from the string of inequalities (2.2) used to derive Castelnuovo's bound, we deduced that an extremal curve lay on a quadric surface. Knowing all curves on such a surface, we were able to characterize geometrically the curves of maximum genus. In the general case we will again use the same string of inequalities to infer that an extremal curve $C \subset \mathbb{P}^r$ must lie on a large number of quadric hypersurfaces. Then, under the hypothesis

$$\deg C = d \geq 2r + 1,$$

we shall argue that C lies on a surface of a very special type, and finally we will be able to characterize extremal curves. Let us then consider an extremal curve $C \subset \mathbb{P}^r$ of degree d and maximal genus $g = \pi(d, r)$. As in the case of space curves, equalities must hold throughout (2.2). In particular,

$$h^0(C, \mathcal{O}(D)) = r + 1,$$

and

$$h^0(C, \mathcal{O}(2D)) = 3r,$$

where, as usual, D denotes a hyperplane section of C. Since

$$h^0(\mathbb{P}^r, \mathcal{O}_{\mathbb{P}^r}(2)) = \binom{r+2}{2},$$

we conclude that C lies on $(r-1)(r-2)/2$ linearly independent quadrics. Of course, in case $r = 3$, we get back the fact that C lies on a unique quadric. The question we must ask is: What is the intersection of the quadrics that contain C? The answer to this question is a consequence of the following beautiful result.

Castelnuovo's Lemma. *Let d and r be integers such that $r \geq 3$ and $d \geq 2r + 1$. Let $\Gamma \subset \mathbb{P}^{r-1}$ be a collection of d points in general position which impose only $2r - 1$ conditions on quadrics. Then Γ lies on a unique rational normal curve X_Γ.*

Castelnuovo's ingenious argument, which actually gives a synthetic construction of X_Γ, is presented in Griffiths–Harris [1, pages 528–531], and we will not reproduce it here.

Consider then our extremal curve $C \subset \mathbb{P}^r$. Let H be a general hyperplane in $\mathbb{P}^r$ and set $\Gamma = C \cap H$. By the general position theorem Γ consists of d distinct points in general position in $H \cong \mathbb{P}^{r-1}$. As we already noticed, $h^0(C, \mathcal{O}(D)) = r + 1$ and $h^0(C, \mathcal{O}(2D)) = 3r$. We can then apply Castelnuovo's lemma and conclude that Γ lies on a unique rational normal curve $X_\Gamma \subset H$. Since

$$d = \deg(\Gamma) \geq 2r + 1 \geq 2 \deg X_\Gamma,$$

every quadric in $\mathbb{P}^r$ containing C, and hence Γ, also contains the rational normal curve X_Γ. Conversely, as we noticed in Section 5 of Chapter II, X_Γ is the intersection of the $(r-1)(r-2)/2$ quadrics containing it, hence the intersection of the quadrics containing C meets H exactly in X_Γ. Thus the intersection of the quadrics containing C is a surface S whose general hyperplane section is a rational normal curve; in particular $\deg S = r - 1$. As a special case of Theorem (5.3) of Chapter II, we recall that, with the exception of the Veronese surface in $\mathbb{P}^5$, the non-degenerate surfaces of minimal degree $r - 1$ in $\mathbb{P}^r$ are the rational normal scrolls.

Let us consider the case of a smooth rational normal scroll S and let us find, on it, extremal curves. The Picard group of S is easy to describe. In fact $\mathrm{Pic}(S)$ is freely generated, over $\mathbb{Z}$, by the classes H of a hyperplane section of S, and L of a line of the ruling. The intersection products are given by

$$H \cdot H = r - 1, \qquad H \cdot L = 1; \qquad L \cdot L = 0.$$

Each of the linear systems $|H|$ and $|L|$ contains smooth rational curves, and the adjunction formula gives

$$0 = p_a(L) = \frac{L \cdot L + K_S \cdot L}{2} + 1,$$

$$0 = p_a(H) = \frac{H \cdot H + K_S \cdot H}{2} + 1.$$

From this we obtain

$$K_S = -2H + (r - 3)L.$$

We may now compute the (arithmetic) genus $p_a(C)$ of any curve $C \in |\alpha H + \beta L|$. Namely,

$$\begin{aligned} p_a(C) &= \frac{C \cdot C + K_S \cdot C}{2} + 1 \\ &= \frac{(\alpha - 1)(\alpha - 2)}{2}(r - 1) + (r - 2 + \beta)(\alpha - 1). \end{aligned}$$

We write the degree d of C as

$$d = m(r - 1) + 1 + \varepsilon,$$

and also as

$$d = (r - 1)\alpha + \beta.$$

Comparing these we see that the maximal genus

$$\pi(d, r) = \frac{m(m - 1)}{2}(r - 1) + m\varepsilon$$

of C is attained exactly when

$$\begin{aligned} &\alpha = m + 1, \qquad \beta = \varepsilon - r + 2 \quad \text{for any } \varepsilon, \\ &\alpha = m, \qquad\quad\ \ \beta = 1 \qquad\qquad\ \ \text{in case } \varepsilon = 0. \end{aligned}$$

A similar analysis may be carried out in the case in which S is a cone over a rational normal curve, and an even easier argument may be given for curves on a Veronese surface. The final conclusion is this:

(2.5) Theorem. *Let d and r be integers such that $r \geq 3$, $d \geq 2r + 1$. Set $m = [(d - 1)/(r - 1)]$ and $d = m(r - 1) + 1 + \varepsilon$. Then extremal curves $C \subset \mathbb{P}^r$ of degree d exist and any such curve is one of the following:*

(i) *The image of a smooth plane curve $C \subset \mathbb{P}^2$ of degree k under the Veronese map $\mathbb{P}^2 \to \mathbb{P}^5$. In this case $r = 5$, $d = 2k$.*
(ii) *A non-singular member of the linear system $|mH + L|$ on a rational normal scroll. In this case $\varepsilon = 0$.*
(iii) *A non-singular element of the linear system $|(m + 1)H - (r - \varepsilon - 2)L|$ on a rational normal scroll.*

Remark that in case (iii), when $\varepsilon = r - 2$, the curve is the complete intersection of a scroll and a hypersurface of degree $m + 1$; in case (iii), for any ε, adding $r - \varepsilon - 2$ lines of the ruling to an extremal curve we get a complete intersection.

The characterization of extremal curves in terms of the surface S yields some corollaries. For example, since the lines of the ruling of a rational normal scroll will cut on C a pencil of degrees m or $m + 1$, we have the

(2.6) Corollary. *If $C \subset \mathbb{P}^r$ is an extremal curve of degree*

$$d = m(r - 1) + 1 + \varepsilon \geq 2r + 1$$

and is not a plane curve, then either:

(i) *C possesses a complete[2] g^1_m, call it $|E|$, satisfying*

$$(r - 2)E \in |K(-(m - 2)H)|,$$

where H is the hyperplane divisor, in case $\varepsilon = 0$; or,
(ii) *C possesses a complete[2] g^1_{m+1}, call it $|E|$, satisfying*

$$(\varepsilon - 1)E \in |K(-(m - 1)H)|,$$

in general; and finally,
(iii) *in case $\varepsilon = 1$, we have*

$$K = \mathcal{O}((m - 1)H),$$

[2] That the g^1_m, or the g^1_{m+1}, is complete, does not follow directly from the above analysis. A proof is outlined in one of the exercises.

i.e., any extremal curve $C \subset \mathbb{P}^r$ of degree

$$d \equiv 2 \quad (\textit{modulo } r - 1), \qquad d \geq 2r$$

is a Klein canonical curve.

We note that one may also use this description of extremal curves to give particularly nice plane models for them; cf. Accola [2].

We close this section by observing that an analysis similar to Castelnuovo's may also be carried out for curves $C \subset \mathbb{P}^r$ whose genus is close to, but not necessarily equal to $\pi(d, r)$. Specifically, we can go back to the inequalities (2.1), (2.2), and (2.3) and ask under what circumstances some may fail; the point is that they do not fail independently (thus, for example, by Castelnuovo's lemma strict inequality in (2.1) for $l = 2$ implies at least strict inequality for all $l < m$). One conclusion we may arrive at is the following: for any d and r, set

$$m_1 = \left[\frac{d-1}{r}\right], \qquad \varepsilon_1 = d - m_1 r - 1,$$

$$\mu_1 = \begin{cases} 1 & \text{if } \varepsilon_1 = r - 1, \\ 0 & \text{if } \varepsilon_1 \neq r - 1, \end{cases}$$

and

$$\pi_1(d, r) = \binom{m_1}{2} r + m_1(\varepsilon_1 + 1) + \mu_1$$

Then we have the statement (cf. Eisenbud–Harris [3]).

(2.7) Theorem. *Any irreducible, reduced, and non-degenerate curve $C \subset \mathbb{P}^r$ of degree $d \geq 2r + 3$ and genus $g > \pi_1(d, r)$ lies on a surface of degree $r - 1$.*

A complete proof of this theorem is beyond our present scope; a proof under the additional hypothesis that $d \gg r$ is outlined in Exercise batch 3J below.

§3. The Enriques–Babbage Theorem and Petri's Analysis of the Canonical Ideal

The statement of Castelnuovo's lemma, which we recalled in the previous section, requires that the set $\Gamma \subset \mathbb{P}^r$ contain at least $2r + 1$ points. This condition cannot be improved: the intersection of three general quadrics in $\mathbb{P}^3$ is a collection of eight points in general position imposing only seven conditions on quadrics, and it does not lie on a twisted cubic curve. Thus our analysis of extremal curves does not apply to canonical curves, which are extremal curves of degree $2g - 2$ in $\mathbb{P}^{g-1}$. In the case of a canonical curve C we can prove the following:

(3.1) Proposition. *If the intersection of the quadrics containing a canonical curve C contains one additional point $p \notin C$, then C lies on either the Veronese surface (in case $g = 6$) or on a rational normal scroll.*

Proof. We first argue that we may assume p does not lie on infinitely many chords of the canonical curve $C \subset \mathbb{P}^{g-1}$; or, equivalently, that the projection of C from p to $\mathbb{P}^{g-2}$ is birational. For if p did lie on infinitely many chords, then every such chord would lie in all the quadrics passing through C. Clearly a general point of the surface S swept out by those chords does not lie on ∞^1 chords; if it did, the union of all the chords to C would be contained in S. We may accordingly replace p by a general point of S and proceed. Thus we may assume that the projection of C from p is birational onto a non-degenerate curve $C' \subset \mathbb{P}^{g-2}$. The hyperplane sections of C' are cut out on C by hyperplanes $H \subset \mathbb{P}^{g-1}$ passing through p. Applying the uniform position theorem we see that the points of $\Gamma = H \cap C$, together with p, form a collection of $2(g - 1) + 1$ points in general position in $H \cong \mathbb{P}^{g-2}$ that impose only $2g - 3$ conditions on quadrics. Castelnuovo's lemma then applies and we conclude that the base S of the linear system of quadrics through C intersects H in a rational normal curve. Thus S is either a Veronese surface on a rational normal scroll. This scroll cannot be a cone over a rational normal curve since C is non-hyperelliptic. Hence such a scroll must be smooth. Q.E.D.

The preceding proposition is the central step in the proof of a nice theorem due to Enriques and Babbage stating that, with only few exceptions, a canonical curve is, set-theoretically, an intersection of quadrics. To characterize those canonical curves that are not intersections of quadrics we use the proposition we just proved and notice that, exactly as in Corollary (2.6), one can easily show that if C is any extremal curve of degree $d = m(r - 1) + 1 + \varepsilon$ in $\mathbb{P}^r$ lying on a scroll, then C possesses a g^1_{m+1}. For canonical curves we have that $r = g - 1, d = 2g - 2$, and $m = 2$. Consequently C is trigonal; i.e., it has a g^1_3. In fact the g^1_3 is the linear system cut out on $C \subset S$ by the ruling of the scroll. Finally, if the canonical curve $C \subset \mathbb{P}^5$ lies on the Veronese surface the curve must be a smooth plane quintic. Summarizing we have proved the

Enriques–Babbage Theorem. *If $C \subset \mathbb{P}^{g-1}$ is a canonical curve, then either C is set-theoretically cut out by quadrics, or C is trigonal or is isomorphic to a plane quintic.*

In a very beautiful and fundamental paper Petri goes far beyond this set-theoretical description. He constructs the first few steps of a resolution of the ideal of a canonical curve by constructing a basis of its homogeneous coordinate ring and a set of generators for the first module of syzygies. In the remaining part of this section we shall carry out Petri's analysis.

Let us consider a smooth non-hyperelliptic curve C of genus g, which we imagine embedded in its canonical space $\mathbb{P}^{g-1}$. The first step in Petri's analysis is to construct an explicit basis for the vector space $H^0(C, K^n)$ of n-canonical differentials on C. Let $p_1, \dots, p_g$ be g general points on C. By the uniform position theorem, which we proved in the first section of this chapter, and by the geometric version of the Riemann–Roch theorem, the dimension of the vector space $H^0(C, K(-p_1 - \cdots - \hat{p}_i - \cdots - p_g))$, $i = 1, \dots, g$, is equal to one. Let ω_i be a generator of this vector space. Thus, as a section of K, ω_i vanishes at all the p_j's, with the sole exception of p_i. It is then clear that $\omega_1, \dots, \omega_g$ is a basis of $H^0(C, K)$; in fact, for any relation $\sum \lambda_i \omega_i = 0$, evaluating at p_j gives $\lambda_j = 0$. Moreover, by general position, we may assume that the divisors (ω_i) consist of $2g - 2$ distinct points and that their supports are pairwise disjoint. We shall adopt the following notation. We let

$$\psi_n: H^0(\mathbb{P}^{g-1}, \mathcal{O}_{\mathbb{P}^{g-1}}(n)) \to H^0(C, K^n)$$

be the natural map. We also let $X_1, \dots, X_g$ be the basis of $H^0(\mathbb{P}^{g-1}, \mathcal{O}_{\mathbb{P}^{g-1}}(1))$ defined by

$$X_i = \psi_1^{-1}(\omega_i), \qquad i = 1, \dots, g.$$

The X_i's can be thought of as homogenous coordinates in $\mathbb{P}^{g-1}$. Given a homogeneous polynomial

$$P = P(X_1, \dots, X_g) \in H^0(\mathbb{P}^{g-1}, \mathcal{O}_{\mathbb{P}^{g-1}}(n)),$$

we set

$$\bar{P} = \psi_n(P),$$

so that, for instance, we have: $\overline{X_1^2 X_3} = \omega_1^2 \omega_3$. We also denote by $I = \bigoplus I_k$ the homogeneous ideal of C and by

$$R(C) = \left(\bigoplus_n H^0(\mathbb{P}^{g-1}, \mathcal{O}_{\mathbb{P}^{g-1}}(n)) \right) / I$$

the homogeneous coordinate ring of C. Now set $D = p_3 + \cdots + p_g$. Again by general position the dimension of the vector space $H^0(C, K(-D))$ is equal to two and its basis is $\{\omega_1, \omega_2\}$. Moreover, by our assumption that the divisors (ω_i) have no point in common, the pencil $|K(-D)|$ is base-point-free. We next consider the tower

$$(3.2) \quad H^0(C, K^n) \supset H^0(C, K^n(-D)) \supset \cdots \supset H^0(C, K^n((-n+1)D)).$$

When $n - 1 \geq s \geq 1$, by the Riemann–Roch theorem we get

$$h^0(C, K^n(-sD)) = (2n - 1)(g - 1) - s(g - 2),$$

so that each vector space in the above filtration has codimension $g - 2$ in the preceding one. The goal is to exhibit, for each s, a set of n-canonical forms in $H^0(C, K^n(-sD))$ which are linearly independent modulo

$$H^0(C, K^n((-s - 1)D))$$

and give a step-by-step construction of a basis for $H^0(C, K^n)$. To find the basis of $H^0(C, K(-sD))$ we proceed inductively by considering, for each n and s such that $n - 1 \geq s \geq 1$, the cup-product mapping

$$(3.3)\quad \phi_{n,s}\colon H^0(C, K^{n-1}((-s + 1)D)) \otimes H^0(C, K(-D)) \to H^0(C, K^n(-sD)).$$

The kernels of these maps can be computed by means of the following lemma.

Base-Point-Free Pencil Trick. *Let C be a smooth curve, L an invertible sheaf on C, and $\mathscr{F}$ a torsion-free $\mathcal{O}_C$-module. Let s_1 and s_2 be linearly independent sections of L, and denote by V the subspace of $H^0(C, L)$ they generate. Then the kernel of the cup-product map*

$$V \otimes H^0(C, \mathscr{F}) \to H^0(C, \mathscr{F} \otimes L)$$

is isomorphic to $H^0(C, \mathscr{F} \otimes L^{-1}(B))$, where B is the base locus of the pencil spanned by s_1 and s_2.

Proof. Let $s_1 \otimes t_2 - s_2 \otimes t_1$ be an element of the kernel; thus $s_1 t_2 - s_2 t_1 = 0$. Write

$$s_1 = r_1 s, \qquad s_2 = r_2 s,$$

where s is a section of $\mathcal{O}(B)$ vanishing on B, and r_1, r_2 are sections of $L(-B)$. Since $\mathscr{F}$ is torsion-free we have that $r_1 t_2 - r_2 t_1 = 0$; as r_1, r_2 have no common zeros, this means that t_i is a section of $\mathscr{F}(-D_i)$, $i = 1, 2$, where D_i is the divisor of zeros of r_i. In other words, there are sections τ_1, τ_2 of $\mathscr{F} \otimes L^{-1}(B)$ such that $t_i = r_i \tau_i$. Plugging this into $r_1 t_2 - r_2 t_1 = 0$ we get, using again the fact that $\mathscr{F}$ is torsion-free, that $\tau_1 = \tau_2 = \tau$. In conclusion

$$t_i = r_i \tau, \qquad i = 1, 2.$$

Conversely, it is clear that any element τ of $H^0(C, \mathscr{F} \otimes L^{-1}(B))$ gives rise, by the above prescription, to a relation $s_1 t_2 - s_2 t_1 = 0$. Q.E.D.

Applying the base-point-free pencil trick to (3.3) yields

$$\operatorname{Ker} \phi_{n,s} \cong H^0(C, K^{n-2}((-s+2)D)).$$

Combining this with the Riemann–Roch theorem and making a straightforward computation, we get the

Lemma. *Let* $n - 1 \geq s \geq 1$. *Then* $\phi_{n,s}$ *is surjective except if* $(n, s) = (3, 2)$. *In this case* $\operatorname{Im} \phi_{3,2}$ *is of codimension one in* $H^0(C, K^3(-2D))$.

Let us now begin Petri's construction.

Basis for $H^0(C, K^2)$. Consider the cup-product mapping

$$\phi_{2,1}: H^0(C, K) \otimes H^0(C, K(-D)) \to H^0(C, K^2(-D)).$$

By the lemma this mapping is surjective and it is immediate to check that the elements

$$\omega_1^2, \omega_1\omega_2, \omega_2^2, \omega_1\omega_i, \omega_2\omega_i, \qquad i = 3, \ldots, g,$$

form a basis of $H^0(C, K^2(-D))$. Now consider the two-step tower

$$H^0(C, K^2) \supset H^0(C, K^2(-D)).$$

It is then clear that $\omega_3^2, \ldots, \omega_g^2$ are differentials in $H^0(C, K^2)$ which are linearly independent modulo $H^0(C, K^2(-D))$. Since the latter space is of codimension $g - 2$ in the former one, we get a basis of $H^0(C, K^2)$:

$$\left.\left.\begin{array}{ll} \omega_1^2, \omega_1\omega_2, \omega_2^2 & \\ \omega_1\omega_i, & i = 3, \ldots, g \\ \omega_2\omega_i, & i = 3, \ldots, g \end{array}\right] \text{basis of } H^0(C, K^2(-D)) \atop \begin{array}{ll} \omega_i^2, & i = 3, \ldots, g \end{array}\right] \text{basis of } H^0(C, K^2).$$

Basis for $H^0(C, K^3)$. This one is, so to speak, the only anomalous case. The difference comes from the fact that, as stated in the lemma, the map

$$\phi_{3,2}: H^0(C, K^2(-2D)) \otimes H^0(C, K(-D)) \to H^0(C, K^3(-2D))$$

is not surjective. We set $\overline{W} = \operatorname{Im} \phi_{3,2}$. By the lemma we know that $\overline{W}$ is of codimension one in $H^0(C, K^3(-2D))$. Let us then consider the filtration

$$H^0(C, K^3) \supset H^0(C, K^3(-D)) \supset H^0(C, K^3(-2D)) \supset \overline{W}.$$

Again, with the exception of $\overline{W}$, each vector space of this tower is of codimension $g - 2$ in the preceding one. We first claim that a basis of $\overline{W}$ is given by

$$\omega_1^3, \omega_1^2\omega_2, \omega_1\omega_2^2, \omega_2^3,$$
$$\omega_1^2\omega_i, \omega_1\omega_2\omega_i, \omega_2^2\omega_i, \qquad i = 3, \ldots, g.$$

Since $\overline{W}$ is $(3g - 2)$-dimensional it suffices to show that the above differentials are linearly independent. Suppose there is a relation

$$\alpha\omega_1^2 + \beta\omega_1\omega_2 + \gamma\omega_2^2 = 0,$$

with $\alpha, \beta, \gamma \in H^0(C, K)$. Setting

$$(\omega_2) = \sum_{\substack{i=1\\ i\neq 2}}^{g} p_i + \sum_{i=1}^{g-1} q_i,$$

and recalling that (ω_1) and (ω_2) have disjoint supports, we see that α belongs to the one-dimensional vector space $H^0(C, K(-p_1 - \sum_{i=1}^{g-1} q_i))$, which is generated by ω_2. Therefore $\alpha = \lambda\omega_2$, and similarly $\gamma = \mu\omega_2$, with $\lambda, \mu \in \mathbb{C}$. We then get

$$\lambda\omega_1^2\omega_2 + \beta\omega_1\omega_2 + \mu\omega_1\omega_2^2 = 0.$$

Writing $\beta = \sum_{i=3}^{g} \beta_i\omega_i$, and looking at the order of zero at p_i, $i = 3, \ldots, g$, we see that $\beta_i = 0$, for all i. This leads to a relation $\lambda\omega_2\omega_1^2 = \mu\omega_1\omega_2^2$, and considering the order of zero at p_1, we see that $\lambda = \mu = 0$. We next try to construct, as canonically as possible, an element of $H^0(C, K^3(-2D))$ which, together with the basis of $\overline{W}$, generates $H^0(C, K^3(-2D))$, or, what is the same, an element of $H^0(C, K^3(-2D)) - \overline{W}$. Actually, we shall exhibit one such element for each index $i = 3, \ldots, g$. For an index i in this range, we consider the unique divisor in the pencil $|K(-D)|$ having a zero of order two at p_i, and let $\bar{\alpha}_i$ be a differential having this as its zero divisor. We claim that $\bar{\alpha}_i\omega_1^2 \in H^0(C, K^3(-2D)) - \overline{W}$. Suppose not; writing $\bar{\alpha}_i = \lambda_i\omega_1 + \mu_i\omega_2$ we then have

$$\bar{\alpha}_i\omega_1^2 = \omega_1^2\psi + \omega_1\omega_2\phi + \omega_2^2\chi,$$

with $\psi, \phi, \chi \in H^0(C, K)$. A moment of reflection shows that, for some $\psi', \phi', \chi' \in H^0(C, K)$, one may write the above relation in the form

$$\bar{\alpha}_i\omega_i^2 = \bar{\alpha}_i\omega_1\psi' + \bar{\alpha}_i\omega_2\phi' + \omega_2^2\chi'.$$

Let $(\bar{\alpha}_i) = p_3 + \cdots + 2p_i + \cdots + p_g + D_i$; since, by hypothesis, $|K(-D)|$ is base-point-free, for each i, the two divisors D_i and (ω_2) are disjoint. The above relation shows then that $\chi' \in H^0(C, K(-D_i))$, which is a one-dimensional vector space generated by $\bar{\alpha}_i$. Our relation reduces then to the form

$$\bar{\alpha}_i \omega_i^2 = \bar{\alpha}_i \left(\sum_{j=1}^{g} c_j \omega_1 \omega_j + d_j \omega_2 \omega_j \right),$$

and looking at the order of vanishing at p_k we easily get a contradiction, proving that, for each $i = 3, \ldots, g$, the differential $\bar{\alpha}_i \omega_i^2$ generates

$$H^0(C, K^3(-2D)) \text{ modulo } \overline{W}.$$

Altering each $\bar{\alpha}_i$ by a suitable non-zero constant, it is possible to write

$$\bar{\alpha}_i \omega_i^2 = \bar{\eta} + \bar{\theta}_i, \qquad \bar{\theta}_i \in \overline{W}, \qquad i = 3, \ldots, g, \tag{3.4}$$

so that $\bar{\eta}$ too generates $H^0(C, K^3(-2D))$ modulo $\overline{W}$. Let us now choose $\bar{\beta}_i \in H^0(C, K(-D))$ in such a way that $\{\bar{\alpha}_i, \bar{\beta}_i\}$ is a basis for $H^0(C, K(-D))$. Using the fact that each step in the filtration

$$H^0(C, K^3) \supset H^0(C, K^3(-D)) \supset H^0(C, K^3(-2D))$$

is of codimension $g - 2$ in the preceding one, we can finally construct a basis of $H^0(C, K^3)$ in the following way:

$$\begin{array}{lll}
\omega_1^3, \omega_1^2\omega_2, \omega_1\omega_2^2, \omega_2^3 & & \\
\omega_1^2\omega_i, & i = 3, \ldots, g, & \\
\omega_1\omega_2\omega_i, & i = 3, \ldots, g, & \\
\omega_2^2\omega_i, & i = 3, \ldots, g, & \big] \text{ basis of } \overline{W} \\
\bar{\eta} & & \big] \text{ basis of } H^0(C, K^3(-2D)) \\
\bar{\beta}_i\omega_i^2, & i = 3, \ldots, g, & \big] \text{ basis of } H^0(C, K^3(-D)) \\
\omega_i^3, & i = 3, \ldots, g, & \big] \text{ basis of } H^0(C, K^3).
\end{array}$$

Basis for $H^0(C, K^n)$, $n \geq 4$. To build a basis for $H^0(C, K^n)$ we use the filtration (3.2). First of all we construct a basis for $H^0(C, K^n((-n+1)D))$ by using the basis of $H^0(C, K^3(-2D))$, which we just constructed, and an induction procedure based on the surjectivity of the maps

$$\phi_{m,m-1}\colon H^0(C, K^{m-1}((-m+2)D)) \otimes H^0(C, K(-D)) \\ \to H^0(C, K^m((-m+1)D))$$

for $m \geq 4$. Specifically one checks that a basis for $H^0(C, K^n((-n+1)D))$ is given by

$$\begin{array}{lll} \omega_1^l \omega_2^m, & l+m=n, & \\ \omega_1^s \omega_2^t \omega_i, & s+t=n-1, & i=3,\ldots,g, \\ \omega_1^h \omega_2^k \bar{\eta}, & h+k=n-2. & \end{array}$$

To get a basis for $H^0(C, K^n)$ we now climb the ladder and notice that a basis of $H^0(C, K^n((-s+1)D))$ modulo $H^0(C, K^n(-sD))$, $s \leq n-1$, is given by the differentials $\bar{\beta}_i^{s-1}\omega_i^{n-s+1}$, $i=3,\ldots,g$. In conclusion, a basis for $H^0(C, K^n)$ is given by

$$\begin{array}{lll} \omega_1^l \omega_2^m, & l+m=n, & \\ \omega_1^s \omega_2^t \omega_i, & s+t=n-1, & \\ & \quad i=3,\ldots,g, & \\ \omega_1^h \omega_2^k \bar{\eta}, & h+k=n-2, & \Big]\text{basis for } H^0(C, K^n((-n+1)D)) \\ \bar{\beta}_i^{n-2}\omega_i^2, & i=3,\ldots,g, & \Big]\text{basis for } H^0(C, K^n((-n+2)D)) \\ \vdots & & \\ \bar{\beta}_i \omega_i^{n-1}, & i=3,\ldots,g, & \Big]\text{basis for } H^0(C, K^n(-D)) \\ \omega_i^n, & i=3,\ldots,g, & \Big]\text{basis for } H^0(C, K^n). \end{array}$$

The second step in Petri's analysis is to construct quadratic and cubic generators for the homogeneous ideal of C. First of all, we notice that for $3 \leq i, k \leq g$, and $i \neq k$ the quadratic differential $\omega_i\omega_k$ belongs to

$$H^0(C, K^2(-D))$$

and vanishes at p_1 and p_2. Looking at the basis of $H^0(C, K^2(-D))$, we see that we can write

$$\omega_i\omega_k = \sum_{s=3}^{g} \bar{a}_{sik}\omega_s + b_{ik}\omega_1\omega_2,$$

where $\bar{a}_{sik} = \lambda_{sik}\omega_1 + \mu_{sik}\omega_2$ and $\lambda_{sik}, \mu_{sik}, b_{ik}$ are complex numbers. It then follows that the quadratic polynomials

$$f_{ik} = X_i X_k - \sum_{s=3}^{g} a_{sik} X_s - b_{ik} X_1 X_2, \qquad 3 \leq i, \quad k \leq g, \quad i \neq k,$$

all vanish on C. We obtain in this way $(g-2)(g-3)/2$ linearly independent elements in I_2. But, by Max Noether's theorem in Section 2, the dimension of

I_2 is exactly $(g-2)(g-3)/2$. Thus the f_{ik}'s form a basis for I_2. Let us now consider cubic relations. Looking at (3.4) we define cubic polynomials G_{kl} by setting

$$G_{kl} = \alpha_k X_k^2 - \alpha_l X_l^2 + \theta_l - \theta_k, \qquad 3 \le k, \quad l \le g, \quad k \ne l.$$

These polynomials obviously belong to I_3 and, moreover, satisfy the cocycle relations

$$(3.5) \qquad \begin{cases} G_{kl} + G_{lk} = 0, \\ G_{kl} + G_{lm} = G_{km}. \end{cases}$$

We can now state

Petri's Theorem. *Let C be a canonical curve of genus $g \ge 4$. Then the ideal of C is generated by the f_{ik}'s, unless C is trigonal or isomorphic to a smooth plane quintic, in which cases it is generated by the f_{ik}'s and the G_{kl}'s.*

To prove Petri's theorem we first show that the f_{ik}'s and the G_{kl}'s generate I. Let us show that they generate I_3. Consider then an element

$$R = \sum \gamma_{ijk} X_i X_j X_k \in I_3.$$

Since the ω_i^3's, $i = 3, \dots, g$, are linearly independent modulo $H^0(C, K^3(-D))$, by restricting R to C we get $\gamma_{iii} = 0$. Using the expression of the f_{ik}'s we can write

$$R = \sum \delta_{lik} X_l f_{ik} + \sum_{i=3}^{g} (\mu_i \alpha_i + v_i \beta_i) X_i^2 + w,$$

where w belongs to the vector space W generated by $x_1^3, x_1^2 x_2, x_1 x_2^2, x_2^2, x_1^2 x_i, x_1 x_2 x_i, x_2^2 x_i$, $i = 3, \dots, g$. Restricting to C and using (3.4), we get $\sum \mu_i = 0$ and $v_i = 0$, for all i. But then we can write

$$R = \sum \sigma_{lik} X_l f_{ik} + \sum \lambda_{kl} G_{kl} + w',$$

with $w' \in W$. Restricting again to C we find that $w' \in I$. Since, by the definition of W, we have $W \cap I = \{0\}$, our claim is proved.

Let us now consider a relation

$$R = \sum \gamma_{ijkl} X_i X_j X_k X_l \in I_4.$$

As above we immediately get $\gamma_{iiii} = 0$ and we can write

$$\begin{aligned} R = \sum_{i,k} P_{ik} f_{ik} &+ \sum (\lambda_i \alpha_i^2 + \mu_i \alpha_i \beta_i + v_i \beta_i^2) X_i^2 \\ &+ \sum (\sigma_i \alpha_i + \tau_i \beta_i) X_i^3 + w, \end{aligned}$$

where w belongs to the subspace W' of $H^0(\mathbb{P}^{g-1}, \mathcal{O}_{\mathbb{P}^{g-1}}(4))$ generated by the monomials $X_1^s X_2^t$, $s + t = 4$, and $X_1^h X_2^k X_i$, $h + k = 3$, $i = 3, \ldots, g$. Again it is immediate to check that $W' \cap I = \{0\}$. Restricting to C and using the explicit basis of $H^0(C, K^4)$, we easily get $\tau_i = \mu_i = 0$. We then write

$$\alpha_i X_i^3 = X_i G_{ij} + \alpha_j X_j^2 X_i + \sum Q_{kl} X_k X_l,$$

where the Q's are quadratic in X_1 and X_2. Using again the expression of the f_{ik}'s we can write

$$R = \sum P'_{ik} f_{ik} + \sum T_{kl} G_{kl} + \sum \lambda'_i \alpha_i^2 X_i^2 \\ + \sum \mu'_i \alpha_i \beta_i X_i^2 + w',$$

with $w' \in W'$. Restricting R to C and using the fact that $\bar{\alpha}_i \omega_i^2 = \bar{\eta} + \bar{\theta}_i$, we get $\sum (\lambda'_i \alpha_i + \mu'_i \beta_i) = 0$, so that

$$R = \sum P'_{ik} f_{ik} + \sum T_{kl} G_{kl} + w'',$$

with $w'' \in W'$. But, as we observed, $W' \cap I = \{0\}$, proving our claim. The general case of I_n is handled in exactly the same way, and the first part of Petri's theorem is proved.

To prove the second half of his theorem Petri exhibits a system of syzygies among the quadratic and cubic relations. First of all we can assume that $g \geq 5$: When $g = 4$ we already know that the canonical curve is the complete intersection of a quadric and a cubic in $\mathbb{P}^3$ and that the rulings of the quadric cut out on C a g_3^1. We also observed that, if C is trigonal or isomorphic to a smooth plane quintic, then its ideal cannot be generated by quadrics only. Before writing down the syzygies among the f_{ik}'s and G_{kl}'s we need the following notation. Consider the relations

$$\omega_i \omega_k = \sum_{s=3}^{g} \bar{a}_{sik} \omega_s + b_{ik} \omega_1 \omega_2;$$

we then see that, when $g \geq 5$, for any triple of distinct integers i, l, k the differential $\bar{a}_{lik}$ vanishes doubly at p_l, so that there are scalars ρ_{ikl} such that

$$a_{lik} = \rho_{ikl} \alpha_l.$$

We can now write Petri's syzygies: for any triple of distinct indices i, k, l such that $3 \leq i, k, l \leq g$, we have

$$f_{ik} X_l - f_{il} X_k = \sum_{s=3}^{g} (a_{sil} f_{sk} - a_{sik} f_{sl}) + \rho_{ikl} G_{kl}, \tag{3.6}$$

where, by convention, $f_{kk} = f_{ll} = 0$. To prove that indeed these relations hold we proceed as follows. First of all, one checks easily that

$$f_{ik}X_l - f_{il}X_k = \sum_{s=3}^{g} (a_{sil}f_{sk} - a_{sik}f_{sl}) + a_{kil}X_k^2 - a_{lik}X_l^2 + w,$$

where $w \in W$. Restricting to C we get the relation

$$\rho_{ilk}\bar{\alpha}_k\omega_k^2 - \rho_{ikl}\bar{\alpha}_l\omega_l^2 + \bar{w} = 0,$$

which can also be written as

$$(\rho_{ilk} - \rho_{ikl})\bar{\eta} + \bar{w}' = 0, \qquad w' \in W.$$

Looking at the basis of $H^0(C, K^3)$ and keeping in mind that $W \cap I = \{0\}$, we obtain that $w' = 0$ and that $\rho_{ikl} = \rho_{ilk}$. This proves that Petri's syzygies hold and that the ρ_{ikl} are *symmetric* in i, k, and l.

To prove his theorem Petri proceeds as follows. Fix an integer k, $3 \le k \le g$, and look at the variety V_k defined by the equations

$$\{f_{ik} = 0\}, \qquad i = 3, \ldots, g, \quad i \neq k.$$

Then write these equations in the following way

$$\left(\delta_{is}X_k - \sum_{\substack{s=3\\ s\neq k}}^{g} a_{sik}\right)X_s = a_{kik}X_k + b_{ik}X_1X_2.$$

Let Δ_k be the determinant of the $(g-3) \times (g-3)$ matrix

$$M_k = (\delta_{is}X_k - \textstyle\sum a_{sik})$$

and let Δ_k^s be the determinant of the matrix obtained by substituting the sth column of M_k with the column ${}^t(a_{kik}X_k + b_{ik}X_1X_2)$, $i = 3, \ldots, g$, $i \neq k$. Both Δ_k and Δ_k^s are homogeneous polynomials in X_1, X_2, and X_k. By Cramer's rule we have

$$\Delta_k X_s = \Delta_k^s, \qquad s = 3, \ldots, g, \quad s \neq k.$$

More geometrically, we can define a (rational) surface F_k whose parametric equations, away from the hypersurface $\Delta_k = 0$, are

$$\text{(3.7)} \qquad \begin{cases} X_1 = X_1, \\ X_2 = X_2, \\ X_k = X_k, \\ X_s = \dfrac{\Delta_k^s}{\Delta_k}, \qquad s = 3, \ldots, g, \quad s \neq k. \end{cases}$$

By construction, the surface F_k is the only component of V_k which is not contained in the hypersurface $\Delta_k = 0$. Now observe that, on one hand C is contained in V_k, and on the other, the homogeneous polynomial Δ_k does not vanish at the point p_k of C. We can therefore conclude that F_k is the only component of V_k containing C. Parenthetically, it could be instructive to notice that, if we denote by C_k the projection of C from the vertex

$$\{X_1 = X_2 = X_k = 0\}$$

into the plane $\{X_s = 0\}$, $s = 3, \ldots, g$, $s \neq k$, then C_k is a degree $g + 1$ plane curve while the curves $X_1\Delta_k = 0$, $X_2\Delta_k = 0$, $X_k\Delta_k = 0$, $\Delta_k^s = 0$, $s = 3, \ldots, g$, $s \neq k$, generate the system of degree $g - 2$ adjoint curves to C_k; so that the surface F_k is nothing but the birational image of $\mathbb{P}^2$ under the map associated to this linear system.

Petri's key observation is the following:

Lemma. *For $3 \leq k, l \leq g$, $k \neq l$, the surface F_k coincides with the surface F_l if and only if $\rho_{ikl} = 0$ for all $i = 3, \ldots, g$, $i \neq k$, $i \neq l$.*

Before proving the lemma let us see how the theorem follows from it. We first assume that all the ρ_{ikl}'s vanish. By the lemma the surfaces $F_3, \ldots, F_g$ all coincide. Writing $F_3 = \cdots = F_g = F$, we have that all the quadrics through C contain F and we may use Proposition (3.1) to conclude that C is either trigonal or isomorphic to a smooth plane quintic. It now remains to show that, if for some triple i, s, t the complex number ρ_{ist} is non-zero, then the ideal I is generated by the f_{ik}'s. Suppose not and let $J \subset I$ be the ideal generated by the f_{ik}'s. Since, as we know, I is generated by the f_{ik}'s and G_{kl}'s, there must be a k and an l such that $G_{kl} \notin J$. Looking at the syzygies (3.6), this implies that $\rho_{ikl} = 0$ for every i, and by the lemma we get $F_k = F_l$. Now, given any j such that $j \neq k$, $j \neq l$, $3 \leq j \leq g$, write

$$G_{kl} = G_{kj} + G_{jl}.$$

Therefore either $G_{kj} \notin J$ or $G_{jl} \notin J$. Arguing exactly as in the case of G_{kl}, this says that either $F_k = F_j$ or $F_l = F_j$. Since $F_k = F_l$ we get $F_3 = \cdots = F_g$. In particular $F_s = F_t$, and by the lemma this implies that $\rho_{ist} = 0$, contrary to our hypothesis.

It now remains to prove the lemma. Suppose first that $\rho_{ikl} = 0$ for every i. Then Petri's syzygies restricted to F_k tell us that

$$\sum_{\substack{s \neq l \\ s \neq k}} (\delta_{is} X_k - a_{sik}) f_{sl} \equiv 0 \pmod{F_k}, \qquad i = 3, \ldots, g, \quad i \neq l, i \neq k.$$

Let Δ_{lk} be the determinant of the $(g-4) \times (g-4)$ matrix $(\delta_{is}X_s - a_{sik})$, $i, s = 3, \ldots, g, i \neq l, k, s \neq l, k$. By Cramer's rule

$$\Delta_{lk} f_{il} \equiv 0 \pmod{F_k}, \qquad i = 3, \ldots, g, \quad i \neq l, \quad i \neq k.$$

Since $\Delta_{lk}(p_k) \neq 0$, the above equations tell us that F_k is contained in V_l. But, as we remarked, F_l is the only component of V_l containing the curve C, so that $F_k = F_l$, proving the first part of the lemma. Suppose, conversely, that $F_k = F_l$, and assume that $\rho_{ikl} \neq 0$ for some index i. Then Petri's syzygies imply that G_{kl} vanishes on F_k. Write G_{kl} in the form

$$\begin{aligned} G_{kl} = {} & \alpha_k X_k^2 - \alpha_l X_l^2 + F(X_i X_1^2, X_i X_1 X_2, X_i X_2^2) \\ & + \lambda X_k X_1^2 + \mu X_k X_1 X_2 + \nu X_k X_2^2, \end{aligned}$$

where F is a linear combination of the monomials $X_i^2 X_1, X_i X_1 X_2, X_i X_2^2$, $i = 3, \ldots, g, i \neq k$. Multiplying by $(\Delta_k)^2$ and using (3.7), we get

$$\begin{aligned} (\Delta_k)^2 G_{kl} = {} & \alpha_k (\Delta_k)^2 X_k^2 - \alpha_l (\Delta_k)^2 X_l^2 \\ & + \Delta_k F(\Delta_k X_i X_1^2, \Delta_k X_i X_1 X_2, \Delta_k X_i X_2^2) \\ & + \lambda (\Delta_k)^2 X_k X_1^2 + \mu (\Delta_k)^2 X_k X_1 X_2 + \nu (\Delta_k)^2 X_k X_2^2, \end{aligned}$$

so that,

$$\begin{aligned} (\Delta_k)^2 G_{kl}|_{F_k} = {} & \alpha_k (\Delta_k)^2 X_k^2 - \alpha_l (\Delta_k^l)^2 \\ & + \Delta_k F(\Delta_k^i X_1^2, \Delta_k^i X_1 X_2, \Delta_k^i X_2^2) \\ & + \lambda (\Delta_k)^2 X_k X_1^2 + \mu (\Delta_k)^2 X_k X_1 X_2 + \nu (\Delta_k)^2 X_k X_2^2. \end{aligned}$$

Since G_{kl} vanishes on F_k, the polynomial on the right-hand side of the last equality is identically zero. Looking at the term $\alpha_k X_k^{2g-4}$ coming from $\alpha_k (\Delta_k)^2 X_k^2$ and recalling that $\alpha_k = \delta X_1 + \gamma X_2$, with $\gamma \neq 0$, $\delta \neq 0$, we get a contradiction. This concludes the proof of Petri's theorem.

Bibliographical Notes

Clifford's theorem is, of course, classical; the original statement is in Clifford [1]. The trilinear algebra set up of Clifford's theorem was used, over $\mathbb{R}$, by H. Hopf to study the question of existence of division algebras in dimension 1, 2, 3, and 8. The argument in Hopf [1] appears to be one of the first applications to geometry of the cup-product in cohomology. The general position theorem is also classical and may be found for instance in Severi [2]. It is perhaps worth remarking that the result is false in characteristic $p \neq 0$. A recent general discussion of monodromy questions is given in Harris [1].

Castelnuovo's bound appears in Castelnuovo [1]. The present discussion of the bound and its consequences is extracted from Griffiths–Harris [1] which contains a proof of Castelnuovo's lemma and a discussion of rational normal scrolls. An analytic proof of Castelnuovo's lemma is given in Griffiths–Harris [2].

The Enriques–Babbage theorem appears in Enriques [1] and Babbage [1]. The ideal theoretical study of the canonical curve was done by K. Petri in [1], and modern versions of this approach can be found in Saint–Donat [1], Arbarello–Sernesi [1], and Mumford [3]. For a somewhat different approach, see Shokurov [1].

Exercises

A. Symmetric Products of $\mathbb{P}^1$

A-1. Show that the dth symmetric product of $\mathbb{P}^1$ is isomorphic to $\mathbb{P}^d$, and that under this identification the locus of divisors of the form dp, $p \in \mathbb{P}^1$, is rational normal curve $C \subset \mathbb{P}^d$.

For the remainder of these exercises we will implicitly make the identifications $(\mathbb{P}^1)_d = \mathbb{P}^d$ and $\{dp\}_{p \in \mathbb{P}^1} = C$.

A-2. Show that the tangent line to C at the point dp is the locus of divisors

$$\{(d-1)p + q\}_{q \in \mathbb{P}^1}.$$

More generally, show that the osculating k-plane to C at dp is the locus of divisors

$$\{(d-k)p + q_1 + \cdots + q_k\}, \qquad q_1 + \cdots + q_k \in (\mathbb{P}^1)_k.$$

(Note that in the case $k = d - 1$ this shows that the locus of osculating hyperplanes to a rational normal curve is a rational normal curve in the dual projective space.)

Observe that if $p_1, \ldots, p_k$ are distinct points of $\mathbb{P}^1$, the span of the points $dp_1, \ldots, dp_k$ is just the $(k-1)$-dimensional linear series spanned by

$$dp_1, \ldots, dp_k$$

(this is tautologous). We will call such a $(k-1)$-plane a *strict secant plane* to C. (For example, a strict secant line is *not* a tangent line.)

A-3. Show that a polynomial $P(X) \in H^0(\mathbb{P}^1, \mathcal{O}(d))$ is expressible as a sum of k dth powers of linear polynomials if and only if the corresponding point in

$$\mathbb{P}^d = \mathbb{P}H^0(\mathbb{P}^1, \mathcal{O}(d))$$

lies on a strict secant $(k-1)$-plane. Conclude the

Theorem. *A general polynomial of degree d is expressible as a sum of k dth powers of linear forms if and only if* $k \geq (d+1)/2$.

Hint: Use the independence of points on a rational normal curve to conclude that the locus of secant k-planes to C contains an open set of dimension $\min(d, 2k+1)$.

A-4. Show that there exist polynomials of degree d which are not expressible as a sum of fewer than d dth powers of linear forms.
Hint: Look at a point on a tangent line to C.

B. Refinements of Clifford's Theorem

In the following series of exercises we will obtain sharper versions of the basic inequality $r(D+D') \geq r(D) + r(D')$ in Section 1 of Chapter III.

One notational convention: if $\mathscr{D} \subset |D|$ and $\mathscr{E} \subset |E|$ are linear series, we will denote by $\mathscr{D} + \mathscr{E}$ the subseries of $|D+E|$ spanned by sums of divisors in $\mathscr{D}$ and divisors in $\mathscr{E}$, and by $\mathscr{E} - \mathscr{D}$ the subseries of $\mathscr{E}$ containing a general divisor D of $\mathscr{D}$. Also, we will write $r(\mathscr{D})$ for the dimension of a linear series $\mathscr{D}$.

B-1. Suppose that $\mathscr{D}, \mathscr{E}$ are linear series on a smooth curve C of genus $g > 0$, and assume that $\mathscr{D} + \mathscr{E}$ is base-point-free and $\phi_{\mathscr{D}+\mathscr{E}}$ is birational. Show that

$$r(\mathscr{D} + \mathscr{E}) \geq r(\mathscr{D}) + r(\mathscr{E}) + 1.$$

(*Hint*: Using the uniform position lemma, show that a general $F \in \mathscr{D} + \mathscr{E}$ is not the sum of a divisor in $\mathscr{D}$ and a divisor in $\mathscr{E}$.)

B-2. Suppose $\mathscr{D}$, $\mathscr{E}$ are base-point-free pencils and that $\phi_{\mathscr{D}} \neq \phi_{\mathscr{E}}$. Show that

$$r(\mathscr{D} + \mathscr{E}) \geq 3.$$

(*Hint*: Let the pencils be given by $\sigma_0, \sigma_1 \in H^0(C, \mathscr{O}(D))$ and $\tau_0, \tau_1 \in H^0(C, \mathscr{O}(E))$ and, under the assumption that $r(\mathscr{D} + \mathscr{E}) = 2$, show that the image of C under the map $[\sigma_0\tau_0, \sigma_0\tau_1, \sigma_1\tau_0, \sigma_1\tau_1]$ lies in a hyperplane section of the quadratic

$$X_0X_3 - X_1X_2 = 0;$$

from this obtain a contradiction to $\phi_{\mathscr{D}} \neq \phi_{\mathscr{E}}$.)

B-3.

(i) As an application of the preceding exercise, show that a smooth, non-hyperelliptic curve of genus $g \geq 5$ cannot have two distinct g^1_3's.
(ii) Show by example that there exist non-hyperelliptic, non-trigonal curves of all genera $g \geq 1$ which possess infinitely many distinct g^1_4's (consider two-sheeted branched covers of elliptic curves).

B-4. Let $\mathscr{D}, \mathscr{E}$ be linear series on a smooth curve and assume that $\mathscr{D}$ is base-point-free. Show that

$$r(\mathscr{D} + \mathscr{E}) \geq 2r(\mathscr{E}) - r(\mathscr{E} - \mathscr{D}).$$

(*Hint*: For disjoint divisors $D_0, D_1 \in \mathscr{D}$, show that the intersection of the linear spaces $D_0 + \mathscr{E}$, $D_1 + \mathscr{E}$ has dimension at most $r(\mathscr{E} - \mathscr{D})$. Alternatively (and, in fact equivalently), look up the base-point-free pencil trick.)

B-5. As an application of the preceding exercise, suppose that C is a smooth, non-hyperelliptic curve having a $g_3^1 = \mathscr{D}$ and that $\mathscr{E}$ is a g_{g-1}^1 on C. Show that either $\mathscr{E}$ or $K - \mathscr{E}$ contains the g_3^1.

(*Hint*: Estimate dim $\mathscr{D} + \mathscr{E}$ and use Riemann–Roch.)

B-6. Let $\mathscr{D}, \mathscr{E}$ be linear series on a smooth curve and assume that $\mathscr{D}$ is base-point-free of degree d and $\phi_{\mathscr{D}}$ is birational. Show that

$$r(\mathscr{D} + \mathscr{E}) \geq \min(r(\mathscr{D}) + 2r(\mathscr{E}) - r(\mathscr{E} - \mathscr{D}), d + r(\mathscr{E})).$$

In particular, if $d \geq r(\mathscr{D}) + r(\mathscr{E}) - r(\mathscr{E} - \mathscr{D})$ then

$$r(\mathscr{D} + \mathscr{E}) \geq r(\mathscr{D}) + 2r(\mathscr{E}) - r(\mathscr{E} - \mathscr{D}).$$

(*Hint*: Use the uniform position lemma to argue that the number of conditions imposed by D on $\mathscr{D} + \mathscr{E}$ is (degree permitting) the number of conditions imposed by D on $\mathscr{D}$ plus the number of conditions imposed by D on $\mathscr{E}$.)

B-7. Using Castelnuovo's theorem, show that if a smooth curve C of genus g has a $g_{2r+\alpha}^r$, say $\mathscr{D}$, with $0 \leq \alpha \leq r - 2$ then either:

(i) $\phi_{\mathscr{D}}$ is not birational and C is a two-sheeted covering of a curve of genus at most $\alpha/2$; or

(ii) $\phi_{\mathscr{D}}$ is birational and either:

(a) $g \leq r + 2\alpha$; or

(b) $g = r + 2\alpha + 1$ and C is trigonal if $\alpha > 0$ while $\mathscr{D} = |K|$ if $\alpha = 0$

(cf. Comessatti [2], Beauville [2]).

C. Complete Intersections

In some of the remaining exercises we will use the following standard facts:

For $X \subset Y$ a smooth divisor on a smooth variety, the *adjunction formula*

$$K_X = K_Y(X) \otimes \mathcal{O}_X$$

holds; an equivalent and perhaps more familiar notation is

$$K_X = K_Y \otimes [X]|_X.$$

It is also well known that

$$(*) \qquad H^i(\mathbb{P}^r, \mathcal{O}(k)) = 0, \qquad 1 \le i \le r - 1 \text{ and any } k;$$

$$(**) \qquad H^i(\mathbb{P}^r, \mathcal{F}(k)) = 0, \qquad i > 0 \text{ and } k \gg 0$$

for any coherent sheaf $\mathcal{F}$ on $\mathbb{P}^r$.

C-1. Suppose that $C = S \cap T$ is the transverse intersection of two smooth surfaces of degrees m, n in $\mathbb{P}^3$ (in particular, C is a smooth *complete intersection*). Show that

$$K_C \cong \mathcal{O}_C(m + n - 4),$$

and that the genus of C is

$$g = \tfrac{1}{2}mn(m + n - 4) + 1.$$

C-2. Suppose that $F_1, \ldots, F_l$ are homogeneous polynomials on $\mathbb{P}^r$ whose locus of common zeros has dimension $r - l$. Let H be any homogeneous polynomial that is locally in the ideal generated by $F_1, \ldots, F_l$ (more precisely, the forms $F_1, \ldots, F_l$ generate a subsheaf $\mathcal{I} \subset \mathcal{O}_{\mathbb{P}^r}$, and H should be a section of $\mathcal{I} \otimes \mathcal{O}_{\mathbb{P}^r}(d)$ where $\deg H = d$). Show that

$$H = \sum_\alpha A_\alpha F_\alpha$$

for suitable homogeneous forms A_α.

(*Hint*: In the local ring at each point of $\mathbb{P}^r$ the localizations of the F_α give a regular sequence, and the Koszul complex (cf. Griffiths–Harris [1]) gives a canonical resolution

$$0 \to \mathcal{O}_{\mathbb{P}^r}(-d_1 - \cdots - d_l) \to \cdots \to \bigoplus_{\alpha<\beta} \mathcal{O}_{\mathbb{P}^r}(-d_\alpha - d_\beta) \to \bigoplus_\alpha \mathcal{O}_{\mathbb{P}^r}(-d_\alpha) \to \mathcal{I} \to 0,$$

where $\deg F_\alpha = d_\alpha$. An argument using (*) then shows that the global sections of this sequence also give an exact sequence.) When $r = l = 2$ this result is *Noether's "AF + BG" theorem.*

C-3. Let $C \subset \mathbb{P}^3$ be a smooth complete intersection of two smooth surfaces of degrees m and n. Using the preceding exercise, compute the number of linearly independent surfaces of degree k containing C. Deduce that the surfaces of degree $m + n - 4$ cut out the complete canonical series on C.

C-4. Show that the curve C in the preceding exercise does not have a g^1_{m+n-3}.

(*Hint*: Show that any set of $d \le m + n - 3$ distinct points in $\mathbb{P}^3$ impose independent conditions on surfaces of degree $m + n - 4$.)

D. Projective Normality (I)

Definitions. A smooth curve $C \subset \mathbb{P}^r$ is *k-normal* if the hypersurfaces of degree k cut out the complete linear series $|\mathcal{O}_C(k)|$. A 1-normal curve is usually said to be *linearly normal*, and $C \subset \mathbb{P}^r$ is *projectively normal* if it is k-normal for every k.

D-1. Show that a smooth irreducible curve $C \subset \mathbb{P}^r$ is k-normal if and only if

$$H^1(\mathbb{P}^r, \mathcal{I}_C(k)) = 0$$

where $\mathcal{I}_C \subset \mathcal{O}_{\mathbb{P}^r}$ is the ideal sheaf of C. Deduce that C is k-normal for $k \gg 0$, and from this infer that every meromorphic function on C is the restriction of a rational function on $\mathbb{P}^r$ (this exercise is a consequence of (*) and (**) in the preceding batch of exercises).

D-2. Let $C \subset \mathbb{P}^r$ be a smooth irreducible curve and $D = H \cdot C$ a hyperplane section. Show that if C is linearly normal then every quadric in $H \cong \mathbb{P}^{r-1}$ containing D is the restriction to H of a quadric in $\mathbb{P}^r$ containing C.

D-3. Keeping the notations of the preceding exercise, show that C is projectively normal if and only if every hypersurface in H containing D is the restriction to H of a hypersurface in $\mathbb{P}^r$ containing C.

D-4. Let $C \subset \mathbb{P}^r$ be a smooth irreducible curve of degree $d < 2r$. Show that C is projectively normal if and only if it is linearly normal.

D-5. Using Exercise B-6, show that if $C \subset \mathbb{P}^r$ is smooth, irreducible, and k-normal, and $\mathcal{O}_C(k-1)$ is non-special, then C is $(k+1)$-normal (and hence l-normal for all $l > k$).

D-6. Let $C \subset \mathbb{P}^r$ be a smooth irreducible curve of degree d. Using the preceding exercise, show that if C is k-normal for $k \leq (d+r)/(r-1)$, then C is projectively normal.

D-7. Let $C \subset Q \subset \mathbb{P}^3$ be a smooth curve of type (m, n) on a smooth quadric surface Q. Show that C is projectively normal if and only if $|m - n| \leq 1$ (i.e., C is as close to a complete intersection as its degree $m + n$ will allow).

D-8. With C as in the preceding exercise, if $m < n$ show that C fails to be k-normal exactly for k satisfying $m \leq k \leq n - 2$.

D-9. More generally, let S and T be smooth surfaces in $\mathbb{P}^3$ whose intersection is the sum of smooth curves C and D. Show that C is projectively normal if and only if D is.

D-10. Let $C \subset \mathbb{P}^r$ be a smooth irreducible curve with defining homogeneous ideal $I \subset S = \mathbb{C}[X_0, \ldots, X_r]$. Show that C is projectively normal if and only if the homogeneous coordinate ring S/I is integrally closed (this is frequently taken as the definition of projective normality).

(*Hint*: Show that the integral closure of S/I is $R = \bigoplus_{n \geq 0} H^0(C, \mathcal{O}(n))$.)

D-11. Show that a smooth complete intersection curve $C \subset \mathbb{P}^r$ is projectively normal.

D-12. Let $Q \subset \mathbb{P}^3$ be a smooth quadric surface, L_1 and L_2 lines from the two distinct rulings, and $\mathcal{O}(m, n) = \mathcal{O}_Q(mL_1 + nL_2)$. Show that

$$h^0(Q, \mathcal{O}(m, n)) = \begin{cases} (m+1)(n+1), & m, n \geq -1, \\ 0, & m \leq -1 \text{ or } n \leq -1, \end{cases}$$

$$h^1(Q, \mathcal{O}(m, n)) = \begin{cases} 0, & m, n \geq -1, \\ -(m+1)(n+1), & m \geq -1, n \leq -2 \text{ or} \\ & m \leq -2, n \geq -1. \end{cases}$$

(*Suggestion*: Use the Künneth formula.) Using these computations give another proof of Exercise D-7.

D-13. Suppose that $|D|$ is any complete linear system on a smooth, linearly normal curve $C \subset \mathbb{P}^r$. Show that, for $\Delta \in |D|$, $\dim \bar{\Delta}$ is independent of the divisor Δ. (*Hint*: Relate $\dim \bar{\Delta}$ to $h^0(C, \mathcal{O}(1)(-\Delta))$.)
This result is false if we do not assume C to be linearly normal (why?).

D-14. Show that the following curves in $\mathbb{P}^3$ (assumed to be smooth) are projectively normal:

(1) A twisted cubic.
(2) A quartic of genus 1.
(3) A quintic of genus 2.
(4) A non-hyperelliptic sextic of genus 3.
(5) A sextic of genus 4.
(6) A septic of genus 5.
(7) A septic of genus 6.

(*Hint*: It may help to observe that, in all the above cases, $\mathcal{O}_C(k)$ is non-special for $k \geq 2$.)

D-15. Show that the curves listed in the preceding exercise are *all* the projectively normal non-degenerate curves of degree ≤ 7 in $\mathbb{P}^3$.

E. Castelnuovo's Bound on k-Normality

This series of exercises will establish the following result due to Castelnuovo:

Theorem. *Let $C \subset \mathbb{P}^r$ be a smooth curve of degree d. Then the hypersurfaces in $\mathbb{P}^r$ of degree $d - 2$ cut out on C a complete linear series.*

Recently, this theorem has been strengthened by Gruson, Lazarsfeld and Peskine, who replace the $d - 2$ in the statement with $d - r + 1$ (assuming C is nondegenerate, of course); cf. Gruson–Lazarsfeld–Peskine [1].

The following notations will be used.

If $r \geq 3$, $\pi: C \to \bar{C} \subset \mathbb{P}^2$ will denote a general projection of C onto a plane curve having only nodes as singularities.

We let $\{r_\alpha\}_{\alpha=1,\ldots,\delta}$ be the set of nodes of $\bar{C}$ and we set

$$\pi^{-1}(r_\alpha) = p_\alpha + q_\alpha;$$

$$\Gamma = \sum_{\alpha=1}^{\delta} (p_\alpha + q_\alpha);$$

$$\Gamma' = \sum_{\alpha=1}^{\delta-1} (p_\alpha + q_\alpha).$$

E-1. Using Appendix A, show that the linear system cut out on C by hypersurfaces of degree $d - 2$ through Γ is the complete series $|\mathcal{O}_C(d-2)(-\Gamma)|$.

E-2. Using Appendix A again, show that there exists a hypersurface of degree $d - 3$ containing Γ' but not p_δ or q_δ, and hence a hypersurface of degree $d - 2$ containing Γ' and p_δ but not q_δ.

E-3. Conclude from the preceding exercise that Γ imposes independent conditions on hypersurfaces of degree $d - 2$ in $\mathbb{P}^r$, and then from this complete the proof of the theorem.

F. Intersections of Quadrics

F-1. Show that a collection $\Gamma = p_1, \ldots, p_{2n}$ of $2n$ points in general position in $\mathbb{P}^n$ is cut out by quadrics.

(*Hint*: If q lies on every quadric containing $p_1, \ldots, p_{2n}$, let $p_1, \ldots, p_k$ be a minimal subset of $p_1, \ldots, p_{2n}$ such that $q \in \overline{p_1, \ldots, p_k}$; consider quadrics of the form

$$Q = \overline{(p_1, \ldots, \widehat{p_i}, \ldots, p_k, p_{\alpha_1}, \ldots, p_{\alpha_{n-k+1}})}$$

$$\cup \overline{(\Gamma - \{p_1, \ldots, \widehat{p_i}, \ldots, p_k, p_{\alpha_1}, \ldots, p_{\alpha_{n-k+1}}\})}.$$

F-2. Let $L \to C$ be a line bundle of degree $d \geq 2g + 2$ over a smooth curve of genus g with corresponding embedding

$$\phi_L: C \to \mathbb{P}^{d-g}.$$

Using the preceding exercise and Exercise D-2, show that $\phi_L(C)$ is the intersection of quadrics.

F-3. Show that the statement of the preceding exercise is false when $d = 2g + 1$. Is there a counterexample with $g \geq 3$?

The remaining exercises are motivated by the discussion in Chapter VI concerning quadrics of minimal rank through a canonical curve, and the reader may wish to postpone them until reading that section.

Let $X \subset \mathbb{P}^n = \mathbb{P}V$ be a quadric hypersurface given by a symmetric bilinear form

$$Q: V \times V \to \mathbb{C},$$

or equivalently

$$\tilde{Q}: V \to V^*.$$

Recall that the rank of X is the rank of $\tilde{Q}$.

F-4. Show that X has rank k if and only if the singular locus of X is an $(n-k)$-plane.
(*Hint*: $X_{\text{sing}} = \mathbb{P}(\ker \tilde{Q})$, and X has rank k exactly when X is a cone over a smooth quadric of dimension $k-1$.)

F-5. Retaining the preceding notations, show that the following are equivalent:

$$\text{rank } X \leq 2m; \quad \text{and}$$

$$X \text{ contains an } (n-m)\text{-plane.}$$

F-6. Let $\mathbb{P}^k \subset \mathbb{P}^n$ be any linear subspace and $X' = X \cap \mathbb{P}^k$. Show that

$$\text{rank } X \geq \text{rank } X' \geq \text{rank } X - 2(n-k).$$

Hint: If $\mathbb{P}^k = \mathbb{P}W$ for $W \subset V$, consider the sequence

$$0 \to W \to V \overset{Q}{\to} V^* \to (V^*/W^{\perp}) \cong W^* \to 0.$$

F-7. By refining the arguments of F-1 and F-2 show that any linearly normal curve of degree $d \geq 2g + 2$ in $\mathbb{P}^r$ is cut out by quadrics of rank 4 (in particular, this applies to curves of degree $d < 2r$).

G. Space Curves of Maximum Genus

In the following series of exercises, taken from Harris [5], we will prove the

Theorem. *Let $C \subset \mathbb{P}^3$ be a smooth curve of genus $g(C)$ and degree $d > k(k-1)$ lying on an irreducible surface of degree k. Then*

$$g(C) \leq \pi_k(d) = \frac{d^2}{2k} + \frac{d(k-4)}{2} - \frac{\varepsilon}{2}\left(k - \varepsilon - 1 + \frac{\varepsilon}{k}\right),$$

where

$$\begin{cases} \varepsilon \equiv -d \text{ modulo } k, \\ 0 \leq \varepsilon \leq k-1. \end{cases}$$

Before doing the general case we will, in the first two exercises, derive the theorem when $k = 3$. In doing both this and the general case we will use the Riemann–Roch theorem for an effective divisor Γ on an irreducible plane curve H (the reason for the notation "H" will appear below), where Γ is supported away from the singularities of H. The statement is

$$(*) \qquad h^0(H, \mathcal{O}_H(\Gamma)) = d - (k-1)(k-2)/2 + 1 + h^0(H, \omega_H(-\Gamma)),$$

where

$$\begin{cases} d = \deg \Gamma, \\ k = \deg H, \\ \omega_H \text{ is the } \textit{dualizing sheaf} \text{ of } H. \end{cases}$$

We remark that

$$\omega_H \cong \mathcal{O}_H(k-3),$$

so that $(*)$ applies just as if H were smooth.

In general, a linear system on any curve H is defined to be $\mathbb{P}V$ where $V \subset H^0(H, L)$ is a linear subspace of the global sections of a line bundle $L \to H$. Any divisor Γ supported on the smooth points of H defines a line bundle $\mathcal{O}_H(\Gamma)$.

G-1. Let $\Gamma \subset \mathbb{P}^2$ be a collection of d distinct points lying in the smooth locus of an irreducible cubic plane curve H, and denote by ρ_l the number of conditions that Γ imposes on curves of degree l. Show that:

(a) if $l < d/3$, then

$$\rho_l = \binom{l+2}{2} - \binom{l-1}{2} = 3l;$$

(b) if $l \geq d/3$, then

$$\begin{cases} \rho_l = d - 1 & \text{if } l = d/3 \text{ and } \Gamma \text{ is a complete intersection}, \\ \rho_l = d & \text{otherwise}. \end{cases}$$

G-2. Let $C_0 \subset \mathbb{P}^3$ be an irreducible curve of degree d lying on an irreducible cubic surface S, and denote by g the genus of the normalization C of C_0. Show that

$$g \leq \pi_1(d, 3) = \begin{cases} \dfrac{d^2}{6} - \dfrac{d}{2} + 1, & d \equiv 0\ (3), \\[2ex] \dfrac{d^2}{6} - \dfrac{d}{2} + \dfrac{1}{3}, & d \not\equiv 0\ (3). \end{cases}$$

(*Suggestion:* Use the preceding exercise applied to a general hyperplane section H of S to estimate the dimension $h^0(C_0, \mathcal{O}_{C_0}(l))$, and then proceed exactly as in the proof of Castelnuovo's bound.)

G-3. Using the preceding exercise and Castelnuovo's bound, show that there does not exist a smooth curve $C \subset \mathbb{P}^3$ of degree 9 and genus 11.

G-4. Give a proof of

Gieseker's Lemma. *Let $V \subset H^0(\mathbb{P}^1, \mathcal{O}(d))$ be any linear subspace and W the image of the map*

$$V \otimes H^0(\mathbb{P}^1, \mathcal{O}_{\mathbb{P}^1}(1)) \to H^0(\mathbb{P}^1, \mathcal{O}_{\mathbb{P}^1}(d+1)).$$

Assume that the linear series $\mathbb{P}V$ has no base points. Then either

$$V = H^0(\mathbb{P}^1, \mathcal{O}_{\mathbb{P}^1}(d)),$$

or

$$\dim W \geq \dim V + 2.$$

(*Hint*: Use the base-point-free pencil trick to deduce an exact sequence

$$0 \to E \to V \otimes \mathcal{O}_{\mathbb{P}^1}(1) \to \mathcal{O}_{\mathbb{P}^1}(d+1) \to 0,$$

defining a vector bundle E, and then estimate $h^0(\mathbb{P}^1, E)$ using the *Grothendieck decomposition*

$$E = \bigoplus_{i=1}^{\dim V - 1} \mathcal{O}_{\mathbb{P}^1}(e_i)$$

of E. Alternatively, show that if $\dim W = \dim V + 1$, then every divisor $D \in \mathbb{P}W$ is the sum of a divisor $D' \in \mathbb{P}V$ and a point $p \in |\mathcal{O}_{\mathbb{P}^1}(1)|$; use a monodromy argument then to conclude that for any $D \in \mathbb{P}W$ and $p \in \operatorname{Supp} D$, $D' = D - p \in \mathbb{P}V$ and hence that $V = H^0(\mathbb{P}^1, \mathcal{O}(d))$.)

G-5. Let $\Gamma \subset \mathbb{P}^2$ be a collection of points in uniform position and let $i_1(\Gamma)$ be the smallest integer i such that there exists a curve of degree i in $\mathbb{P}^2$ containing Γ. Show that every curve of degree $i_1(\Gamma)$ containing Γ is irreducible.

(*Hint*: Show that if $C = C_1 \cup C_2$ is a plane curve of degree i_1 containing Γ, then $\Gamma \cap C_\alpha$ fails to impose independent conditions on curves of degree $\deg C_\alpha$ for $\alpha = 1$ or 2; then apply the uniform position statement.)

G-6. Continuing with the notations of the preceding exercises, let H be a curve of degree $i_1(\Gamma)$ containing Γ, let $i_2(\Gamma)$ be the smallest integer i such that there exists a curve of degree i containing Γ but not containing H, and let

$$\sigma_l = \rho_l - \rho_{l-1}.$$

Show that

(a)
$$\sum_l \sigma_l = d;$$

(b)
$$\sigma_l = l + 1, \qquad 0 \leq l \leq i_1(\Gamma) - 1;$$

(c)
$$\sigma_l = i_1(\Gamma), \qquad i_1(\Gamma) - 1 \leq l \leq i_2(\Gamma) - 1.$$

Having done this, define m to be the least positive integer such that

(d)
$$\sigma_l = 0, \qquad l \geq m.$$

G-7. By applying Gieseker's lemma to the linear series $\mathbb{P}V_l$ cut out on a general line $L \subset \mathbb{P}^2$ by $|\mathscr{I}_\Gamma \otimes \mathcal{O}_{\mathbb{P}^2}(l)|$, show that

(e)
$$\sigma_l < \sigma_{l-1}, \qquad i_2(\Gamma) \leq l \leq m.$$

G-8.

(i) Compute the sequences $\rho_l = \rho_l(i_1, d)$ and $\sigma_l = \sigma_l(i_1, d)$ in case Γ is the complete intersection of a pair of curves of degrees i_1 and i_2.

(ii) Similarly, compute $\rho_l = \rho_l(i_1, d)$ and $\sigma_l = \sigma_l(i_1, d)$ in case Γ is residual to $i_1 i_2 - d$ collinear points in an intersection of curves of degrees i_1 and i_2.

(*Suggestions* : One way to do (i) is by Noether's $AF + BG$ theorem. One way to do (ii) is by appealing to the Riemann–Roch formula (*).)

G-9. Now fix i_1 and $d > i_1(i_1 - 1)$. Show that, if $\{\sigma_l\}$ is *any* sequence satisfying (a)–(e) above, then

$$\rho_l \geq \rho_l(i_1, d).$$

Moreover, if equality holds for a sequence $\{\sigma_l\}$ arising from Γ as in G-5 then Γ is residual to $i_1 i_2 - d$ collinear points in an intersection of two curves of degrees i_1 and i_2.

G-10. Finally, let $C_0 \subset \mathbb{P}^3$ be an irreducible curve of genus $d > k(k - 1)$ lying on an irreducible surface S of degree k. Applying the preceding two exercises to a general plane section H, estimate the differences

$$h^0(C_0, \mathcal{O}_{C_0}(l)) - h^0(C_0, \mathcal{O}_{C_0}(l - 1))$$

and deduce the bound in the theorem.

Remark. Using G-9 it can be shown that, if equality holds in the estimate in the theorem, then C_0 is residual to a plane curve in a complete intersection of S with a surface of degree $[(d - 1)/k] + 1$.

H. G. Gherardelli's Theorem

In these three exercises we will retain the notation of the preceding batch and prove the

Theorem (G. Gherardelli). *Let $C \subset \mathbb{P}^3$ be a projectively normal curve such that $K_C \cong \mathcal{O}_C(m)$ for some m. Then C is a complete intersection of two surfaces.*

H-1. Setting

$$\alpha_l = h^0(C, \mathcal{O}_C(l)),$$

$$\beta_l = \alpha_l - \alpha_{l-1},$$

and

$$\gamma_l = \beta_l - \beta_{l-1},$$

show that

$$\beta_l + \beta_{m-l+1} = d,$$

$$\gamma_l = \gamma_{m-l+2},$$

and

$$\sum_i \gamma_i = d = \deg(C).$$

H-2. Let k be the smallest degree of a surface $S \subset \mathbb{P}^3$ containing C, and let n be the smallest degree of a surface $T \subset \mathbb{P}^3$ containing C but not S. Show that

$$\gamma_l = l + 1, \qquad 0 \leq l \leq k - 1,$$

$$\gamma_l = k, \qquad k \leq l \leq n - 1.$$

H-3. Conclude that

$$\gamma_l \geq k - l + n - 1, \qquad n \leq l \leq n + k - 1.$$

and hence that

$$\sum_i \gamma_i \geq nk.$$

This implies the theorem.

I. Extremal Curves

I-1. Let $X \subset \mathbb{P}^r$ be a smooth surface scroll, and denote by L and H the divisors of a line of the ruling and a hyperplane section, respectively. Show that

$$K_X = \mathcal{O}_X(-2H + (r - 3)L).$$

I-2. Let $C \in |\alpha H + \beta L|$ be a smooth curve. Show that

$$g(C) = \frac{\alpha(\alpha - 1)}{2}(r - 1) + (\alpha - 1)(\beta - 1).$$

I-3. Show that

$$H^1(X, \mathcal{O}_X(\alpha H + \beta L)) = 0 \quad \text{for } \alpha > 0, \beta \geq -\alpha.$$

By using the Riemann–Roch theorem for surfaces

$$\chi(\mathcal{O}_X(D)) = \chi(\mathcal{O}_X) + \tfrac{1}{2}(D \cdot D - K_X \cdot D),$$

where D is a divisor on X, deduce that

$$h^0(X, \mathcal{O}_X(\alpha H + \beta L)) = \frac{\alpha(\alpha + 1)}{2}(r - 1) + (\alpha + 1)(\beta + 1).$$

(*Hint*: Use the exact sequence

$$0 \to \mathcal{O}_X(\alpha H + (\beta - 1)L) \to \mathcal{O}_X(\alpha H + \beta L) \to \mathcal{O}_L(\alpha) \to 0.)$$

I-4. Show that a smooth curve $C \in |\alpha H + \beta L|$ is projectively normal if and only if $-(r - 1) \leq \beta \leq 1$.

I-5. Let $C \subset \mathbb{P}^r$ be an extremal curve of degree

$$(*) \qquad d = m(r - 1) + 1 + \varepsilon > 2r,$$

and let $I \subset \mathbb{C}[X_0, \ldots, X_r]$ be the homogeneous ideal defining C. Show that a minimal set of generators for I consists of $(r - 1)(r - 2)/2$ quadrics and $r - 1 - \varepsilon$ polynomials of degree $m + 1$.

(*Sketch*: In case $C \in |(m + 1)H - (r - \varepsilon - 2)L|$ show that:

(i) Every hypersurface of degree $l \leq m$ containing C contains X.
(ii) Modulo the ideal of X there are exactly

$$h^0(X, \mathcal{I}_C(m + 1)) = h^0(X, \mathcal{O}((r - \varepsilon - 2)L)) = r - \varepsilon - 1$$

linearly independent hypersurfaces of degree $m + 1$ containing C.
(iii) $H^0(X, \mathcal{O}((r - \varepsilon - 2)L)) \otimes H^0(X, \mathcal{O}(\alpha H)) \to H^0(X, \mathcal{O}(\alpha H + (r - \varepsilon - 2)L))$ is surjective.)

Perform a similar analysis in case $C \sim mH + L$ or in case C lies on a Veronese surface.

I-6. Show that the ideal of a trigonal canonical curve of genus g is generated by the $(g - 2)(g - 3)/2$ quadrics containing it plus $g - 3$ additional cubics.

I-7. Let $C \in |\alpha H + \beta L|$ be a smooth curve on a scroll X with $-(r-2) \leq \beta$. Show that the linear system $|\mathcal{O}_X((\alpha - 2)H + (r - 3 + \beta)L)|$ cuts out the complete canonical series on C.

(*Hint*: Use I-1, I-2, I-3, the adjunction sequence

$$0 \to K_X \to K_X(C) \to K_C \to 0,$$

and the fact that

$$\begin{aligned} h^1(X, K_X) &= h^1(X, \mathcal{O}) \quad \text{by Kodaira–Serre duality,} \\ &= 0 \quad\quad\quad\quad \text{(since } X \text{ is rational).)} \end{aligned}$$

I-8. Let $C \subset \mathbb{P}^r$ be an extremal curve of degree $d > 2r$ given by (*) above. Show that any $m - 1$ points of C impose independent conditions on $|K_C|$, and m points fail to do so only if they are collinear. Conclude that C possesses no g^1_{m-1}, and that the g^1_m cut out by the lines on X is unique. (Use the preceding exercise.)

J. Nearly Castelnuovo Curves

The following series of exercises will give a proof of theorem (2.7) of Chapter III under the additional assumptions:

$$(*) \qquad \begin{cases} \text{(i)} \quad C = C_0 \text{ is smooth;} \\ \text{(ii)} \quad \deg C > 2^{r-1}. \end{cases}$$

The first assumption is primarily a technical convenience. However, the second assumption allows a much simpler argument than is required in the general case; for example, it is not too difficult to deduce Castelnuovo's lemma on quadrics from general theorems when $d > 2^{r-1}$ (cf. Exercise J-4 below).

J-1. This exercise is combinatorial. Setting

$$\rho^1_l = \begin{cases} lr, & lr < d, \\ lr - 1, & lr = d, \\ d, & lr > d, \end{cases}$$

show that

$$\pi_1(d, r) = ld - \sum_{i=1}^{l} \rho^1_i,$$

where $\pi_1(d, r)$ is defined as in (2.7).

In the next two exercises we let $X \subset \mathbb{P}^r$ be an irreducible non-degenerate variety of dimension k and codimension $c = r - k$, Y be a general hyperplane section of X, and

$$\rho_l(X) = h^0(\mathbb{P}^r, \mathcal{O}(l)) - h^0(\mathbb{P}^r, \mathcal{I}_X(l)).$$

J-2. Show that

$$\rho_l(X) \geq \rho_{l-1}(X) + \rho_l(Y).$$

J-3. By repeatedly applying the preceding exercise, deduce that

(i)
$$\rho_l(X) \geq \binom{l+k-1}{k}c + \binom{l+k}{k},$$

(ii) for $0 \leq \alpha \leq c$, if $d \geq c + \alpha + 1$ then

$$\rho_l(X) \geq \binom{l+k-1}{k}c + \binom{l+k-2}{k}\alpha + \binom{l+k}{k}.$$

The next exercise will use the following general result (cf. Fulton [2]).

Theorem. *If $W_i \subset \mathbb{P}^r$ are irreducible varieties of degree d_i and $X = \bigcap W_i$, then the sum of the degrees of the irreducible components of X is at most $\prod_i d_i$.*

J-4. Suppose that $C \subset \mathbb{P}^r$ is a non-degenerate curve of degree $d > 2^{r-1}$ and Γ is a general hyperplane section of C. Suppose that

$$\rho_2(\Gamma) \leq 2r - 1 + \alpha$$

for some α with $0 \leq \alpha < r - 2$. Show that Γ lies on a curve of degree at most $r - 1 + \alpha$.

(*Hint*: If X is the intersection of the quadrics containing Γ use $d > 2^{r-1}$ to show that $\dim X \geq 1$; $\rho_2(\Gamma) = \rho_2(X) < 3r - 3$ and uniform position to show $\dim X = 1$ and estimate its degree.)

J-5. Suppose now that C is smooth of genus g, and show that for l sufficiently large

$$g \leq ld - \sum_{i=1}^{l} \rho_i(\Gamma).$$

J-6. Suppose now that Γ lies on an elliptic normal curve H, that is, an irreducible, non-degenerate curve of arithmetic genus 1 and degree r in $\mathbb{P}^{r-1}$. With the notations of *J*-1 show that

$$\rho_l(\Gamma) \geq \rho_l^1$$

with equality holding if and only if $d \not\equiv 0\ (r)$ or Γ is a complete intersection of H with a hypersurface in $\mathbb{P}^{r-1}$.

J-7. Using the inequalities in J-3, assuming that

$$\rho_2(\Gamma) > 2r - 1,$$

show that

$$\rho_l(\Gamma) \geq \rho_l^1$$

for all l.

J-8. Suppose that Γ lies on a smooth rational normal curve in $\mathbb{P}^{r-1}$. Show that

$$\rho_2(\Gamma) = 2r - 1.$$

J-9. Combining Exercises J-1 and J-5, show that if Γ does *not* lie on a rational normal curve, then $\rho_l(\Gamma) \geq \rho_l^1$ for all l, with equality holding only if Γ lies on an elliptic normal curve.

J-10. With Castelnuovo's number $\pi(d, r)$ as defined in Section 2 of Chapter III, show that

$$\pi_1(d, r) > \pi(d, r + 1).$$

Using the above exercises show that if

$$g \geq \pi_1(d, r),$$

then C is linearly normal.

Now complete the proof of the theorem under the additional assumptions $(*)$.

K. Castelnuovo's Theorem

In this sequence of exercises we will prove the following generalizations of Max Noether's theorem.

Theorem. *Let $|D|$ be a complete base-point-free linear series of dimension $r = r(D) \geq 3$ on a smooth curve C, and assume that the mapping*

$$\phi_D : C \to \mathbb{P}^r$$

is birational onto its image. Then the natural map

$$\operatorname{Sym}^l H^0(C, \mathcal{O}(D)) \otimes H^0(C, K) \to H^0(C, K(lD))$$

is surjective for $l \geq 0$.

Before beginning we recall

Castelnuovo's Lemma. *Let $L \to C$ be a line bundle such that*

$$H^1(C, L(-D)) = 0,$$

and let $\mathcal{D} \subset |D|$ be a base-point-free pencil given by a two-dimensional linear subspace $V \subset H^0(C, \mathcal{O}(D))$. Then the natural map

$$V \otimes H^0(C, L) \to H^0(C, L(D))$$

is surjective.

This follows immediately from the base-point-free pencil trick which gives the exact sheaf sequence

$$0 \to L(-D) \to V \otimes L \to L(D) \to 0.$$

K-1. Let $\mathscr{D} \subset |D|$ be a base-point-free pencil corresponding to

$$V \subset H^0(C, \mathcal{O}(D)).$$

Show that the image of the natural map

$$\operatorname{Sym}^l V \otimes H^0(C, K) \to H^0(C, K(lD))$$

is of codimension $lr - 1$ in $H^0(C, K(lD))$.

(*Hint*: Do first the cases $l = 1, 2$ using the base-point-free pencil trick and the Riemann–Roch theorem. By Castelnuovo's lemma applied to $L = K(lD)$ deduce that every element in $H^0(C, K(lD))$ is, for $l \geq 3$, of the form

$$\sum_\alpha P_\alpha \omega_\alpha + Q\eta,$$

where $\omega_1, \ldots, \omega_g$ is a basis for $H^0(C, K)$, $\eta \in H^0(C, K(2D))$, $P_\alpha \in \operatorname{Sym}^l V$, and $Q \in \operatorname{Sym}^{l-2} V$. Then argue by induction on l (the case $l = 2$ shows you what to do).)

K-2. Suppose that $p_1 + \cdots + p_d$ is a general divisor in $|D|$ and set $E = p_1 + \cdots + p_{r-2}$. Show that $|D - E|$ is a base-point-free pencil, and that in the exact cohomology sequence of

$$0 \to \mathcal{O}(D - E) \to \mathcal{O}(D) \to \mathcal{O}(D) \otimes \mathcal{O}_E \to 0$$

the map

$$H^0(C, \mathcal{O}(D)) \to H^0(\mathcal{O}(D) \otimes \mathcal{O}_E)$$

is surjective (use the general position theorem).

K-3. By tensoring the exact cohomology sequence of the preceding exercise with $H^0(C, K((l - 1)D)$ and arguing by induction on l, complete the proof of the theorem.

L. Secant Planes

L-1. Using the general position theorem, show that a curve $C \subset \mathbb{P}^r$ cannot have ∞^d $(d + 1)$-secant $(d - 1)$-planes spanned by their intersection with C, for any $d < r$.

Chapter IV

The Varieties of Special Linear Series on a Curve

As we have said, our approach to the study of curves is to analyze their projective realizations, and to speak of these is to speak of linear series. A more in-depth reflection leads to the appreciation of the fact that it is not only the single linear series which are important, but rather the configuration of all linear series of given type that a curve carries. To make precise what these "configurations" are we shall introduce three main kinds of varieties, which we now describe set-theoretically.

The first one, C_d^r, is the subvariety of the d-fold symmetric product C_d parametrizing effective divisors of degree d on C moving in a linear series of dimension at least r:

$$C_d^r = \{D \in C_d : \dim |D| \geq r\}.$$

The second, $W_d^r(C)$, is the subvariety of $\operatorname{Pic}^d(C)$ parametrizing *complete* linear series of degree d and dimension at least r:

$$W_d^r(C) = \{|D| : \deg D = d, r(D) \geq r\}.$$

The third one, $G_d^r(C)$, parametrizes linear series of degree d and dimension *exactly* r on C:

$$G_d^r(C) = \{g_d^r\text{'s on } C\}.$$

The link between C_d^r and $W_d^r(C)$ is given by the abelian sum mapping

$$u: C_d \to \operatorname{Pic}^d(C);$$

in fact

$$u(C_d^r) = W_d^r(C).$$

An essential feature of these subvarieties is that, on one hand, C_d^r and $W_d^r(C)$ carry natural determinantal structures while, on the other, $G_d^r(C)$ turns out to be the canonical blow-up of the determinantal variety $W^r_d(C)$ as defined in Chapter II, Section 4.

The reader should be aware of the fact that in the next two chapters we shall not need the results of this one in their full generality; a more naive approach would have been sufficient. At the cost of disregarding history and of rendering the chapter technically rather heavy, we have chosen to collect in a unique body all the various constructions and foundational results concerning the varieties of special divisors, in order to avoid repetitions and possible confusion in the sequel. In our opinion the sections that should get the most attention are Section 1, which is the closest in spirit to the origin of the subject; Section 4, which introduces one of the most widely used technical tools of the book; and Section 5, which contains a few practical applications. On a first reading a more superficial look at the remaining sections should be sufficient.

§1. The Brill–Noether Matrix and the Variety C^r_d

Let C be a smooth genus g curve. As usual we denote by C_d its d-fold symmetric product. Our goal is to define a subvariety C^r_d of C_d parametrizing effective divisors of degree d on C that vary in a linear series of dimension at least r. In symbols, we would like to have

$$\operatorname{Supp}(C^r_d) = \{D \in C_d : r(D) \geq r\}.$$

As with so much of the theory, the key tool to define C^r_d as a variety is the Brill–Noether matrix which we are now going to introduce. Let $D = \sum p_i$ be an effective degree d divisor on C. Choose a local coordinate z on C in a neighborhood of $\{p_1, \ldots, p_d\}$ and write

$$\omega_\alpha = f_\alpha\, dz, \qquad \alpha = 1, \ldots, g,$$

where $\omega_1, \ldots, \omega_g$ form a basis for the vector space of holomorphic differentials on C and the f_α's are holomorphic functions in a neighborhood of $\{p_1, \ldots, p_d\}$. Assuming that $p_1, \ldots, p_d$ are distinct, it follows from the Riemann–Roch theorem that $r(D) \geq r$ if and only if

$$\operatorname{rank}\begin{pmatrix} f_1(z(p_1)) & \cdots & f_1(z(p_d)) \\ \vdots & & \vdots \\ f_g(z(p_1)) & \cdots & f_g(z(p_d)) \end{pmatrix} \leq d - r.$$

The above matrix, which is usually written

$$\begin{pmatrix} \omega_1(p_1) & \cdots & \omega_1(p_d) \\ \vdots & & \vdots \\ \omega_g(p_1) & \cdots & \omega_g(p_d) \end{pmatrix},$$

is called the *Brill–Noether* matrix: the important role it plays in the fundamental paper [1] by Brill and Noether justifies this terminology. The fact that $r(D) \geq r$ if and only if the rank of the Brill–Noether matrix does not exceed $d - r$ shows that, in a neighborhood of D in C_d, the subset

$$\{D' \in C_d: r(D') \geq r\}$$

is the common zero locus of the $(d - r + 1) \times (d - r + 1)$ minors of this matrix. Clearly, the ideal generated by these minors does not depend on the choice of local coordinate z and of the differentials $\omega_1, \ldots, \omega_g$. In this way we may define C_d^r determinantally, at least away from the diagonals.

We now wish to extend the above determinantal description by means of an intrinsic interpretation of the Brill–Noether matrix which holds regardless of whether D consists of distinct points or not. There is really nothing to this. Just recall the basic abelian sum mapping

$$u: C_d \to J(C)$$

defined by

$$u(D) = (\ldots, u_\alpha(D), \ldots),$$

where

$$u_\alpha(D) = \sum_i \int_{p_0}^{p_i} \omega_\alpha, \qquad D = \sum_{i=1}^{d} p_i,$$

and p_0 is a base point, chosen once and for all. It is apparent that, in case D is the sum of d distinct points, the Brill–Noether matrix is just the Jacobian matrix of u relative to the coordinates u_α on $J(C)$ and the coordinates

$$\zeta_i(D) = z(p_i), \qquad i = 1, \ldots, d,$$

on C_d. Thus the Brill–Noether matrix can be thought of as giving a local coordinate representation of the homomorphism of locally free sheaves

$$u_*: \Theta_{C_d} \to u^*\Theta_{J(C)}.$$

We now define C_d^r as the $(d - r)$th determinantal variety attached to u_*, the one whose support is

$$\{D \in C_d: \operatorname{rank}((u_*)_D) \leq d - r\}.$$

As we just remarked, *away from the diagonals*, we have

$$\operatorname{supp}(C_d^r) = \{D \in C_d: r(D) \geq r\},$$

as desired. We now show that this is the case *everywhere* by proving the following fundamental

(1.1) Lemma. *$r(D) \geq r$ if and only if the rank of u_* at D does not exceed $d - r$. In particular, the support of C_d^r is precisely the set $\{D \in C_d: r(D) \geq r\}$, and the fibers of u are linear series (scheme-theoretically).*

Here we shall proceed by direct computation; a more intrinsic proof will be found in the third section of this chapter. One implication is trivial: since $u^{-1}(u(D)) = |D|$ is a projective space of dimension $r(D)$, the rank of u_* at D does not exceed $d - r(D)$. The other implication has already been proved when D consists of d distinct points.

Before treating the general case we shall recall some elementary properties of symmetric functions. Consider the elementary symmetric functions in h variables:

$$\sigma_j(x_1, \ldots, x_h) = \sum_{i_1 < \cdots < i_j} x_{i_1} x_{i_2} \cdots x_{i_j}, \qquad j = 1, \ldots, h.$$

It is well known that these functions generate the ring of all symmetric functions in h variables. For our particular purposes we shall need another set of generators: the Newton functions. These are symmetric functions defined by

$$v_i(x_1, \ldots, x_h) = x_1^i + \cdots + x_h^i, \qquad i = 1, 2, \ldots$$

We shall show that the first h Newton functions generate the ring of all symmetric functions, by showing that each σ_i, $i = 1, \ldots, h$, can be expressed as a polynomial in the v_i's, $i = 1, \ldots, h$. We have

$$1 + \sum_{j=1}^{h} \sigma_j t^j = \prod_{j=1}^{h} (1 + x_j t).$$

Hence

$$\log\left(1 + \sum_{j=1}^{h} \sigma_j t^j\right) = \sum_{j=1}^{h} \sum_{i=1}^{\infty} (-1)^{i+1} \frac{x_j^i}{i} t^i$$
$$= \sum_{i=1}^{\infty} (-1)^{i+1} \frac{v_i}{i} t^i.$$

Therefore

$$1 + \sum_{j=1}^{h} \sigma_j t^j = \exp\left(\sum_{i=1}^{\infty} (-1)^{i+1} \frac{v_i}{i} t^i\right). \tag{1.2}$$

Let us now use the following notation. Given a set y of infinitely many variables

$$y = (y_1, y_2, y_3, \ldots),$$

define functions $\Delta_j(y) = \Delta_j(y_1, y_2, \ldots, y_j)$ by setting

$$\Delta_j(y) = \sum_{h_1 + 2h_2 + \cdots + jh_j = j} \frac{1}{h_1! \ldots h_j!} y_1^{h_1} \cdots y_j^{h_j}.$$

A straightforward formal computation shows that

$$\exp\left(\sum_{i=1}^{\infty} y_i t^i\right) = \sum_{j \geq 0} \Delta_j(y) t^j.$$

Looking at (1.2) we can then conclude that

$$\sigma_j = \Delta_j\left(v_1, \frac{-v_2}{2}, \ldots, (-1)^{j+1} \frac{v_j}{j}\right), \qquad j = 1, \ldots, h,$$

proving that each σ_j can be expressed as a polynomial in $v_1, \ldots, v_h$.

Let us make a further observation. Consider the Newton functions $v_n = v_n(x_1, \ldots, x_h)$ for $n \geq h + 1$. By what we just proved these are polynomials in $v_1, \ldots, v_h$. We claim that these polynomials contain no linear term. In symbols

$$(1.3) \qquad v_n \in (v_1, \ldots, v_h)^2, \qquad n \geq h + 1.$$

By recursion it really suffices to show that v_n belongs to $(v_1, \ldots, v_{n-1})^2$, for $n \geq h + 1$. Looking at the coefficient of t^n in (1.2) we get

$$0 = \Delta_n\left(v_1, \ldots, (-1)^{n+1} \frac{v_n}{n}\right).$$

On the other hand, the very definition of Δ_n tells us that

$$\Delta_n\left(v_1, \ldots, (-1)^{n+1} \frac{v_n}{n}\right) - \frac{1}{n!}(-1)^{n+1} \frac{v_n}{n} \in (v_1, \ldots, v_{n-1})^2,$$

proving our claim.

We shall finally need the following remark.

Lemma. *Let $g(w)$ be a holomorphic function in a neighborhood of $0 \in \mathbb{C}$. Consider the Newton functions*

$$v_i = v_i(w_1, \dots, w_h) = w_1^i + \cdots + w_h^i, \qquad i = 1, \dots, h.$$

Write $\gamma(v_1, \dots, v_h) = \sum_{i=1}^h g(w_i)$. Then

$$\gamma(v_1, \dots, v_h) = ng(0) + \sum_{i=1}^{h} \frac{1}{i!}\frac{d^ig}{dw^i}(0)v_i + O((v_1, \dots, v_h)^2).$$

Proof. We have

$$g(w_j) = g(0) + \sum_{i=1}^{\infty} \frac{1}{i!}\frac{d^ig}{dw^i}(0)w_j^i,$$

so that

$$\gamma(v_1, \dots, v_h) = ng(0) + \sum_{i=1}^{\infty} \frac{1}{i!}\frac{d^ig}{dw^i}(0)v_i,$$

and the lemma follows from (1.3). Q.E.D.

We now go back to the proof of Lemma (1.1) and prove that the rank of u_* at $D \in C_d$ equals $d - r(D)$. Write $D = \sum h_i p_i$, where $p_1, \dots, p_n$ are distinct. For each i, let z_i be a local coordinate centered at p_i. We take as local coordinates on C_d, in a neighborhood of D, the functions

$$\begin{aligned}
&\zeta_i(\textstyle\sum q_j) = v_i(z_1(q_1), \dots, z_1(q_{h_1})), \quad i = 1, \dots, h_1,\\
&\zeta_i(\textstyle\sum q_j) = v_{i-h_1}(z_2(q_{h_1+1}), \dots, z_2(q_{h_1+h_2})), \qquad i = h_1 + 1, \dots, h_1 + h_2,\\
&\quad\vdots \qquad\qquad\qquad \vdots
\end{aligned}$$

where v_i stands for the ith Newton function. Write

$$\omega_\alpha = f_{\alpha i}\, dz_i,$$

where $f_{\alpha i}$ is a holomorphic function in a neighborhood of p_i. We now compute the Jacobian matrix $(\partial u_\alpha/\partial\zeta_i)$ at D, applying the lemma to the abelian sums

$$u_\alpha\left(\sum_j q_j\right) = \sum_j u_\alpha(q_j),$$

viewed as functions of $\zeta_1, \ldots, \zeta_d$. Since near p_i the derivative of $u_\alpha(q)$ with respect to z_i is $f_{\alpha i}$, the Jacobian $(\partial u_\alpha / \partial \zeta_i)$ at D is

$$\begin{pmatrix} \underbrace{f_{11}(0) \quad f_{11}^{(1)}(0) \quad \cdots \quad \frac{1}{(h_1-1)!} f_{11}^{(h_1-1)}(0)}_{h_1} & \underbrace{f_{12}(0) \quad f_{12}^{(1)}(0) \quad \cdots \quad \frac{1}{(h_2-1)!} f_{12}^{(h_2-1)}(0)}_{h_2} & \cdots \\ \vdots & \vdots & \vdots \\ f_{g1}(0) \quad f_{g1}^{(1)}(0) \quad \cdots \quad \frac{1}{(h_1-1)!} f_{g1}^{(h_1-1)}(0) & f_{g2}(0) \quad f_{g2}^{(1)}(0) \quad \cdots \quad \frac{1}{(h_2-1)!} f_{g2}^{(h_2-1)}(0) & \cdots \end{pmatrix}.$$

To say that the rank of this matrix is $d - r(D)$ is just to say that there are exactly $g - d + r(D)$ independent relations amongst its rows, that is, there are exactly $g - d + r(D)$ linearly independent holomorphic differentials vanishing at D: this is the Riemann–Roch theorem.

The above matrix will also be called the *Brill–Noether matrix*. Sometimes we shall improperly use this name to denote the sheaf homomorphism u_* as well.

The determinantal description of the variety C^r_d provides a lower bound on its dimension. Since C^r_d is locally defined by the simultaneous vanishing of all $(d - r + 1) \times (d - r + 1)$ minors of a $d \times g$ matrix of holomorphic functions, as we observed in Chapter II, Section 4, the codimension of C^r_d can be at most

$$[d - (d - r)] \cdot [g - (d - r)] = r(g - d + r).$$

Thus every component of C^r_d has dimension at least

$$d - r(g - d + r) = r + g - (r + 1)(g - d + r).$$

The number

$$\rho(g, r, d) = g - (r + 1)(g - d + r),$$

which is of fundamental importance in the study of special divisors, is called the *Brill–Noether number* and will usually be denoted simply by ρ, unless confusion is likely. Thus we can say that *every component of C^r_d has dimension at least equal to the "expected" dimension $\rho + r$, where ρ is the Brill–Noether number.*

It is now instructive to determine the Zariski tangent space to C^r_d at a divisor D not belonging to C^{r+1}_d, for this computation brings to the foreground again the basic cup-product mapping μ_0, which we already encountered in the proof of Clifford's theorem. Here we shall give only an elementary argument covering the case $D = \sum p_i$, where the p_i's are distinct;

the whole question will be reconsidered, from a slightly less elementary, but more natural, point of view, in Section 4 of this chapter.

We first choose local coordinates $z_1, \ldots, z_d$ on C, centered at $p_1, \ldots, p_d$, respectively. The z_i's can also be viewed as local coordinates on C_d, centered at D. Next, when $\sum q_i$ is close to D, we denote by $\phi(\sum q_i)$ the transpose of the Brill–Noether matrix at $\sum q_i$, i.e., we set

$$\phi(\sum q_i) = \begin{pmatrix} \omega_1(q_1) & \cdots & \omega_g(q_1) \\ \vdots & & \vdots \\ \omega_1(q_d) & \cdots & \omega_g(q_d) \end{pmatrix}.$$

Denoting by M the variety of $d \times g$ matrices, we shall view ϕ as an M-valued mapping defined in a neighborhood of D. By definition, in this neighborhood, C_d^r is just the pull-back, via ϕ, of the determinantal subvariety M_{d-r} of all matrices of rank not exceeding $d - r$. In particular, the Zariski tangent space to C_d^r at D is the pull-back, under the differential of ϕ, of the tangent space to M_{d-r} at $A = \phi(D)$. This has been described in Chapter II, Section 2, where it is shown that

$$T_A(M_{d-r}) = \{B \in M : B \cdot \ker A \subset A \cdot \mathbb{C}^g\},$$

under the assumption that A does not belong to M_{d-r-1}. This is true in our case, since we are assuming that D does not belong to C_d^{r+1}. We then have to describe, in intrinsic terms,

$$\begin{aligned} T_D(C_d^r) &= \phi_*^{-1}(T_A(M_{d-r})) \\ &= \{v \in T_D(C_d) : \phi_*(v) \cdot \ker A \subset \text{Image } A\}, \end{aligned}$$

where $A = \phi(D)$. To begin with, let $v = \sum v_i \, \partial/\partial z_i$ be a tangent vector to C_d at D. To v we may associate

$$\sum_i v_i \frac{1}{z_i},$$

viewed as a section of $\mathcal{O}_D(D)$. This sets up an isomorphism

$$T_D(C_d) \cong H^0(C, \mathcal{O}_D(D)).$$

The important fact, which can be checked by a trivial calculation, is that this isomorphism is *intrinsic*, i.e., does not depend on the choice of coordinates. Actually $T_D(C_d)$ and $H^0(C, \mathcal{O}_D(D)$ are intrinsically isomorphic even if D contains multiple points. In keeping with our "naive" approach, we shall postpone a proof of this, together with a global description of the tangent bundle to C_d, to a later section of this chapter. If $v = \sum v_i\, \partial/\partial z_i$ is as above,

$$\phi_*(v) = \begin{pmatrix} v_1\dot{\omega}_1(p_1) & \cdots & v_1\dot{\omega}_g(p_1) \\ \vdots & & \vdots \\ v_d\dot{\omega}_1(p_d) & \cdots & v_d\dot{\omega}_g(p_d) \end{pmatrix},$$

where $\dot{\omega}_i(p_j)$ stands for the derivative of ω_i/dz_j at p_j with respect to z_j. On the other hand, if we identify the $\mathbb{C}^g$ on which A operates with $H^0(C, K)$, it is clear that A "is" the restriction mapping

$$\alpha\colon H^0(C, K) \to H^0(C, K \otimes \mathcal{O}_D).$$

Thus, recalling the identification given above between $T_D(C_d)$ and $H^0(C, \mathcal{O}_D(D))$, we may identify $\phi_*(v) \cdot \ker A$ with the image, under the cup-product homomorphism

$$\beta\colon H^0(C, \mathcal{O}_D(D)) \otimes H^0(C, K(-D)) \to H^0(C, K \otimes \mathcal{O}_D)$$

of

$$v \otimes H^0(C, K(-D)).$$

It follows that

$$\phi_*(v) \cdot \ker A \subset \mathrm{Image}(A)$$

if and only if for any ω belonging to $H^0(C, K(-D))$,

$$\beta(v \otimes \omega) = \alpha(\omega')$$

for some ω' in $H^0(C, K)$. In other terms, if we let

$$\beta'\colon H^0(C, \mathcal{O}_D(D)) \otimes H^0(C, K(-D)) \to H^1(C, K(-D))$$

be the composition of β with the coboundary mapping

$$H^0(C, K \otimes \mathcal{O}_D) \to H^1(C, K(-D)),$$

v belongs to $T_D(C_d^r)$ if and only if

$$\beta'(v \otimes \omega) = 0 \quad \text{for any} \quad \omega \in H^0(C, K(-D)).$$

This condition is most conveniently expressed by saying that

$$\langle s, \beta'(v \otimes \omega) \rangle = 0$$

for any ω belonging to $H^0(C, K(-D))$ and any s belonging to $H^0(C, \mathcal{O}(D))$, where $\langle \ , \ \rangle$ stands for the Serre duality pairing. Now look at the commutative diagram

$$\begin{array}{ccc} H^0(C, \mathcal{O}_D(D)) \otimes H^0(C, K(-D)) & \xrightarrow{\beta} & H^0(C, K \otimes \mathcal{O}_D) \\ \downarrow{\scriptstyle \delta \otimes 1} & & \downarrow{\scriptstyle \delta} \\ H^1(C, \mathcal{O}) \otimes H^0(C, K(-D)) & \xrightarrow{\beta''} & H^1(C, K(-D)), \end{array}$$

where the horizontal rows are cup-product homomorphisms. Clearly,

$$\begin{aligned} \langle s, \beta'(v \otimes \omega) \rangle &= \langle s, \beta''(\delta v \otimes \omega) \rangle \\ &= \langle \mu_0(s \otimes \omega), \delta v \rangle \\ &= \langle \alpha\mu_0(s \otimes \omega), v \rangle, \end{aligned}$$

where

$$\mu_0 \colon H^0(C, \mathcal{O}(D)) \otimes H^0(C, K(-D)) \to H^0(C, K) \tag{1.4}$$

is the cup-product mapping and

$$\alpha \colon H^0(C, K) \to H^0(C, K \otimes \mathcal{O}_D)$$

is the restriction mapping. Putting everything together, we conclude

(1.5) Lemma. *If D belongs to C_d^r but not to C_d^{r+1}, the tangent space to C_d^r at D is*

$$T_D(C_d^r) = (\mathrm{Image}(\alpha\mu_0))^{\perp},$$

where μ_0 is the basic cup-product mapping (1.4) *and α is the restriction mapping from $H^0(C, K)$ to*

$$H^0(C, K \otimes \mathcal{O}_D) \cong T_D(C_d)^*.$$

Of course, this lemma has been proved, so far, only under the additional assumption that D has no multiple points; it is, however, valid in general.

It is now very easy to say when D is a smooth point of C_d^r. Notice that the kernel of α is just $H^0(C, K(-D))$, so that, in particular, it is contained in the image of μ_0. Thus

$$\begin{aligned}\dim T_D(C_d^r) &= d - \dim(\text{Image } \alpha\mu_0)\\ &= d - \dim(\text{Image } \mu_0) + h^0(C, K(-D))\\ &= d - (r+1)(g-d+r) + \dim(\ker \mu_0) + (g-d+r)\\ &= \rho + r + \dim(\ker \mu_0).\end{aligned}$$

As we already observed that every component of C_d^r has dimension at least $\rho + r$, we reach the following conclusion.

(1.6) Lemma. *The variety C_d^r is smooth and has the "expected" dimension*

$$\rho + r = g - (r+1)(g-d+r) + r$$

at $D \in C_d^r - C_d^{r+1}$ if and only if the cup-product homomorphism

$$\mu_0 \colon H^0(C, \mathcal{O}(D)) \otimes H^0(C, K(-D)) \to H^0(C, K)$$

is injective.

At this point, one might ask under what circumstances is C_d^r different from C_d^{r+1}. The answer is that this is always the case, unless the contrary is forced "by the numbers." More exactly we have the simple but basic

(1.7) Lemma. *Assume $g - d + r \geq 0$. Then no component of C_d^r is entirely contained in C_d^{r+1}.*

Proof. We shall argue by contradiction. Let D be a general point of a component of C_d^r and assume $r(D) > r$. If q is a point of C there is an effective divisor E such that D is linearly equivalent to $E + q$. Moreover, if q is general (more exactly, if q is not a base point of $|D|$),

$$r(E) = r(D) - 1 \geq r,$$

and, by the Riemann–Roch theorem,

$$i(E) = i(D) = g - d + r(D) \geq 1.$$

Thus, if p is a general point of C, p is not a base point of $|K(-E)|$, and hence

$$\begin{aligned}i(E+p) &= i(E) - 1,\\ r(E+p) &= r(E) = r(D) - 1 \geq r,\end{aligned}$$

by the Riemann–Roch theorem. On the other hand, by the generality of p and q, $E + p$ belongs to the same component of C^r_d as D. This contradicts the generality of D. Q.E.D.

§2. The Universal Divisor and the Poincaré Line Bundles

In the previous section we have introduced the Brill–Noether matrix and explained, in elementary terms, its significance in the study of special divisors. Now the time has come to give more solid foundations to our study of the symmetric products of a curve and their relation to the Jacobian variety.

The first new concept we wish to introduce is the one of universal divisor. As usual, we let C be a fixed smooth curve of genus g. The *universal effective divisor of degree d on C* is the divisor

$$\Delta \subset C \times C_d,$$

which, for any D in C_d, cuts on

$$C \cong C \times \{D\}$$

exactly the divisor D. To convince ourselves that this prescription really defines a divisor, and for later use, we shall now give explicit local equations for Δ. Let $D = \sum_{i=1}^d p_i$ be a point of C_d. Choose a coordinate ζ on C in a neighborhood U of $\{p_1, \dots, p_d\}$. In a neighborhood V of D there are coordinates

$$\sigma = (\sigma_1, \dots, \sigma_d)$$

defined by the property that the natural morphism

$$\gamma: C^d \to C_d$$

is given, in a neighborhood of $(p_1, \dots, p_d)$, by

$$\sigma_i(\gamma(q_1, \dots, q_d)) = i\text{th symmetric function of } (\zeta(q_1), \dots, \zeta(q_d)).$$

Then Δ is described, in $U \times V$, by the equation

$$F(\zeta, \sigma) = 0,$$

where

$$F(\zeta, \sigma) = \zeta^d - \sigma_1 \zeta^{d-1} + \sigma_2 \zeta^{d-2} - \cdots + (-1)^d \sigma_d.$$

We would now like to explain the adjective "universal" attached to the divisor Δ. First of all, we recall what is meant by a *relative divisor in* $C \times S$ *over* an analytic space S. This is simply an effective (Cartier) divisor on $C \times S$ which does not contain fibers of the projection onto S. We shall say that such a divisor has degree d if it cuts a degree d divisor on each fiber of this projection. Then the universal property of Δ is the following:

(2.1) Lemma. *For any analytic space S, any relative degree d divisor in $C \times S$ over S is the pull-back $(1_C \times f)^*(\Delta)$ of the universal divisor for an unique morphism*

$$f: S \to C_d.$$

In other words, C_d represents the functor

$$G: \{\text{analytic spaces}\} \to \{\text{sets}\}$$

defined by

$$G(S) = \{\text{relative degree } d \text{ divisors in } C \times S \text{ over } S\}.$$

The proof of the lemma is very simple. Let D be a relative degree d divisor in $C \times S$ over S. First of all, since the problem is local on S, we may assume that D is supported in $U \times S$, where U is a coordinate open subset of C, and that D is defined, in $U \times S$, by a global equation ϕ. Next, if ξ is a coordinate in U, using the Weierstrass preparation theorem we may write, in a *unique* way,

$$\phi = hg,$$

where h and $1/h$ are holomorphic and

$$g = \xi^d - \tau_1 \xi^{d-1} + \cdots + (-1)^d \tau_d$$

is a Weierstrass polynomial in ξ. It is then clear that the only map

$$f: S \to C_d$$

such that

$$(1_C \times f)^*(\Delta) = D$$

is defined, in the local coordinates σ_i on C_d introduced in the definition of Δ, by

$$\sigma_i = \tau_i, \qquad i = 1, \ldots, d.$$

This completes the proof of the lemma.

As we know $\mathrm{Pic}^d(C)$ parametrizes degree d line bundles on C. The first use of the universal divisor Δ will be in showing that actually there is, on $C \times \mathrm{Pic}^d(C)$, a universal line bundle, in a sense which we shall now make precise. By a *Poincaré line bundle of degree d for C* we shall mean a line bundle $\mathscr{L}$ on $C \times \mathrm{Pic}^d(C)$ which, for each L in $\mathrm{Pic}^d(C)$, restricts exactly to L on

$$C \cong C \times \{L\}.$$

If $\mathscr{L}$, $\mathscr{L}'$ are two Poincaré line bundles, we may write, in a unique way,

$$\mathscr{L}' = \mathscr{L} \otimes v^*\mathscr{R},$$

where

$$v\colon C \times \mathrm{Pic}^d(C) \to \mathrm{Pic}^d(C)$$

is the projection, and $\mathscr{R}$ is a line bundle on $\mathrm{Pic}^d(C)$. It suffices to take

$$\mathscr{R} = v_*(\mathscr{L}' \otimes \mathscr{L}^{-1}).$$

In fact, by the very definition of Poincaré line bundle, $\mathscr{L}' \otimes \mathscr{L}^{-1}$ is trivial on each fiber of v, hence $\mathscr{R}$ is a line bundle. Moreover,

$$v_*(\mathscr{L} \otimes \mathscr{L}'^{-1} \otimes v^*\mathscr{R})$$

is trivial. Hence

$$\mathscr{L} \otimes \mathscr{L}'^{-1} \otimes v^*\mathscr{R}$$

has a global section which does not vanish identically on any fiber of v. Since the above line bundle restricts to the trivial bundle on every fiber of v, this global section never vanishes. This shows that $\mathscr{L} \otimes \mathscr{L}'^{-1} \otimes v^*\mathscr{R}$ is trivial, as desired.

Of course, it is not clear, at this point, that Poincaré line bundles exist. To show that this is the case is our next task. We begin by noticing that it suffices to construct a Poincaré line bundle for one d. In fact, let $\mathscr{L}$ be one such, and let e be an integer. Fix a line bundle L_0 on C of degree $d - e$. We get an identification

$$a\colon \mathrm{Pic}^e(C) \to \mathrm{Pic}^d(C)$$

by setting

$$a(L) = L \otimes L_0.$$

Then, clearly,

$$(1_C \times a)^*\mathscr{L} \otimes \nu'^*(L_0^{-1})$$

is a Poincaré line bundle of degree e, where

$$\nu' : C \times \operatorname{Pic}^e(C) \to C$$

is the projection.

In constructing Poincaré line bundles, we may therefore limit our attention to large values of d. In particular, we shall assume that $d \geq 2g - 1$, so that the fibers of

$$u: C_d \to \operatorname{Pic}^d(C)$$

are all projective spaces of dimension $d - g$. For any point q in C we consider the divisor in C_d:

$$X_q = \{q + D' : D' \in C_{d-1}\}.$$

We then set

$$\mathscr{L} = (1_C \times u)_*(\mathscr{O}(\Delta - \pi^* X_q)),$$

where

$$\pi: C \times C_d \to C_d$$

is the projection. We claim that $\mathscr{L}$ is a Poincaré line bundle of degree d. To see this, fix a point, say $\mathscr{O}(D)$, in $\operatorname{Pic}^d(C)$; all we have to do is show that $\mathscr{O}(\Delta - \pi^* X_q)$ restricts, on $C \times |D|$, to the pull-back of $\mathscr{O}(D)$ from C. But this is clear, since on one hand, for every q' in C, Δ and $\pi^* X_q$ cut on the projective space

$$|D| \cong \{q'\} \times |D|$$

the divisors $X_{q'} \cap |D|$ and $X_q \cap |D|$, which are both hyperplanes, while $\mathscr{O}(\Delta - \pi^* X_q)$ restricts to $\mathscr{O}(D)$ on $C \times \{D\}$.

Having established that there exist Poincaré bundles, we may notice, in passing, that for any q in C there is a unique degree d Poincaré line bundle whose restriction to $\{q\} \times \operatorname{Pic}^d(C)$ is trivial.

Poincaré line bundles have a universal property, which is expressed by the following

(2.2) Lemma. *Let $\mathscr{L}_d$ be a degree d Poincaré line bundle for the smooth curve C. Let S be an analytic space and let $\mathscr{L}$ be a line bundle over $C \times S$ such that, for each s in S, $\mathscr{L}|_{C \times \{s\}}$ has degree d. Then there exists a unique map*

$$f\colon S \to \mathrm{Pic}^d(C)$$

such that the pull-back, via $1_C \times f$, of $\mathscr{L}_d$ is of the form $\mathscr{L} \otimes \phi^*\mathscr{R}$, where $\mathscr{R}$ is a line bundle on S and ϕ is the projection of $C \times S$ onto S.

Proof. As in the construction of the Poincaré line bundle, we may and will assume that $d \geq 2g - 1$, so that C_d is fibered over $\mathrm{Pic}^d(C)$ in projective spaces of dimension $d - g$. The problem is local on S. In fact, assume the lemma holds for any "sufficiently small" S; we can find covers $\{S_i\}$ of S by arbitrarily small open subsets such that $\mathrm{Pic}(S_i)$ is trivial. Then if we can find morphisms

$$f_i\colon S_i \to \mathrm{Pic}^d(C)$$

and isomorphisms

$$\xi_i\colon (1_C \times f_i)^*\mathscr{L}_d \to \mathscr{L}|_{C \times S_i},$$

then the f_i's patch together into a morphism f from S to $\mathrm{Pic}^d(C)$, by uniqueness, and moreover, since C is complete,

$$\xi_i\xi_j^{-1}\colon L|_{C \times (S_i \cap S_j)} \to L|_{C \times (S_i \cap S_j)}$$

is a never vanishing holomorphic function g_{ij} on $S_i \cap S_j$ *alone*. Furthermore,

$$g_{ij}g_{jk} = g_{ik},$$

so that the g_{ij}'s are the transition functions of a line bundle $\mathscr{R}$ on S. The ξ_i's can then be viewed as local expressions of a global isomorphism

$$\xi\colon (1_C \times f)^*\mathscr{L}_d \to \mathscr{L} \otimes \phi^*\mathscr{R}.$$

We may then assume S to be "small," and, in particular, that $\mathrm{Pic}(S)$ is trivial. The existence of f is then clear, for we may assume that $\mathscr{L}$ has a global section s which does not vanish identically on any fiber of ϕ. Then s defines a relative degree d divisor, which corresponds, by the universal property of the universal divisor Δ, to a morphism of S into C_d. Composing this morphism with u gives f.

As for the uniqueness of f, this is clear over S_{red} by the very definition of Poincaré bundle. Thus if f' is another map of S into $\mathrm{Pic}^d(C)$ under which $\mathscr{L}_d$ pulls back to $\mathscr{L}$, we may assume that both f and f' map into an open subset U of $\operatorname{Pic}^d(C)$ over which C_d is a trivial $\mathbb{P}^{d-g}$-bundle. Let g, g' be liftings of f, f' to maps into C_d. Also, let D, D' be the corresponding liftings of $\mathscr{L}$ to relative divisors on $C \times S$. Now, to give a lifting of either f or $\mathscr{L}$ corresponds to choosing a section over S of

$$\mathbb{P}(\phi_* \mathscr{L}) \cong \mathbb{P}^{d-g} \times S.$$

Thus there is a lifting g'' of f whose corresponding relative divisor is D'. But then, by the universal property of the universal divisor, $g'' = g'$, and hence $f = f'$. Q.E.D.

By a *family of degree d line bundles over C parametrized by* an analytic space S we shall simply mean a line bundle on $C \times S$ which restricts to a degree d line bundle on $C \times \{s\}$, for any s in S. We shall say that two families $\mathscr{L}$ and $\mathscr{L}'$ of degree d line bundles on C parametrized by S are *equivalent* if there is a line bundle $\mathscr{R}$ on S such that

$$\mathscr{L}' \cong \mathscr{L} \otimes \phi^* \mathscr{R},$$

where ϕ stands for the projection of $C \times S$ onto S. In this language the lemma we just proved simply says that $\operatorname{Pic}^d(C)$ represents the functor

$$F: \{\text{analytic spaces}\} \to \{\text{sets}\},$$

$$F(S) = \{\text{equivalence classes of families of degree } d \text{ line bundles on } C \text{ parametrized by } S\}.$$

We would finally like to give an intrinsic description of the tangent bundle to C_d and of the Brill–Noether homomorphism. The first step is to make explicit the canonical identification between the tangent spaces to $\operatorname{Pic}^d(C)$ and $H^1(C, \mathcal{O})$ provided by the (non-canonical) isomorphism between $\operatorname{Pic}^d(C)$ and $H^1(C, \mathcal{O})/H^1(C, \mathbb{Z})$.

Let L be an element of $\operatorname{Pic}^d(C)$. Recall that $T_L(\operatorname{Pic}^d(C))$ can be described as the set of morphisms of pointed analytic spaces

$$\operatorname{Spec} \mathbb{C}[\varepsilon] = S \to (\operatorname{Pic}^d(C), L),$$

where $\mathbb{C}[\varepsilon] = \mathbb{C}[T]/(T^2)$ is the ring of dual numbers. In other words

$$T_L(\operatorname{Pic}^d(C)) \cong \operatorname{Hom}(S, (\operatorname{Pic}^d(C), L)).$$

Now let us define a *first-order deformation of* L to be a family $\mathscr{L}$ of line bundles on C parametrized by $S = \operatorname{Spec} \mathbb{C}[\varepsilon]$ such that

$$j^*(\mathscr{L}) = L,$$

where j is the inclusion of C in $C \times S$. By the universal property of the Poincaré bundle there is a natural 1–1 correspondence between $T_L(\mathrm{Pic}^d(C))$ and the set of equivalence classes of first-order deformations of L.

Now suppose that L is given by transition data $(\{U_\alpha\}, \{g_{\alpha\beta}\})$, where

$$\mathscr{U} = \{U_\alpha\}$$

is an open cover of C and the functions

$$g_{\alpha\beta} \in \Gamma(U_\alpha \cap U_\beta, \mathcal{O}_C^*)$$

are transition functions for L. Thus, in triple intersections the cocycle rule

$$g_{\alpha\beta} g_{\beta\gamma} = g_{\alpha\gamma}$$

holds.

We may now write a first-order deformation $\mathscr{L}$ of L as being given by transition functions

$$\tilde{g}_{\alpha\beta} = g_{\alpha\beta}(1 + \varepsilon\phi_{\alpha\beta}),$$

defined in $(U_\alpha \cap U_\beta) \times S$, where $\phi_{\alpha\beta}$ is holomorphic on $U_\alpha \cap U_\beta$. Again we have the cocycle rule

$$\tilde{g}_{\alpha\beta}\tilde{g}_{\beta\gamma} = \tilde{g}_{\alpha\gamma}.$$

It is easy to see that this cocycle rule is equivalent to

$$g_{\alpha\beta} g_{\beta\gamma} = g_{\alpha\gamma},$$

$$\phi_{\alpha\beta} + \phi_{\beta\gamma} = \phi_{\alpha\gamma}.$$

This just says that the 1-cochain $\{\phi_{\alpha\beta}\}$ is in fact a cocycle, and hence defines a class

$$\phi \in H^1(C, \mathcal{O}).$$

Next, recall that the identification between $\mathrm{Pic}^d(C)$ and $H^1(C, \mathcal{O})/H^1(C, \mathbb{Z})$ is provided by the exponential sequence

$$0 \to \mathbb{Z} \to \mathcal{O} \xrightarrow{\exp(2\pi\sqrt{-1})} \mathcal{O}^* \to 1.$$

It follows that the element of $H^1(C, \mathcal{O})$ associated to the tangent vector corresponding to $\mathscr{L}$ is simply the class of the logarithmic derivatives, with respect to ε, of the transition data of $\mathscr{L}$, divided by $2\pi\sqrt{-1}$. But

$$\frac{1}{2\pi\sqrt{-1}}\frac{d}{d\varepsilon}\log\tilde{g}_{\alpha\beta} = \frac{1}{2\pi\sqrt{-1}}g_{\alpha\beta}^{-1}\frac{d}{d\varepsilon}(g_{\alpha\beta}(1+\varepsilon\phi_{\alpha\beta}))$$
$$= \frac{1}{2\pi\sqrt{-1}}\phi_{\alpha\beta}.$$

Thus the element of $H^1(C, \mathcal{O})$ corresponding to the tangent vector defined by $\mathscr{L}$ is $\phi/2\pi\sqrt{-1}$. In what follows, however, we will find it more convenient to associate to this tangent vector the class ϕ, which will be called the *Kodaira–Spencer class of the first-order deformation* $\mathscr{L}$.

We have finally assembled all the elements which are necessary to describe the Brill–Noether homomorphism in global, intrinsic terms. As usual, we denote by

$$\pi\colon C \times C_d \to C_d$$

the projection and by Δ the universal divisor in $C \times C_d$. Consider the exact sequence of coherent sheaves on $C \times C_d$

$$0 \to \mathcal{O} \to \mathcal{O}(\Delta) \to \mathcal{O}_\Delta(\Delta) \to 0,$$

and the corresponding exact sequence of direct image sheaves

$$\cdots \to \pi_*\mathcal{O}_\Delta(\Delta) \xrightarrow{\delta} R^1\pi_*\mathcal{O} \to \cdots.$$

We then have the following result, the embryo of which we encountered in the first section of this chapter.

(2.3) Lemma. *There are isomorphisms, ϕ and ψ, making the following diagram commutative*

$$\begin{array}{ccc} \Theta_{C_d} & \xrightarrow{u_*} & u^*\Theta_{\mathrm{Pic}^d(C)} \\ \psi\downarrow\cong & & \phi\downarrow\cong \\ \pi_*\mathcal{O}_\Delta(\Delta) & \xrightarrow{\delta} & R^1\pi_*\mathcal{O} \end{array}$$

so that the Brill–Noether homomorphism may be identified with the coboundary homomorphism δ.

Proof. The identification ϕ between $u^*\Theta_{\mathrm{Pic}^d(C)}$ and $R^1\pi_*\mathcal{O}$ is given by the Kodaira–Spencer class. As for ψ, let X be a vector field on an open subset

U of C_d. The field X lifts, via π, to a vector field $\tilde{X}$ on $C \times U$. Now, if $F_1 = 0$, $F_2 = 0$ are local equations for Δ in an open subset of $C \times U$, then

$$F_1 = GF_2,$$

where G is a non-vanishing function. Thus

$$\frac{X(F_1)}{F_1} = \frac{X(F_2)}{F_2} + (\text{a holomorphic function}). \tag{2.4}$$

Therefore, if $\{V_\alpha\}$ is an open cover of $C \times U$ and, for each α, $F_\alpha = 0$ is a local equation for Δ in V_α, the cocycle

$$\left\{\frac{X(F_\alpha)}{F_\alpha}\right\} \in C^0(\{V_\alpha\}, \mathcal{O}(\Delta)|_{C \times U})$$

determines, by restriction to Δ, a section

$$\psi_U(X) \in \Gamma(C \times U, \mathcal{O}_\Delta(\Delta)).$$

Formula (2.4) also shows that the collection of homomorphisms $\{\psi_U\}$ defines a sheaf homomorphism

$$\psi \colon \Theta_{C_d} \to \pi_* \mathcal{O}_\Delta(\Delta).$$

We now show that $\phi u_* = \delta\psi$. Let $D = \sum_{i=1}^d p_i$ be a point of C_d, and let

$$F = F(\zeta, \sigma) = \zeta^d - \sigma_1 \zeta^{d-1} + \cdots + (-1)^d \sigma_d$$

be a local equation for Δ near $\{p_1, \ldots, p_d\} \times \{D\}$, as in the formal definition of Δ. Consider a tangent vector

$$X = \sum a_i \frac{\partial}{\partial \sigma_i}$$

to C_d at D. This corresponds to a morphism

$$\xi \colon \operatorname{Spec} \mathbb{C}[\varepsilon] \to (C_d, D).$$

By definition, $\phi u_*(X)$ is the Kodaira–Spencer class of the pull-back bundle $(1_C \times \xi)^* \mathcal{O}(\Delta)$. To compute this class consider the pull-back of F via $1_C \times \xi$, namely

$$F(\zeta, \sigma(D) + \varepsilon(a_1, \ldots, a_d)) = G(\zeta, \varepsilon),$$

let U be a small neighborhood of $\{p_1, \ldots, p_d\}$, and set $V = C - \{p_1, \ldots, p_d\}$. Then G is the transition function of the bundle $(1_C \times \xi)^* \mathcal{O}(\Delta)$ relative to the cover $\{U', V'\}$, where $U' = U \times \operatorname{Spec} \mathbb{C}[\varepsilon]$, $V' = V \times \operatorname{Spec} \mathbb{C}[\varepsilon]$. Therefore the Kodaira–Spencer class $\phi u_*(X)$ of $(1_C \times \xi)^*\mathcal{O}(\Delta)$ is the class of

$$\left\{G^{-1}\frac{\partial G}{\partial \varepsilon}\bigg|_{U\cap V}\right\} = \left\{F^{-1}\tilde{X}(F)\bigg|_{U\cap V}\right\} \in C^1(\{U, V\}, \mathcal{O}).$$

On the other hand, it follows directly from the definitions that

$$\delta\psi(X) = \{F^{-1}\tilde{X}(F)|_{U\cap V}\}.$$

This proves the commutativity of our diagram.

It remains to show that ψ is an isomorphism. Let g be a holomorphic function in a neighborhood of $\Delta \cap (C \times U)$, where U is an open set in C_d containing D. Using Weierstrass' division theorem, we may write

$$g = hF + R,$$

where h is holomorphic and

$$R = \sum_{i=1}^{d} (-1)^i b_i(\sigma)\zeta^{d-i}.$$

Then the section of $\pi_*\mathcal{O}_\Delta(\Delta)$ over U determined by g/F is equal to $\psi(Y)$, where

$$Y = \sum b_i(\sigma)\frac{\partial}{\partial \sigma_i}.$$

This shows that ψ is onto. Conversely, if Y is as above,

$$\frac{\tilde{Y}(F)}{F} = \frac{\sum_{i=1}^{d}(-1)^i b_i(\sigma)\zeta^{d-i}}{F}$$

can be holomorphic only when all the b_i's vanish. This completes the proof of the lemma, achieving the description of the Brill–Noether homomorphism in cohomological terms. Q.E.D.

A final and important comment is in order. Lemma (2.3) shows that the cokernel of u_* is just $R^1\pi_*\mathcal{O}(\Delta)$. In fact, since $R^1\pi_*\mathcal{O}_\Delta(\Delta)$ is zero, a piece of the exact cohomology sequence of

$$0 \to \mathcal{O} \to \mathcal{O}(\Delta) \to \mathcal{O}_\Delta(\Delta) \to 0$$

is

$$\pi_*\mathcal{O}_\Delta(\Delta) \to R^1\pi_*\mathcal{O} \to R^1\pi_*\mathcal{O}(\Delta) \to 0. \tag{2.5}$$

This provides a global presentation of $R^1\pi_*\mathcal{O}(\Delta)$, i.e., exhibits $R^1\pi_*\mathcal{O}(\Delta)$ as the cokernel of a homomorphism of vector bundles. It is crucial to notice that the above presentation is *functorial* in the following sense. Let

$$f: S \to C_d$$

be a morphism, let Δ' be the corresponding relative divisor in $C \times S$ and denote by ϕ the projection of $C \times S$ onto S. Then we have an analogue of (2.5), namely

$$\phi_*\mathcal{O}_{\Delta'}(\Delta') \xrightarrow{\delta'} R^1\phi_*\mathcal{O} \to R^1\phi_*\mathcal{O}(\Delta') \to 0.$$

Using the Weierstrass preparation theorem, as we did for the universal divisor Δ, it is easy to show that $\phi_*\mathcal{O}_{\Delta'}(\Delta')$ is locally free and that, for any s in S,

$$\phi_*\mathcal{O}_{\Delta'}(\Delta') \otimes k(s) \cong H^0(C, \mathcal{O}_{\Delta'_s}(\Delta'_s)),$$

where Δ'_s is the divisor cut on $C \cong C \times \{s\}$ by Δ'. On the other hand, by pull-back, we get another presentation

$$f^*\pi_*\mathcal{O}_\Delta(\Delta) \xrightarrow{f^*(\delta)} f^*R^1\pi_*\mathcal{O} \to f^*R^1\pi_*\mathcal{O}(\Delta) \to 0.$$

Clearly the natural morphism

$$f^*R^1\pi_*\mathcal{O} \to R^1\phi_*\mathcal{O}$$

is an isomorphism. As for

$$f^*\pi_*\mathcal{O}_\Delta(\Delta) \to \phi_*\mathcal{O}_{\Delta'}(\Delta'),$$

it is also an isomorphism, since the two sheaves in question are locally free and, for any s in S,

$$f^*\pi_*\mathcal{O}_\Delta(\Delta) \otimes k(s) \cong \phi_*\mathcal{O}_{\Delta'}(\Delta') \otimes k(s).$$

Summarizing, we have natural identifications

$$f^*\pi_*\mathcal{O}_\Delta(\Delta) \cong \phi_*\mathcal{O}_{\Delta'}(\Delta'),$$

$$f^*R^1\pi_*\mathcal{O} \cong R^1\phi_*\mathcal{O},$$

such that $\delta' = f^*(\delta)$. It follows, in particular, that

$$f^*R^1\pi_*\mathcal{O}(\Delta) \cong R^1\phi_*\mathcal{O}(\Delta').$$

An important special case is the one in which f is the inclusion in C_d of a point D. What we have just said means that $H^1(C, \mathcal{O}(D))$ is the cokernel of the homomorphism

$$T_D(C_d) \xrightarrow{u_{*,D}} T_{\mathcal{O}(D)}(\mathrm{Pic}^d(C)).$$

This provides another proof of the statement that the rank of u_* at D does not exceed $d - r$ if and only if

$$i(D) \geq g - d + r,$$

or, which is the same, if and only if

$$r(D) \geq r$$

(cf. Section 1 of this chapter).

To put the functorial presentation (2.5) in proper perspective, we recall the basic result about base change in cohomology

(2.6) Theorem. *Let $f: X \to Y$ be a proper morphism of analytic spaces, and let $\mathcal{F}$ be a coherent $\mathcal{O}_X$-module. Suppose $\mathcal{F}$ is $\mathcal{O}_Y$-flat. Then, for every point of Y there are a neighborhood U and a finite complex*

$$0 \to K^0 \to K^1 \to \cdots$$

of free finitely generated $\mathcal{O}_U$-modules, with the following property. For every morphism of analytic spaces $g: V \to U$, there are functorial isomorphisms

$$R^i\beta_*\alpha^*(\mathcal{F}) \cong \mathcal{H}^i(g^*K^\cdot),$$

where β is the projection of $f^{-1}(U) \times_U V$ onto V and α is the composition of the projection of $f^{-1}(U) \times_U V$ onto $f^{-1}(U)$ with the inclusion of $f^{-1}(U)$ in X. Moreover, if the dimension of all the fibers of f does not exceed n, we may assume that $K^m = 0$ when $m > n$.

It is important to notice that Theorem (2.6) is still true (and has, in fact, a much simpler proof) if we replace the words "analytic space" with "algebraic variety," and "neighborhood" with "Zariski neighborhood."

Applying Theorem (2.6) to the morphism

$$\pi: C \times C_d \to C_d$$

and to the sheaf $\mathcal{O}(\Delta)$, we get, locally on C_d, a functorial presentation of $R^1\pi_*\mathcal{O}(\Delta)$ (and one whose kernel is $\pi_*\mathcal{O}(\Delta)$, as well). In a sense, then, at least for what concerns $R^1\pi_*\mathcal{O}(\Delta)$, the functorial presentation (2.5) is a global and explicit version of the functorial cohomology complex whose existence is asserted by (2.6).

§3. The Varieties $W_d^r(C)$ and $G_d^r(C)$ Parametrizing Special Linear Series on a Curve

Our first goal, in this section, is easily stated. We fix a smooth genus g curve C and two non-negative integers r, d; we would then like to construct a subvariety $W_d^r(C)$ of $\mathrm{Pic}^d(C)$ whose support is the set of complete linear series of degree d and dimension at least r. In other words, we would like to have

$$\mathrm{Supp}(W_d^r(C)) = \{L \in \mathrm{Pic}^d(C): h^0(C, L) \geq r + 1\}.$$

This is certainly possible since the above set is the image of C_d^r under the proper map

$$u: C_d \to \mathrm{Pic}^d(C);$$

the problem is to choose the "right" scheme structure.

We shall denote by $\mathscr{L}$ a fixed Poincaré line bundle of degree d for C, and by v the projection

$$v: C \times \mathrm{Pic}^d(C) \to \mathrm{Pic}^d(C).$$

Also, we choose, once and for all, an effective divisor E of degree

$$m = \deg(E) \geq 2g - d - 1$$

on C, and let

$$\Gamma = E \times \mathrm{Pic}^d(C)$$

be the product divisor in $C \times \mathrm{Pic}^d(C)$. By Theorem (2.6), the hypothesis on the degree of E guarantees that

$$R^1 v_* \mathscr{L}(\Gamma) = 0$$

and that $v_* \mathscr{L}(\Gamma)$ is locally free of rank

$$n = \mathrm{rank}(v_* \mathscr{L}(\Gamma)) = d + m - g + 1.$$

Thus the exact higher direct image sequence of

$$0 \to \mathscr{L} \to \mathscr{L}(\Gamma) \to \mathscr{L}(\Gamma)/\mathscr{L} \to 0$$

is

$$0 \to v_* \mathscr{L} \to v_* \mathscr{L}(\Gamma) \xrightarrow{\gamma} v_*(\mathscr{L}(\Gamma)/\mathscr{L}) \to \mathrm{R}^1 v_* \mathscr{L} \to 0.$$

The two middle terms are locally free of ranks n and m, and it is natural to define $W_d^r(C)$ as the locus "where the fiber of $R^1\nu_*\mathscr{L}$ has dimension at least $g - d + r$," or, more precisely, as the $(m + d - g - r)$th determinantal variety attached to γ. In the language of Chapter II, Section 4, and setting, for brevity

$$X = \operatorname{Pic}^d(C),$$

we have

$$W_d^r(C) = X_{m+d-g-r}(\gamma).$$

We shall normally write $W_d(C)$ instead of $W_d^0(C)$. We shall later see that this $W_d(C)$ is the same as the one already defined in Chapter I. We also let

$$G_d^r(C) = \tilde{X}_{m+d-g-r}(\gamma)$$

be the canonical blow-up of $W_d^r(C)$, which was also defined in Section 4 of Chapter II.

We shall return to $G_d^r(C)$ later in this section. For the moment, we have to justify our definition of $W_d^r(C)$, in the sense that the support of $W_d^r(C)$ is precisely the set of those L in $\operatorname{Pic}^d(C)$ such that $h^0(C, L)$ is at least $r + 1$. We can do much better. For brevity, we set

$$K^0 = \nu_*\mathscr{L}(\Gamma); \qquad K^1 = \nu_*(\mathscr{L}(\Gamma)/\mathscr{L}).$$

We have seen that the kernel and cokernel of

$$\gamma\colon K^0 \to K^1$$

are $\nu_*(\mathscr{L})$ and $R^1\nu_*\mathscr{L}$, respectively. The analogue of this is true for any family of degree d line bundles on C. Namely, let S be any analytic space and let L be a family of line bundles on C parametrized by S. Denote by ϕ the projection of $C \times S$ onto S, and set $\Gamma' = E \times S$. As before, there is an exact sequence

$$0 \to \phi_* L \to \phi_* L(\Gamma') \xrightarrow{\gamma'} \phi_*(L(\Gamma')/L) \to R^1\phi_* L \to 0.$$

Now, by the universal property of Poincaré bundles, there are a unique map

$$f\colon S \to \operatorname{Pic}^d(C)$$

and a unique line bundle $\mathscr{R}$ on S such that

$$(1_C \times f)^*\mathscr{L} \cong L \otimes \phi^*\mathscr{R}.$$

We then have the following

(3.1) Lemma. *There are canonical isomorphisms between the kernel and cokernel of*

$$f^*(\gamma): f^*(K^0) \to f^*(K^1)$$

and $\phi_*(L) \otimes \mathscr{R}$, $R^1\phi_*(L) \otimes \mathscr{R}$, *respectively. In particular*

$$R^1\phi_*(L) \otimes \mathscr{R} \cong f^*(R^1 v_* \mathscr{L}).$$

The proof of this lemma is straightforward. Applying Theorem (2.6) to the projection

$$v: C \times \operatorname{Pic}^d(C) \to \operatorname{Pic}^d(C)$$

and to the sheaf $\mathscr{L}(\Gamma)$ shows that the natural homomorphism from $f^*(K^0)$ to

$$\phi_*(1_C \times f)^*(\mathscr{L}(\Gamma)) \cong \phi_*(L(\Gamma') \otimes \phi^*\mathscr{R}) \cong \phi_*(L(\Gamma')) \otimes \mathscr{R}$$

is in fact an isomorphism. On the other hand, it is clear (or, for skeptics, follows from (2.6)) that the natural homomorphism from $f^*(K^1)$ to

$$\phi_*(1_C \times f)^*(\mathscr{L}(\Gamma)/\mathscr{L}) = \phi_*(L(\Gamma')/L \otimes \phi^*\mathscr{R}) \cong \phi_*(L(\Gamma')/L) \otimes \mathscr{R}$$

is also an isomorphism. We then have a commutative diagram

$$\begin{array}{ccc} f^*(K^0) & \xrightarrow{f^*(\gamma)} & f^*(K^1) \\ \downarrow & & \downarrow \\ \phi_*(L(\Gamma')) \otimes \mathscr{R} & \xrightarrow{\gamma' \otimes 1_{\mathscr{R}}} & \phi_*(L(\Gamma')/L) \otimes \mathscr{R}. \end{array}$$

Since the kernel and cokernel of $\gamma' \otimes 1_{\mathscr{R}}$ are, respectively, $\phi_*(L) \otimes \mathscr{R}$ and $R^1\phi_*(L) \otimes \mathscr{R}$, the lemma follows.

As a special case, we look at the one when S is a single point. Thus L is simply a degree d line bundle on C, and f maps S to the corresponding point of $\operatorname{Pic}^d(C)$. Then the lemma simply says that

$$h^1(C, L) = m - \operatorname{rank}_L(\gamma).$$

Therefore we have

$$\begin{aligned} \operatorname{Supp}(W^r_d(C)) &= \{L \in \operatorname{Pic}^d(C): \operatorname{rank}_L(\gamma) \le m + d - g - r\} \\ &= \{L \in \operatorname{Pic}^d(C): h^1(C, L) \ge g - d + r\} \\ &= \{L \in \operatorname{Pic}^d(C): h^0(C, L) \ge r + 1\}, \end{aligned}$$

as desired.

There seems to be a certain degree of arbitrariness in the definition of $W^r_d(C)$, coming from the fact that we had to choose both a Poincaré line bundle and an effective divisor of high degree on C. This is not a serious problem, however. Changing the Poincaré line bundle has the effect of tensoring both K^0 and K^1 with the same line bundle. This clearly does not change the determinantal varieties attached to γ. As for our second choice, let q be a point of C, set $E' = E + q$, $\Gamma' = E' \times \mathrm{Pic}^d(C)$, and look at the homomorphism

$$\gamma' \colon \nu_* \mathscr{L}(\Gamma') \to \nu_*(\mathscr{L}(\Gamma')/\mathscr{L}).$$

Choose a local frame for $\nu_* \mathscr{L}(\Gamma')$ by adding to a frame for $\nu_* \mathscr{L}(\Gamma)$ a section s of $\nu_* \mathscr{L}(\Gamma')$ which induces a nowhere vanishing section of $\nu_* \mathscr{L}(\Gamma')/\nu_* \mathscr{L}(\Gamma)$. Then choose a frame for $\nu_*(\mathscr{L}(\Gamma')/\mathscr{L})$ by adding to a frame for $\nu_*(\mathscr{L}(\Gamma)/\mathscr{L})$ the image of s under γ'. Relative to these frames, the homomorphism γ' is represented by a matrix of holomorphic functions

$$A' = \begin{pmatrix} 1 & B \\ 0 & A \end{pmatrix},$$

where A represents γ. Thus the minors of size $m + d - g - r + 1$ of A are minors of size $(m + 1) + d - g - r + 1$ of A'. Conversely, every minor of size $(m + 1) + d - g - r + 1$ of A' is a linear combination, with holomorphic coefficients, of minors of size $m + d - g - r + 1$ of A. Thus the varieties $W^r_d(C)$ defined by choosing, as fixed effective divisor, E or E', are the same. Adding one point at a time, we can thus show that, for any effective divisor E'' of degree at least $2g - d - 1$, we get the same $W^r_d(C)$ if we choose, as fixed divisor, E or $E + E''$. Reversing the roles of E and E'' we conclude that the definition of $W^r_d(C)$ is independent of the choice of E.

(3.2) Remark. The independence of $W^r_d(C)$ of the choices made, which we checked by direct computations, also follows from general properties of Fitting ideals. We shall describe them here, without proofs, although they will not be strictly necessary in the sequel. For more details the reader is referred to Fitting [1], Northcott [1].

Let X be an analytic space and let $\mathscr{F}$ be a coherent analytic sheaf on X. A *presentation* of $\mathscr{F}$ is an exact sequence

$$\mathcal{O}_X^n \xrightarrow{\alpha} \mathcal{O}_X^m \to \mathscr{F} \to 0.$$

The homomorphism α is represented by an $m \times n$ matrix A whose entries are holomorphic functions on X. The hth *Fitting ideal of* the above presentation is the ideal in $\mathcal{O}_X$ generated by the $(m - h + 1) \times (m - h + 1)$ minors of A in case $h \leq m$, and is defined to be $\mathcal{O}_X$ when $h > m$.

The basic property of Fitting ideals is that *any two presentations of $\mathscr{F}$ have the same Fitting ideals*. This will enable us to speak of the Fitting ideals of $\mathscr{F}$ without reference to any particular presentation, and also makes it possible to define the Fitting ideals in the absence of a presentation for $\mathscr{F}$. In fact, $\mathscr{F}$ will always have local presentations, and the hth Fitting ideals of these patch together to yield a global ideal in $\mathcal{O}_X$, by the basic property of Fitting ideals. This ideal will be called the hth Fitting ideal of $\mathscr{F}$.

An easy property of Fitting ideals is that they are compatible with base change. This means that, whenever,

$$f: Y \to X$$

is a morphism of analytic spaces, the hth Fitting ideal of $f^*\mathscr{F}$ is generated, as an $\mathcal{O}_Y$-module, by the hth Fitting ideal of $\mathscr{F}$. This is simply a reflection of the fact that the pull-back of a presentation of $\mathscr{F}$ is a presentation of $f^*(\mathscr{F})$, since the tensor product is a right exact functor.

In the language of Fitting ideals, then, $W^r_d(C)$ is simply the subvariety of $\operatorname{Pic}^d(C)$ defined by the $(g - d + r)$th Fitting ideal of $R^1\nu_*\mathscr{L}$.

Let $\mathscr{F}$ be a coherent analytic sheaf over the analytic space X. The *Fitting rank* of $\mathscr{F}$ is defined to be the largest integer h such that the hth Fitting ideal of $\mathscr{F}$ vanishes. A moment of reflection will then convince the reader that Lemma (3.1) essentially says that $W^r_d(C)$ represents the functor

$$S \to \left\{\begin{matrix}\text{equivalence classes of families } L \text{ of degree } d \text{ line bundles on} \\ C \times S \xrightarrow{\phi} S \text{ such that the Fitting rank of } R^1\phi_* L \text{ is at least} \\ g - d + r\end{matrix}\right\}.$$

The variety $W^r_d(C)$ has been defined as the $(m + d - g - r)$th determinantal variety attached to a homomorphism between suitable vector bundles of ranks $m + d - g + 1$ and m, respectively, over $\operatorname{Pic}^d(C)$. Thus (cf. Chapter II, Section 4) every component of $W^r_d(C)$ has dimension at least equal to

$$\begin{aligned} g - [(m + d - g + 1) - (m + d - g - r)][m - (m + d - g - r)] \\ = g - (r + 1)(g - d + r), \end{aligned}$$

provided, of course, that

$$r \geq d - g.$$

We then see that the "expected" dimension of $W^r_d(C)$ is exactly the Brill–Noether number

$$\rho = g - (r + 1)(g - d + r),$$

which we first met in the previous section. Because of its importance, we express this as a formal lemma.

(3.3) Lemma. *Suppose $r \geq d - g$. Then every component of $W_d^r(C)$ has dimension greater or equal to the Brill–Noether number*

$$\rho(g, r, d).$$

We remark, in passing, that, when L belongs to $W_d^r(C)$, but not to $W_d^{r+1}(C)$, the Brill–Noether number is just

$$g - l(L)i(L) = g - h^0(C, L)h^0(C, KL^{-1}).$$

In a sense, then, the Brill–Noether number is invariant under the operation of passing from a linear series to the residual one.

We may now ask what is the precise relation between C_d^r and $W_d^r(C)$. Of course, *as sets*, the second is just the image of the first under the abelian sum mapping

$$u: C_d \to \mathrm{Pic}^d(C).$$

The question then is, what can we say if we also take the scheme structures of C_d^r and $W_d^r(C)$ into account? The following result clarifies the situation.

(3.4) Proposition. $u^{-1}(W_d^r(C)) = C_d^r$.

Proof. We let π be the projection of $C \times C_d$ onto the second factor, and let $\mathscr{L}$ be a Poincaré line bundle of degree d for C. We set

$$L = (1_C \times u)^*(\mathscr{L}).$$

By Remark (3.2) and Lemma (3.1), the ideal sheaf of $u^{-1}(W_d^r(C))$ is the $(g - d + r)$th Fitting ideal of $R^1\pi_*(L)$. Now let Δ be the universal divisor in $C \times C_d$. By the definition of Poincaré line bundle, $L \otimes \mathcal{O}(-\Delta)$ is trivial on the fibers of π, and hence there is a line bundle $\mathscr{R}$ on C_d such that

$$R^1\pi_*\mathcal{O}(\Delta) = R^1\pi_*(L) \otimes \mathscr{R}.$$

Since tensoring with a line bundle obviously has no effect on the Fitting ideals, this in turn means that the ideal sheaf of $u^{-1}(W_d^r(C))$ is the $(g - d + r)$th Fitting ideal of $R^1\pi_*\mathcal{O}(\Delta)$. On the other hand, the discussion at the end of Section 2 shows that this is precisely the ideal of C_d^r (see, in particular, the presentation (2.5)). Q.E.D.

The preceding result shows, in particular, that the restriction of u to C_d^r factors through the inclusion of $W_d^r(C)$ in $\mathrm{Pic}^d(C)$.

Lemma (1.6) has an exact analogue for $W_d^r(C)$.

(3.5) Lemma. *Suppose $g - d + r \geq 0$. Then no component of $W_d^r(C)$ is entirely contained in $W_d^{r+1}(C)$.*

This is just a corollary of (1.6). In fact, if a component X of $W_d^r(C)$ were contained in $W_d^{r+1}(C)$, $u^{-1}(X)$ would contain a component of C_d^r, which would then be contained in $u^{-1}(W_d^{r+1}(C)) = C_d^{r+1}$.

We now turn to the variety $G_d^r(C)$ which we introduced, along with $W_d^r(C)$, as the canonical blow-up of $W_d^r(C)$ itself. It will turn out that $G_d^r(C)$ parametrizes linear series of degree d and *fixed* dimension r on C, so that we will find it much more manageable than $W_d^r(C)$ in studying mappings from C to $\mathbb{P}^r$.

To see that the points of $G_d^r(C)$ are, in fact, the g_d^r's on C, recall the two-stage complex

$$\gamma \colon K^0 \to K^1$$

of locally free sheaves over $\mathrm{Pic}^d(C)$ used in the definition of $W_d^r(C)$ and $G_d^r(C)$. The points of $G_d^r(C)$ are the couples (x, V), where x belongs to $\mathrm{Pic}^d(C)$, and V is an $(r + 1)$-dimensional subspace of the kernel of

$$\gamma_x \colon K^0 \otimes k(x) \to K^1 \otimes k(x).$$

Now, x corresponds to a degree d line bundle L on C, while, by Lemma (3.1), the kernel of γ_x is canonically isomorphic to $H^0(C, L)$. Thus $G_d^r(C)$ parametrizes couples (L, V), where L is a degree d line bundle on C and V is an $(r + 1)$-dimensional vector subspace of $H^0(C, L)$. In symbols

$$\mathrm{Supp}(G_d^r(C)) = \{(L, V) \colon L \in \mathrm{Pic}^d(C),\ V \in G(r + 1, H^0(C, L))\}.$$

In other words, as we announced above $G_d^r(C)$ *parametrizes the* g_d^r's (*complete or not*) *on* C.

We now wish to show how the above description of the points of $G_d^r(C)$ has an exact analogue for S-valued points of $G_d^r(C)$, for any analytic space S. To this end, we introduce the cumbersome but useful concept of *family of* g_d^r's *on* C (*parametrized by an analytic space* S). By this we mean the datum of

(i) A family L of degree d line bundles on C, parametrized by S.
(ii) A locally free, rank$(r + 1)$ subsheaf F of $\phi_* L$, where ϕ is the projection of $C \times S$ onto S, with the property that, for each $s \in S$, the homomorphism

$$F \otimes k(s) \to H^0(\phi^{-1}(s), L \otimes \mathcal{O}_{\phi^{-1}(s)})$$

is injective.

Condition (ii) means that, writing

$$L \otimes \phi^*\mathscr{R} = (1_C \times f)^*(\mathscr{L}),$$

where $\mathscr{L}$ is a degree d Poincaré line bundle and f is a suitable morphism of S into $\mathrm{Pic}^d(C)$, we require that $F \otimes \mathscr{R}$ be a *vector subbundle* of $f^*(K^0)$.

At least when S is reduced, a family of g^r_d's on C parametrized by S can be thought of as a holomorphically varying family

$$L_s \to C, \qquad s \in S,$$

of degree d line bundles on C, together with a holomorphically varying family $\mathscr{D}_s$ of g^r_d's

$$\mathscr{D}_s \subset |L_s|, \qquad s \in S.$$

If we are given a family of g^r_d's, $\mathscr{G} = (L, F)$ on C parametrized by S, and a morphism

$$f: T \to S,$$

we can define the pull-back

$$f^*(\mathscr{G}) = ((1_C \times f)^*L, f^*(F)),$$

which is a family of g^r_d's on C parametrized by T.

Two families (L, F), (L', F') of g^r_d's on C parametrized by S are said to be *equivalent* if there exist a line bundle $\mathscr{R}$ on S and an isomorphism

$$L' \cong L \otimes \phi^*\mathscr{R}$$

such that F' is identified with $F \otimes \mathscr{R}$. We are now going to construct a universal family of g^r_d's on C parametrized by $G^r_d(C)$. We recall that $G^r_d(C)$ is naturally a subvariety of the Grassman bundle $G(r + 1, K^0)$ over $\mathrm{Pic}^d(C)$. We denote by

$$c: G^r_d(C) \to \mathrm{Pic}^d(C)$$

the restriction of the projection from $G(r + 1, K^0)$. Then the universal family of g^r_d's on C parametrized by $G^r_d(C)$ is $(c^*(\mathscr{L}), \mathscr{F})$, where $\mathscr{L}$ is a Poincaré line bundle of degree d and $\mathscr{F}$ is the restriction to $G^r_d(C)$ of the universal subbundle on $G(r + 1, K^0)$ (of course, to say "the" universal family is somewhat improper, for there is one for each Poincaré line bundle, but we shall ignore this). We are now going to explain in what precise sense the family we just constructed is universal.

(3.6) Theorem. *For any analytic space S and any family $\mathscr{G}$ of g^r_d's on C parametrized by S, there is a unique morphism from S to $G^r_d(C)$ such that the pull-back of the universal family parametrized by $G^r_d(C)$ is equivalent to $\mathscr{G}$.*

Proof. Write $\mathscr{G} = (L, F)$. By the universal property of the Poincaré line bundle, there are a unique morphism

$$f: S \to \mathrm{Pic}^d(C)$$

and a line bundle $\mathscr{R}$ on S such that, denoting by

$$\phi: C \times S \to S$$

the projection,

$$(1_C \times f)^*(\mathscr{L}) \cong L \otimes \phi^*\mathscr{R}.$$

As we already noticed, $F \otimes \mathscr{R}$ can be viewed as a vector subbundle of $f^*(K^0)$ contained in

$$\phi_* L \otimes \mathscr{R} \cong \ker(f^*(\gamma)),$$

where

$$\gamma: K^0 \to K^1$$

is the functorial complex constructed at the beginning of this section. By the universal property of Grassmann bundles, $F \otimes \mathscr{R}$ is the pull-back of the universal subbundle via a unique section of

$$G(r+1, f^*(K^0)) = G(r+1, K^0) \times_{\mathrm{Pic}^d(C)} S$$

over S. This section corresponds to a unique morphism of analytic spaces over $\mathrm{Pic}^d(C)$

$$S \to G(r+1, K^0),$$

which factors through the inclusion

$$G^r_d(C) \subset G(r+1, K^0),$$

since $F \otimes \mathscr{R}$ is annihilated by $f^*(\gamma)$. Q.E.D.

A moment of reflection will convince the reader that to speak of families of g^0_d's, up to equivalence, is just to speak of relative divisors. Thus a first consequence of the theorem we just proved (and of the universal property of the universal divisor) is that $G^0_d(C)$ is nothing but C_d.

§4. The Zariski Tangent Spaces to $G^r_d(C)$ and $W^r_d(C)$

We shall now begin a study of the infinitesimal structure of $G^r_d(C)$ and $W^r_d(C)$ by describing their tangent spaces. Throughout the discussion, C will stand for a fixed smooth genus g curve.

Let w be a point of $G^r_d(C)$, corresponding to a degree d line bundle L on C plus an $(r + 1)$-dimensional subspace W of $H^0(C, L)$. The Zariski tangent space to $G^r_d(C)$ at w is the set of morphisms of analytic spaces from $\mathrm{Spec}(\mathbb{C}[\varepsilon])$ to $G^r_d(C)$ sending the unique point of $\mathrm{Spec}(\mathbb{C}[\varepsilon])$ to w. In symbols

$$T_w(G^r_d(C)) = \mathrm{Hom}(\mathrm{Spec}(\mathbb{C}[\varepsilon]), (G^r_d(C), w)).$$

The universal property of $G^r_d(C)$ expressed by Theorem (3.6) implies that *the tangent space to $G^r_d(C)$ at w is just the set of isomorphism classes of families of g^r_d's on C parametrized by* $\mathrm{Spec}(\mathbb{C}[\varepsilon])$ *which restrict to* (L, W) on C. Such a family will be called a *first-order deformation of* (L, W).

Our task is then to describe the isomorphism classes of first-order deformations of (L, W). This could be done directly, but we choose instead an indirect approach, relying in part on the explicit construction of $G^r_d(C)$. This was based on the functorial cohomology complex

$$\gamma\colon K^0 \to K^1$$

constructed in the previous section, and $G^r_d(C)$ was described as a subvariety of the Grassmann bundle $G(r + 1, K^0)$ over $\mathrm{Pic}^d(C)$. As we did there, we let

$$c\colon G^r_d(C) \to \mathrm{Pic}^d(C)$$

be the restriction of the projection of $G(r + 1, K^0)$ onto $\mathrm{Pic}^d(C)$. We then have an exact sequence

$$0 \to T_w(c^{-1}(L)) \to T_w(G^r_d(C)) \xrightarrow{c_*} T_L(\mathrm{Pic}^d(C)).$$

By the universal property of $G^r_d(C)$, $c^{-1}(L)$ parametrizes g^r_d's in $|L|$, i.e., $(r + 1)$-dimensional subspaces of $H^0(C, L)$. In symbols,

$$c^{-1}(L) = G(r + 1, H^0(C, L)).$$

But then

$$\begin{aligned} T_w(c^{-1}(L)) &= T_W(G(r + 1, H^0(C, L))) \\ &= \mathrm{Hom}(W, H^0(C, L)/W). \end{aligned}$$

Having described the kernel of c_*, we now turn to its image. This requires an explicit computation. Suppose L is defined by the transition data $\{U_\alpha\}$, $\{g_{\alpha\beta}\}$. Let L' be a first-order deformation of L, defined by the transition data

$$\tilde{g}_{\alpha\beta} = g_{\alpha\beta}(1 + \varepsilon\phi_{\alpha\beta}),$$

where

$$\phi_{\alpha\beta} + \phi_{\beta\gamma} = \phi_{\alpha\gamma}.$$

A holomorphic section $s \in H^0(C, L)$ is given by a collection $\{s_\alpha\}$ of functions,

$$s_\alpha \in \Gamma(U_\alpha, \mathcal{O}_C)$$

satisfying

$$s_\alpha = g_{\alpha\beta} s_\beta \quad \text{in} \quad U_\alpha \cap U_\beta.$$

An extension of s to a section of L' is given locally by

$$\tilde{s}_\alpha = s_\alpha + \varepsilon s'_\alpha,$$

where $\tilde{s}_\alpha = \tilde{g}_{\alpha\beta}\tilde{s}_\beta$. But this is equivalent to saying, again, that

$$s_\alpha = g_{\alpha\beta} s_\beta \quad \text{in} \quad U_\alpha \cap U_\beta$$

and that, moreover,

$$\phi_{\alpha\beta} s_\alpha = s'_\alpha - g_{\alpha\beta} s'_\beta \quad \text{in} \quad U_\alpha \cap U_\beta.$$

We observe that the left-hand side of the above inequality is a cocycle representing the cup-product

$$\phi \cdot s \in H^1(C, L)$$

under the natural pairing

$$H^1(C, \mathcal{O}) \otimes H^0(C, L) \to H^1(C, L);$$

here, of course, ϕ stands for the cohomology class of $\{\phi_{\alpha\beta}\}$, i.e., it is the Kodaira–Spencer class of the first-order deformation L'. The right-hand side is the coboundary $\delta s'$, where $s' = \{s'_\alpha\}$ belongs to $C^0(\{U_\alpha\}, L)$. Our computation establishes the following

Lemma. *Given L in* $\mathrm{Pic}^d(C)$ *and a section s of L, the set of tangent vectors*

$$\phi \in T_L(\mathrm{Pic}^d(C)) \cong H^1(C, \mathcal{O})$$

such that s may be extended to a section of the corresponding first-order deformation of L is given by

$$\{\phi \in H^1(C, \mathcal{O}) \colon \phi \cdot s = 0 \text{ in } H^1(C, L)\}.$$

It is important to remark that, when $\phi \cdot s = 0$, the extensions of s correspond to the different ways of writing the cocycle representing $\phi \cdot s$ as a coboundary.

Returning to the problem of computing the image of c_*, notice that the image of the tangent vector to $G^r_d(C)$ represented by the first-order deformation (L', W') of (L, W) is represented by the first-order deformation L' of L. Now, letting L'' be a first-order deformation of L with Kodaira–Spencer class ϕ, clearly (L, W) has a first-order deformation (L'', W'') if and only if every section of W extends to a section of L'', i.e., by the preceding lemma, if and only if $\phi \cdot W = 0$. In other terms the image of c_* is the set of those ϕ in $H^1(C, \mathcal{O})$ such that $\phi \cdot W = 0$. The result of our computations is incorporated in the following, more general, proposition.

(4.1) Proposition. (i) *Every component of $G^r_d(C)$ has dimension at least equal to the Brill–Noether number*

$$\rho = g - (r + 1)(g - d + r).$$

(ii) *Let w be a point of $G^r_d(C)$, corresponding to a line bundle L and an $(r + 1)$-dimensional vector subspace W of $H^0(C, L)$. Then the tangent space to $G^r_d(C)$ at w fits into an exact sequence*

$$0 \to \mathrm{Hom}(W, H^0(C, L)/W) \to T_w(G^r_d(C)) \xrightarrow{c_*} T_L(\mathrm{Pic}^d(C)),$$

where

$$\mathrm{Image}(c_*) = \{\phi \in H^1(C, \mathcal{O}) = T_L(\mathrm{Pic}^d(C)) \colon \phi \cdot W = 0\}.$$

Dually

$$\mathrm{Image}(c_*) = (\mathrm{Image}\ \mu_{0,W})^{\perp},$$

where

$$\mu_{0,W} \colon W \otimes H^0(C, KL^{-1}) \to H^0(C, K) = H^1(C, \mathcal{O})^*$$

is the cup-product homomorphism.

(iii) *We have*

$$\dim T_w(G^r_d(C)) = \rho + \dim(\ker \mu_{0,W}).$$

In particular, $G^r_d(C)$ is smooth of dimension ρ at w if and only if

$$\ker \mu_{0,W} = 0.$$

Part (ii) has just been proved, except for the dual description of the image of c_*. Here the basic cup-product mapping

$$\mu_0 \colon H^0(C, L) \otimes H^0(C, KL^{-1}) \to H^0(C, K)$$

encountered in the proof of Clifford's theorem and in the first section of this chapter appears again. In fact $\mu_{0,W}$ is just its restriction to $W \otimes H^0(C, KL^{-1})$. From now on $\mu_{0,W}$ will often be written μ_0, unless confusion is likely. To show that

$$\mathrm{Image}(c_*) = (\mathrm{Image}\ \mu_{0,W})^{\perp},$$

it suffices to notice that, writing $\langle\ ,\ \rangle$ for the Serre duality pairing, $\phi \cdot W = 0$ if and only if

$$\langle \phi, \mu_0(s \otimes r)\rangle = \langle \phi \cdot s, r\rangle = 0$$

for every s in W and every r in $H^0(C, KL^{-1})$.

Part (i) of the proposition follows from Lemma (3.3) when $r \geq d - g$. When $r < d - g$, $G^r_d(C)$ maps onto $\mathrm{Pic}^d(C)$ and the fiber over a point L of $\mathrm{Pic}^d(C)$ is the Grassmannian of $(r + 1)$-planes in $H^0(C, L)$ which, by the Riemann–Roch theorem, has dimension at least $(r + 1)(d - g - r)$. Thus

$$\dim G^r_d(C) \geq g + (r + 1)(d - g - r) = \rho.$$

Part (iii) of the proposition follows from part (ii) by a trivial computation. In fact we have

$$\begin{aligned}\dim T_w(G^r_d(C)) &= (r + 1)(h^0(C, L) - r - 1) + g - \dim(\mathrm{Image}\ \mu_{0,W}) \\ &= g + (r + 1)(h^0(C, L) - r - 1) - (r + 1)h^0(C, KL^{-1}) \\ &\qquad + \dim(\ker \mu_{0,W}) \\ &= \rho + \dim(\ker \mu_{0,W}).\end{aligned}$$

This ends the proof of Proposition (4.1).

It is now very easy to deduce from (4.1) a description of the tangent spaces to $W^r_d(C)$.

(4.2) Proposition. (i) *Let L be a point of $W_d^r(C)$ not belonging to $W_d^{r+1}(C)$ (thus $r \geq d - g$). The tangent space to $W_d^r(C)$ at L is*

$$T_L(W_d^r(C)) = (\text{Image } \mu_0)^\perp,$$

where

$$\mu_0: H^0(C, L) \otimes H^0(C, KL^{-1}) \to H^0(C, K)$$

is the cup-product mapping. Thus $W_d^r(C)$ is smooth of dimension ρ at L if and only if μ_0 is injective.

(ii) *Let L be a point of $W_d^{r+1}(C)$. Then*

$$T_L(W_d^r(C)) = T_L(\text{Pic}^d(C)).$$

In particular, if $W_d^r(C)$ has the expected dimension ρ and $r > d - g$ (i.e., $\rho < g$), L is a singular point of $W_d^r(C)$.

There is very little, here, that needs proof. Part (i) follows from part (ii) of (4.1) once we notice that

$$c: G_d^r(C) \to W_d^r(C)$$

is biregular off $W_d^{r+1}(C)$. To prove (ii), first recall that, for suitable integers n, m, k, $W_d^r(C)$ is locally the pull-back, via a mapping ϕ from an open subset of $\text{Pic}^d(C)$ to the variety $M(n, m)$ of $n \times m$ matrices, of the subvariety $M_k(n, m)$ of matrices of rank at most k. Moreover, to say that L belongs to $W_d^{r+1}(C)$ means that $\phi(L)$ belongs to $M_{k-1}(n, m)$. On the other hand, we showed in Chapter II, Section 2 that the tangent space to $M_k(n, m)$ at a point of $M_{k-1}(n, m)$ is the tangent space to all of $M(n, m)$. Thus

$$\begin{aligned} T_L(W_d^r(C)) &= \phi_*^{-1}(T_{\phi(L)}(M_k(n, m))) \\ &= \phi_*^{-1}(T_{\phi(L)}(M(n, m))) \\ &= T_L(\text{Pic}^d(C)), \end{aligned}$$

which finishes the proof of (4.2).

To give a more geometrical description of the tangent spaces to $W_d^r(C) - W_d^{r+1}(C)$, consider the canonical map

$$\phi_K: C \to \mathbb{P}H^0(C, K)^* = \mathbb{P}^{g-1}.$$

Given L in $W_d^r(C) - W_d^{r+1}(C)$, Proposition (4.2) states that the Zariski tangent space $T_L(W_d^r(C))$ is defined by the simultaneous vanishing of the differentials in Image μ_0. On the other hand, given any non-zero section s in $H^0(C, L)$, the image of

$$s \otimes H^0(C, KL^{-1}) \to H^0(C, K)$$

is equal to $H^0(C, K(-D))$, where $D = (s)$, and the linear subspace of $\mathbb{P}^{g-1}$ defined by the simultaneous vanishing of the forms in $H^0(C, K(-D))$ is, by definition, $\overline{\phi_K(D)}$. We therefore have the following geometric description of the projectivized Zariski tangent space to $W_d^r(C)$ at L:

$$\mathbb{P}T_L(W_d^r(C)) = \bigcap_{D \in |L|} \overline{\phi_K(D)} \subset \mathbb{P}^{g-1}. \tag{4.3}$$

The subvariety $W_d^r(C)$ of $\mathrm{Pic}^d(C)$, as a set, is the image of $G_d^r(C)$ via the mapping c. It is natural to ask whether this is also an equality of schemes. We shall now settle the question in the affirmative when $G_d^r(C)$ is smooth and has the "correct" dimension.

(4.4) Proposition. *Suppose $G_d^r(C)$ is smooth of dimension*

$$\rho = g - (r + 1)(g - d + r).$$

Then the variety $W_d^r(C)$ is Cohen–Macaulay, reduced, and normal. If $d < g + r$ the singular locus of $W_d^r(C)$ is $W_d^{r+1}(C)$.

Proof. We can limit ourselves to the case $d < g + r$. The variety $W_d^r(C) - W_d^{r+1}(C)$ is isomorphic to $G_d^r(C) - c^{-1}(W_d^{r+1}(C))$, and hence smooth. Moreover Lemma (3.5) shows that it has non-empty intersection with every component of $W_d^r(C)$. In particular, to show that $W_d^r(C)$ is reduced, it suffices to show that it has no embedded components. This is true since $W_d^r(C)$, being a determinantal variety of the "correct" dimension ρ, is Cohen–Macaulay. It remains to show that it is normal. To this end, notice, first of all, that every component of C_d^{r+1} has codimension at least one in C_d^r, which is of pure dimension $\rho + r$, and, secondly, that a general fiber of

$$C_d^{r+1} \to W_d^{r+1}(C)$$

is $(r + 1)$-dimensional. Thus $W_d^{r+1}(C)$ has codimension at least two in $W_d^r(C)$. Since $W_d^r(C)$ is Cohen–Macaulay, this suffices to show that it is normal (cf. Proposition (5.4) in Chapter II).

Since $G_d^0(C) = C_d$ is smooth and connected of dimension $d = \rho(g, 0, d)$, we find, as a consequence of the result we just proved, the following

(4.5) Corollary. *The variety $W_d(C)$ is reduced, irreducible, normal and Cohen–Macaulay. If $d < g$ the singular locus of $W_d(C)$ is $W_d^1(C)$.*

To finish the section, we shall reconsider, in the light of the results proved in later sections, the naive computation of the tangent space to C^r_d at a point D not belonging to C^{r+1}_d, which we gave in the first section (Lemma (1.5)). First of all, by Proposition (3.4), the Zariski tangent space to C^r_d at D is the pull-back, via

$$u_*: T_D(C_d) \to T_{\mathcal{O}(D)}(\mathrm{Pic}^d(C))$$

of $T_{\mathcal{O}(D)}(W^r_d(C))$. Secondly, by the basic Lemma (2.3), u_* can be identified with the coboundary mapping

$$H^0(C, \mathcal{O}_D(D)) \to H^1(C, \mathcal{O})$$

gotten from the exact sheaf sequence

$$0 \to \mathcal{O} \to \mathcal{O}(D) \to \mathcal{O}_D(D) \to 0.$$

Thus, by Serre duality, the transpose of u_* is the restriction homomorphism

$$\alpha: H^0(C, K) \to H^0(C, K \otimes \mathcal{O}_D).$$

Combining the above remarks, and letting

$$\mu_0: H^0(C, \mathcal{O}(D)) \otimes H^0(C, K(-D)) \to H^0(C, K)$$

be the cup-product mapping, we find

$$\begin{aligned} T_D(C^r_d) &= u_*^{-1}(T_{\mathcal{O}(D)}(W^r_d(C))) \\ &= u_*^{-1}((\text{Image } \mu_0)^\perp) \\ &= \text{Image}(\alpha\mu_0)^\perp. \end{aligned}$$

This is precisely the statement of Lemma (1.5).

§5. First Consequences of the Infinitesimal Study of $G^r_d(C)$ and $W^r_d(C)$

As an application of the study of Zariski tangent spaces to $G^r_d(C)$ and $W^r_d(C)$ that we did in the previous section, we shall present Saint-Donat's proof and Mumford's subsequent refinement of a nice theorem of Martens which provides an upper bound for the dimension of $W^r_d(C)$.

(5.1) Theorem (Martens). *Let C be a smooth curve of genus $g \geq 3$. Let d be an integer such that $2 \leq d \leq g - 1$ and let r be an integer such that $0 < 2r \leq d$.*

Then if C is not hyperelliptic, every component of $W^r_d(C)$ has dimension at most equal to $d - 2r - 1$; in symbols

$$\dim W^r_d(C) \leq d - 2r - 1.$$

If C is hyperelliptic

$$\dim W^r_d(C) = d - 2r.$$

Proof. By the remark immediately following the proof of the geometric version of the Riemann–Roch theorem, any complete g^r_d on a hyperelliptic C is of the form

$$rg^1_2 + p_1 + \cdots + p_{d-2r}.$$

This takes care of the hyperelliptic case. Now assume C to be non-hyperelliptic and suppose there is a d, $2 \leq d \leq g - 1$, such that $W^r_d(C)$ is not empty and

$$\dim W^r_d(C) \geq d - 2r.$$

Choose the minimum d such that this holds. Notice that, by Clifford's theorem, the number $d - 2r$ must be positive. Let L be a point belonging to a component of $W^r_d(C)$ of maximal dimension. We may assume that $h^0(C, L) = r + 1 = l$. Set $i = h^1(C, L)$. Since d is minimum we may also assume that $|L|$ has no base points. By Proposition (4.2) we know that

$$\dim T_L(W^r_d(C)) = g - li + \dim(\operatorname{Ker} \mu_0).$$

Therefore, by our assumption, we get

$$\dim(\operatorname{Ker} \mu_0) \geq li - l - i + 1.$$

Let $s_1, \ldots, s_l$ be a basis for $H^0(C, L)$. Since $|L|$ has no base points we may assume that s_1, s_2 have no common zeros. Set

$$W_h = \operatorname{Span}(s_1, \ldots, s_h), \qquad h = 1, \ldots, l.$$

Obviously

$$\dim(\operatorname{Ker} \mu_{0, W_h}) - \dim(\operatorname{Ker} \mu_{0, W_{h-1}}) \leq i,$$

and therefore the last two inequalities and the base-point-free pencil trick give us

$$h^0(C, KL^{-2}) = \dim(\operatorname{Ker} \mu_{0, W_2}) \geq i - l + 1 = g - d,$$

so that

$$h^0(C, L^2) \geq g - d + 2d - g + 1 = d + 1.$$

If $d \leq g - 2$, then $4 \leq 2d \leq 2g - 4$, so, by Clifford's theorem, C is hyperelliptic, a contradiction. Suppose now that $d = g - 1$. Since $W_d^r(C)$ has positive dimension, we may assume that L is not a theta characteristic, i.e., that $L^2 \ncong K$. This contradicts the last inequality. Q.E.D.

Let us now turn to Mumford's refinement of Martens' theorem. We shall say that a smooth curve C is *bi-elliptic* if it can be represented as a ramified double covering of an elliptic curve.

(5.2) Theorem (Mumford). *Let C be a smooth non-hyperelliptic curve of genus $g \geq 4$. Suppose that there exist integers r and d such that $2 \leq d \leq g - 2$, $d \geq 2r > 0$ and a component X of $W_d^r(C)$ with*

$$\dim X = d - 2r - 1.$$

Then C is either trigonal, bi-elliptic or a smooth plane quintic.

Proof (Mumford). In what follows, to simplify the notation we shall always write the symbol $W_d^r(C)$ to mean "a component of $W_d^r(C)$." Assume then that

$$\dim W_d^r(C) = d - 2r - 1.$$

Notice that by imposing $r - 1$ general base points to the series belonging to $W_d^r(C)$ we get

$$\dim W_{d-r+1}^1(C) \geq (d - r + 1) - 3.$$

On the other hand, since C is not hyperelliptic, equality holds by Martens' theorem. We may therefore limit ourselves to the case $r = 1$. Let d be the minimum integer such that

$$\dim W_d^1(C) = d - 3.$$

Arguing as in the proof of Martens' theorem we conclude that

$$h^0(C, L^2) \geq d,$$

and therefore

$$\dim W_{2d}^{d-1}(C) \geq d - 3. \tag{5.3}$$

We want to apply Martens' theorem to this situation. There are three possible cases:

(a) $2d \leq g - 1$. Then

$$\dim W_{2d}^{d-1}(C) \leq 2d - 2(d - 1) - 1,$$

and hence, by (5.3),

$$d \leq 4.$$

(b) $g - 1 < 2d < 2g - 4$. Then, passing to the residual series, (5.3) gives

$$\dim W^{g-d-2}_{2g-2d-2}(C) \geq d - 3.$$

Again by Martens' theorem, we get

$$d \leq 4.$$

(c) $2d = 2g - 4$. Then, passing to the residual series, (5.3) gives

$$\dim W^0_2(C) \geq d - 3,$$

and therefore

$$d \leq 5, \qquad g = d + 2.$$

In conclusion, in every case $d \leq 5$, and if $d = 5$, $g = 7$. If $d = 3$, C is trigonal. Assume now that C is not trigonal and $d = 4$: in particular, then, $g \geq 6$. Our hypothesis that the dimension of $W^1_d(C)$ is equal to $d - 3$ gives

$$\dim W^1_4(C) = 1.$$

Let L and L' be two distinct points of $W^1_4(C)$. We notice that

$$h^0(C, L \otimes L') = 4. \tag{5.4}$$

In fact, by Clifford's theorem, $h^0(C, L \otimes L')$ is less than or equal to 4. On the other hand, since L has no base points, by the base-point-free pencil trick the kernel of the multiplication map

$$v\colon H^0(C, L) \otimes H^0(C, L') \to H^0(C, L \otimes L')$$

is isomorphic to $H^0(C, L^{-1} \otimes L')$. Since L and L' are distinct points of $W^1_4(C)$ this implies that v is injective, proving (5.4). By duality

$$h^0(C, KL^{-1}L'^{-1}) = g - 5.$$

If $p_1, \ldots, p_{g-6}$ are general points on C,

$$|KL^{-1}L'^{-1}(-\textstyle\sum p_i)| \neq \varnothing,$$

therefore $|L'|$ is contained in

$$|KL^{-1}(-\textstyle\sum p_i)| = |M|.$$

On the other hand, $\dim |M| = 2$. Let

$$\phi: C \to \mathbb{P}^2$$

be the morphism defined by $|M|$. The morphism

$$\phi_{L'}: C \to \mathbb{P}^1$$

defined by $|L'|$ is obtained by composing ϕ with a projection from a point of $\phi(C)$. Since $W^1_4(C)$ has positive dimension, we may assume that the center of projection is a smooth point of $\phi(C)$. Therefore

$$4 = d = (\deg \phi) \cdot (\deg \phi(C) - 1).$$

Two cases are possible. Either $\deg \phi = 1$, in which case $\phi(C)$ is a plane quintic, which has to be smooth since $g \geq 6$, or else C is a double covering of a plane cubic, which has to be smooth since C is not hyperelliptic.

There remains to examine the case $d = 5$, $g = 7$. We wish to show that this cannot occur. If it did, on C there would be a two-dimensional family of degree five line bundles L such that $h^0(C, L) \geq 2$ and $|KL^{-2}| \neq \emptyset$: in other terms

$$KL^{-2} \cong \mathcal{O}(p + q) \tag{5.5}$$

for some points $p, q \in C$. Moreover, by our choice of d, for a general L the linear series $|L|$ would have no base points. But now, for any divisor $D \in |L|$ not containing p or q, we would have:

$$\overline{\phi_K(D)} \supset \mathbb{P}T_L(W^1_5(C)) = \mathbb{P}T_{KL^{-2}}(W_2(C)) = \overline{\phi_K(p + q)}.$$

By the Riemann–Roch theorem, then, $|L(p + q)|$ would be a g^3_7 and $|KL^{-1}(-p - q)|$ a g^2_5, which cannot exist on a non-hyperelliptic curve of genus 7.

Bibliographical Notes

The Brill–Noether number and the Brill–Noether matrix first appear in Brill–Noether [1]. A standard reference for Poincaré line bundles, both in the case at hand and in the one of a family of curves is Grothendieck's lecture [1]. The basic theorems about base change in cohomology in an analytic setting are to be found in Grauert [1]; for a lucid account of the algebraic case we strongly recommend Mumford's book [1, Section 5].

Our definition of the loci $W^r_d(C)$ as the determinantal varieties associated to a functorial cohomology complex for the Poincaré line bundle is as in

Kleiman–Laksov [1], or Kempf [1]. The variety $G^r_d(C)$ is introduced in Arbarello–Cornalba [1].

For the basic properties of Fitting ideals we refer to Northcott [1], as well as to Fitting's original paper [1].

The first part of (4.2), in a more general context, can be found in Griffiths [1]. Corollary (4.5) is in fact part of Kempf's singularity theorem, to be proved in Chapter VI. The original proof is in Kempf [2].

Martens' theorem first appeared in Martens [2]. Our proof is adapted from Saint-Donat [2]. Mumford's refinement of Martens' theorem is taken from Mumford's paper on Prym varieties [2].

Exercises

A. Elementary Exercises on μ_0

A-1. Let C be a curve of genus g, L a line bundle of degree $g-1$ on C such that $|L|$ is a base-point-free pencil. Show that the map

$$\mu_0: H^0(C, L) \otimes H^0(C, KL^{-1}) \to H^0(C, K)$$

will fail to be injective if and only if $L^2 \cong K$. (More generally, suppose that $|L|$ is a pencil with base divisor D. Show that μ_0 will fail to be injective if and only if $h^0(C, KL^{-2}(D)) > 0$.)

A-2. In general, suppose that L is any theta characteristic; that is, a line bundle such that $L^2 = K$. In terms of the corresponding identification $H^0(C, L) = H^0(C, KL^{-1})$ we may then write

$$H^0(C, L) \otimes H^0(C, KL^{-1}) = \Lambda^2 H^0(C, L) \oplus \operatorname{Sym}^2 H^0(C, L).$$

Show that $\mu_0(\Lambda^2 H^0(C, L)) = 0$, and conclude that if $r = r(L)$

$$\dim T_L(W^r_{g-1}(C)) \geq g - \binom{r+1}{2}.$$

A-3. Suppose now that C is a non-hyperelliptic curve of genus 4. The canonical model of C is then the intersection of a quadric Q and a cubic F in $\mathbb{P}^3$. Show directly:

(i) If Q is smooth, $W^1_3(C)$ consists of two reduced points.
(ii) If Q is singular, $W^1_3(C)$ consists of one non-reduced point, whose tangent space is one-dimensional. Identify this tangent space.

A-4. With C as in A-3 above, let $D \in C^1_3$ be any point. Identify explicitly the tangent spaces to C^1_3 at D in case (i) Q is smooth, and (ii) Q is singular.

B. An Interesting Identification

Let $D \in C_d$. In this chapter we encountered the identification

$$(*) \qquad H^0(C, \mathcal{O}_D(D)) \underset{\alpha}{\tilde{\to}} \begin{Bmatrix} \text{first-order deformations of} \\ D \text{ as a divisor on } C \end{Bmatrix}.$$

Recall that the right-hand side are relative divisors on $C \times S$ ($S = \operatorname{Spec} \mathbb{C}[\varepsilon]/(\varepsilon^2)$) whose restriction to $C \times \operatorname{Spec} \mathbb{C}$ is D.

B-1. Make the mapping α explicit. For example, if $D = dp$ where $p \in C$ and if z is a local coordinate on C centered at p, then $v \in H^0(C, \mathcal{O}_D(D))$ is

$$v = \frac{a_d}{z^d} + \cdots + \frac{a_1}{z}.$$

What is the corresponding divisor on $C \times S$?

B-2. We also encountered the identification

$$\{\text{first-order deformations of } D \text{ as a divisor on } C\} \underset{\beta}{\tilde{\to}} T_D(C_d).$$

Combining this with $(*)$ gives

$$H^0(C, \mathcal{O}_D(D)) \underset{\beta \circ \alpha}{\tilde{\to}} T_D(C_d)$$

(this is ψ^{-1} in (2.3)). Using B-1 you may make $\beta \circ \alpha$ explicit. For example, what is

$$(\beta \circ \alpha)^{-1} \text{ (tangent space to the diagonal)?}$$

B-3. Use these explicit identifications to verify the commutativity of the diagram in Lemma (2.3).

C. Tangent Spaces to $W_1(C)$

In the following three exercises C will be a non-hyperelliptic curve.

C-1. Show that for $p \in C$ a general point there do not exist an integer m and a point $q \neq p$ such that $mq \sim mp$.

(*Hint*: Use a tangent space computation to show that

$$mW_1(C) \neq nW_1(C) + v, \qquad v \in J(C),$$

unless $m = n$ and $v = 0$. (Here $mW_1(C) = \{mu(p), p \in C\}$.))

C-2. Let $\Lambda \subset C_2$ be an algebraic curve giving a one-dimensional family $\{D_\lambda\}$ of divisors of degree 2 on C, and assume that

$$u(\Lambda) = -W_1(C) + v.$$

Show that C has a g_3^1, denoted by $|D|$, and $\Lambda = \{D - p\}_{p \in C}$ consists of all divisors of degree 2 subordinate to the g_3^1 (again use tangent spaces).

C-3. Retaining the assumptions of the preceding exercise, assume that

$$u(\Lambda) = W_1(C) + v,$$

and show that $\Lambda = \{p + q\}_{q \in C}$ for some fixed $p \in C$.

D. Mumford's Theorem for g_d^2's

The proof of Mumford's theorem proceeds by first reducing to the case $r = 1$. As we will see in the following exercises, this means that it is not necessarily sharp in case $r > 1$.

D-1. Show that if C is bi-elliptic and $d \leq g - 2$, then $\dim W_d^2(C) = d - 5$.

D-2. Show that if C is trigonal and $d \leq g - 2$, then $\dim W_d^2(C) = d - 6$.

D-3. Show that if C is a two-sheeted cover of a curve of genus 2 and $d \leq g - 2$, then $\dim W_d^2(C) = d - 6$.

D-4. Use Mumford's theorem and Exercise D-2 above to conclude that for any C, if $d \leq g - 2$ and $\dim W_d^2(C) = d - 5$, then C is bi-elliptic.

D-5. More generally, show that if C is bi-elliptic (resp., trigonal; a two-sheeted cover of a curve of genus 2) and $d \leq g - 2$, then $\dim W_d^r(C) = d - 2r - 1$ (resp., $d - 3r$; $d - 2r - 2$). Use this to give a sharpened form of Mumford's theorem when $r \geq 2$.

D-6. Can you prove this result directly, by giving an argument that uses g_d^r's rather that first reducing to g_d^1's? (We don't know how this might go.)

E. Martens–Mumford Theorem for Birational Morphisms

E-1. Let C be any curve of genus g, $L = \mathcal{O}(D)$ a special line bundle on C of degree d with $h^0(C, L) = r + 1$. Show that

$$d \geq g - d + 2r + h^0(K_C(-2D)).$$

E-2. Now assume that ϕ_L is birational. Use Exercise B-6 of Chapter III and the estimate in the preceding excercise to conclude that

$$\begin{aligned}\operatorname{rank}(\mu_0: H^0(C, L) \otimes H^0(C, K_C \otimes L^*) \to H^0(C, K_C)) \\ \geq 2g - 2d + 3r - 1 - h^0(C, K_C(-2D))\end{aligned}$$

and hence prove the

Theorem. *If $|L|$ is a g^r_d on C with ϕ_L birational, then in a neighborhood of $u(L) \in J(C)$,*

$$\dim W^r_d(C) \leq h^0(C, \mathcal{O}(2D)) - 3r.$$

Conclude in particular that if $d < g - 1$ or $d = g - 1$ and $L^2 \neq K_C$, then

$$\dim W^r_d(C) \leq d - 3r.$$

F. Linear Series on Some Complete Intersections

In these exercises we will examine the varieties $W^r_d(C)$ (written W^r_d, for brevity) for two special types of curves, both complete intersections $C = S \cap T$ in $\mathbb{P}^3$. In each case we will use the geometry of the surfaces containing C, the fact that $K_C \cong \mathcal{O}_C(2)$ (because in both examples $\deg S + \deg T = 6$), and the following superabundance lemma:

F-1. Let $\Gamma \subset \mathbb{P}^3$ be a collection of $d \leq 7$ points. Assuming that Γ fails to impose independent conditions on quadrics, show that either:

(i) Γ contains four collinear points;
(ii) Γ contains six points on a conic; or
(iii) Γ contains seven coplanar points.

F-2. Suppose that $C = S \cap T$ is the smooth complete intersection of a smooth quadric S and a smooth quartic T (thus C is a curve of type (4, 4) on S). Using the preceding exercise show that set-theoretically:

(i) $W^1_3 = \emptyset$;
(ii) W^1_4 consists of the two points corresponding to the two pencils cut out by the rulings of S (in particular, these g^1_4's are distinct);
(iii) $W^1_5 = W^1_4 + W_1$ (i.e., every g^1_5 on C has a base point), and that W^1_5 consists of two disjoint copies of C;
(iv) $W^3_8 = \{\mathcal{O}_C(1)\}$ (i.e., the hyperplane series is the unique g^3_8);
(v) $W^1_6 = (W^1_4 + W_2) \cup (W^3_8 - W_2)$. Conclude that W^1_6 consists of three copies of C_2, with the two components of $W^1_4 + W_2$ being disjoint, and each component meeting $W^3_8 - W_2$ in a curve;
(vi) $W^2_7 = W^3_8 - W_1$ (thus, W^2_7 is set-theoretically isomorphic to C);
(vii) $W^1_7 = (W^1_4 + W_3) \cup (W^3_8 - W_2 + W_1)$; and
(viii) finally, show that W^1_8 is *not* contained in

$$(W^1_4 + W_4) \cup (W^3_8 + W_2 - W_2).$$

(*Hint*: $\dim W^1_{g-1}(C) \geq g - 4$ for any curve C.)

F-3. In parts (ii), (iii), and (iv) of the preceding exercise, show that all schemes are reduced, and thus all set-theoretic equalities are in fact scheme-theoretic equalities.

F-4. Suppose that $C = S \cap T$ is the smooth intersection of two smooth cubic surfaces. We consider C as the base locus of the pencil $\{S_\lambda\}_{\lambda \in \mathbb{P}}$ of cubic surfaces spanned by S and T. Using the analysis of cubic surfaces show that:

(i) $W_5^1 = \emptyset$;
(ii) $\dim W_6^1 = 1$ (even better, show that set-theoretically $W_6^1 \cong \{(L, \lambda): L \subset S_\lambda\} \subset G(2, 4) \times \mathbb{P}^1$;
(iii) $W_8^3 = \emptyset$ and W_9^3 is scheme-theoretically one point (i.e., the hyperplane series is the unique g_9^3 on C);
(iv) $W_7^2 = \emptyset$ and scheme-theoretically $W_8^2 = W_9^3 - W_1$;
(v) $W_7^1 = (W_6^1 + W_1) \cup (W_9^3 - W_2)$. Show that W_7^1 thus reduces to a union of two surfaces meeting along a curve; and
(vi) finally, show that W_8^1 is *not* contained in

$$(W_6^1 + W_2) \cup (W_9^3 - W_2 + W_1).$$

In both these examples, the g_{g-1}^1's are the only special linear series not arising from linear subspaces of $\mathbb{P}^3$.

G. Keem's Theorems

In this batch of exercises we will look at one of several extensions of Mumford's theorem due to Keem.

(∗) **Theorem.** *Let C be a smooth algebraic curve of genus $g \geq 11$, and suppose that for some integers d and r satisfying $d \leq g + r - 4, r \geq 1$, we have*

$$\dim W_d^r(C) \geq d - 2r - 2.$$

Then C possesses a g_4^1.

For further results along these lines, see Keem [1].

G-1. Suppose C satisfies the conditions of Theorem (∗) above, but does not possess a g_4^1. Show that for some e such that $5 \leq e \leq g - 3$, we have

$$\dim W_e^1(C) \geq e - 4,$$

and a general element of a component of maximal dimension in $W_e^1(C)$ has no base points.

G-2. Show that with e as in the preceding exercise, we must have

$$\dim W_{2e}^{e-2}(C) \geq e - 4,$$

and, applying Mumford's theorem, deduce that we must have either

$$e = 5 \text{ or } 6,$$

or

$$e = 8 \quad \text{and} \quad g = 11.$$

In the following exercises, we consider the cases $e = 5$, $e = 6$, and $e = 8$ allowed above.

G-3. Suppose that we have $e = 5$, and let L_0 and L be general members of $W_5^1(C)$. Show that the map $\phi_{L_0^2}$ associated to the complete series $|L_0^2|$ is a birational embedding of C in $\mathbb{P}^3$.

G-4. Use the base-point-free pencil trick in the situation of the preceding exercise to conclude that

$$h^0(C, L_0^2 \otimes L) = 7,$$

so that

$$h^0(C, KL_0^{-2}L^{-1}) = g - 9,$$

and

$$h^0(C, KL_0^{-2}) = g - 7.$$

Now apply Exercise L-1 of Chapter III to $\phi_{KL_0^{-2}}$ to obtain a contradiction.

G-5. Suppose now that in the situation of Exercise G-1 we have $e = 6$. Show that for L_0, L general points of $W_6^1(C)$, we have

$$h^0(C, L_0^2 L) \leq 8,$$

and hence, by the base-point-free pencil trick, that

$$h^0(C, L_0^2 L^{-1}) \geq 2.$$

G-6. Using the preceding exercise and Exercise L-1 of Chapter III, show that the map $\phi_{L_0^2}: C \to \mathbb{P}^4$ is not birational, and derive a contradiction.

G-7. Finally, suppose that $e = 8$ and $g = 11$. Show that for general $L \in W_8^1(C)$ we have

$$L^2 \cong K(-p - q - r - s)$$

for some $p, q, r, s \in C$.

G-8. Let $J(C) \xrightarrow{m} J(C)$ be the map given by multiplication by m. Show that $m^{-1}(W_k(C))$ is still irreducible. Conclude that in the circumstances of the preceding exercise, we must have

$$h^0(C, L) \geq 2$$

for every line bundle L such that $L^2 \cong K(-p - q - r - s)$.

G-9. Now let M be any line bundle on C with $M^2 \cong K$. Show that $h^0(C, M) \leq 3$, and conclude that for general $p, q \in C$, $h^0(C, M(-p - q)) \leq 1$ to obtain a contradiction with the preceding exercise.

Chapter V

The Basic Results of the Brill–Noether Theory

As we recalled in Chapter I, a genus g curve depends on $3g - 3$ parameters, describing the so-called moduli of the curve. Our goal in this chapter is to describe how the projective realizations of a curve vary with its moduli, and what it means, from this point of view, to say that a curve is "general" or "special." Accordingly, we would like to know, first of all, what linear series can we expect to find on a general curve and, secondly, what the subvarieties of the moduli space corresponding to curves possessing linear series of specified type look like. In the previous chapter we introduced the varieties

$$W_d^r(C) = \{L \in \mathrm{Pic}^d(C): h^0(C, L) \geq r + 1\}$$

$$G_d^r(C) = \{g_d^r\text{'s on } C\},$$

referring to them as "configurations of linear series on C," and it is on these that we will focus our attention. A natural question is, how can we tell one curve from another by looking at these configurations, or more precisely, what do these look like in general, and how—and where—can they degenerate? Clearly, from the point of view of non-special linear series, all curves look alike. It is at the level of *special* linear series that differences appear: this is what was classically called the problem of special linear series. In this chapter we shall deal only with one aspect of the problem; namely, we shall state and discuss a few fundamental results describing what happens on a *general* curve (and, in some instances, on every curve). All of these results will be proved in Chapters VII and XI of this book.

In describing the configurations of linear series of given dimension r, the essential step consists in studying the complete ones; in symbols, this amounts to dealing with

$$W_d^r(C) - W_d^{r+1}(C).$$

As we already pointed out, the really interesting case is the one of special linear series or, more precisely, the case of exceptional linear series; these are the complete, degree d, series $|D|$ for which

$$r = r(D) > 0, \qquad g - d + r = i(D) > 0.$$

The cases in which these inequalities are not satisfied are very easily dealt with. Consider, in fact, the abelian sum mapping

$$u: C_d \to \mathrm{Pic}^d(C).$$

If $r = 0$, by Abel's theorem, u is an isomorphism between an open set of C_d and $W_d(C) - W_d^1(C)$. If, on the other hand, $r > 0$ and $g - d + r = 0$, then, by the Riemann–Roch theorem, u maps C_d onto $W_d^r(C)$ and, again by Abel's theorem, the general fiber of u is a $(d - g)$-dimensional projective space, showing that, in this case, $W_d^r(C)$ and $\mathrm{Pic}^d(C)$ coincide.

The study of complete (exceptional) special linear series can be slightly simplified by means of residuation. In fact, passing to the residual series provides an isomorphism

$$W_d^r(C) \to W_{2g-2-d}^{g-d+r-1}(C),$$

given by

$$|L| \mapsto |KL^{-1}|.$$

Thus we can restrict our attention to complete linear series of degree less than g. A final consideration comes from Clifford's theorem. According to this, any linear series $|D|$ of degree d less than $2g - 1$ satisfies the inequality

$$d \geq 2r(D),$$

and equality holds only in a few cases: when $D = 0$, $D = K$ or else when the curve C is hyperelliptic and $|D|$ is a multiple of the hyperelliptic involution on C. Therefore, when $0 < d < 2g - 2$, $W_d^{d/2}(C)$ either consists of a single point, when C is hyperelliptic, or else is empty.

In the light of these considerations we emphasize that the results that we shall discuss and illustrate in this chapter, although stated in general, really acquire a non-trivial meaning when thought of as theorems about complete linear series g_d^r such that

$$r > 0, \qquad g > d > 2r.$$

In particular, these limitations already take care of the cases for which $g \leq 3$.

For a fixed value of the genus it is useful to represent on a diagram the values of d and r that we are interested in:

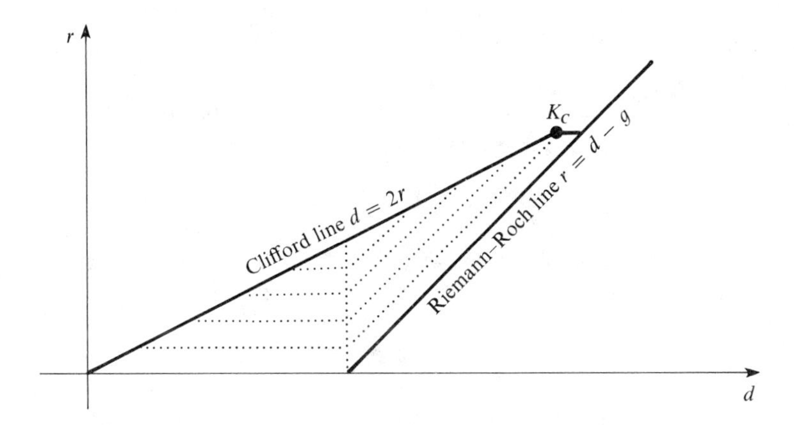

The union of the two shaded parts, with the exclusion of the lines $r = 0$ and $r = d - g$, constitutes the region of "exceptional special divisors." The two shaded parts are divided by the vertical line $d = g$ about which residuation produces a diffracted reflection, and the region we are interested in is the open triangle: $r > 0$, $g > d > 2r$.

Let us now bring into the picture the Brill–Noether number

$$\rho = g - (r + 1)(g - d + r).$$

In the diagram below, the shaded area is that part of the region of exceptional special divisors for which $\rho \geq 0$.

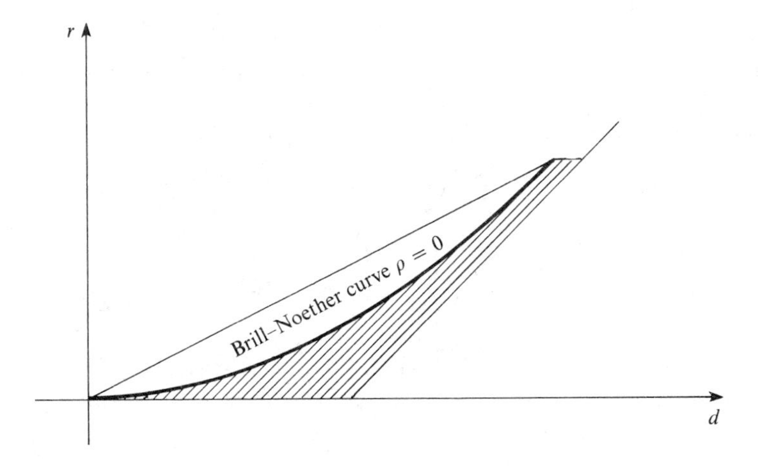

As we explained in Chapter IV, Section 3, the varieties $W^r_d(C)$ and $G^r_d(C)$, if non-empty, have dimension at least equal to ρ. The first natural question is whether the condition $\rho \geq 0$ suffices to conclude that $W^r_d(C)$ and $G^r_d(C)$ are non-empty. This is answered by the following existence theorem, independently proved by Kempf and Kleiman–Laksov.

(1.1) Existence Theorem. *Let C be a smooth curve of genus g. Let d, r, be integers such that $d \geq 1$, $r \geq 0$; then if*

$$\rho = g - (r + 1)(g - d + r) \geq 0,$$

$G^r_d(C)$, and hence $W^r_d(C)$, are non-empty. Furthermore, every component of $G^r_d(C)$ has dimension at least equal to ρ; the same is true for $W^r_d(C)$ provided $r \geq d - g$.

We now wish to offer, as an illustration, a direct analysis of the varieties of special divisors for curves of genus $g \leq 6$. We shall not use the Existence Theorem, but will rely on ad hoc arguments. As we already explained, we shall limit ourselves to the case of complete linear series g^r_d for which

$$r > 0, \qquad g > d > 2r.$$

The first case is $g = 4$. Here the only values of d and r to be considered are $d = 3$ and $r = 1$, and for these the Brill–Noether number ρ is equal to zero. Two possibilities are to be examined. First assume that the curve C is hyperelliptic. Then, as we observed in Chapter I, Section 2, any g^1_3 on C is obtained by adding an arbitrary point to the g^1_2. Thus, in this case, $W^1_3(C)$ (or at least its support) is isomorphic to C. Suppose now that C is not hyperelliptic. Then its canonical model, Γ, is a smooth complete intersection of a quadric and a cubic in $\mathbb{P}^3$. Consider a g^1_3 on C. By the geometric version of the Riemann–Roch theorem, the linear span in $\mathbb{P}^3$ of a divisor D belonging to the g^1_3 is a line, which is thus a trisecant to Γ. By Bezout's theorem this line lies in the quadric Q containing Γ. This shows that a g^1_3 on C is cut out by the lines of a ruling of Q. Two cases can then occur. Either Q is smooth, hence doubly ruled, and then $W^1_3(C)$ consists of two distinct points, or else Q is a cone and then $W^1_3(C)$ consists of a single point.

The next case is $g = 5$. Here the values of d and r to be considered are

$$d = 3, \qquad r = 1 \qquad (\rho < 0),$$

$$d = 4, \qquad r = 1 \qquad (\rho = 1).$$

When $d = 3$ and $r = 1$ the Brill–Noether number ρ is negative. In this case the existence theorem doesn't give any information, but we still try and describe $W^1_3(C)$, when non-empty. First of all, if C is hyperelliptic, then, as before, any g^1_3 is obtained by adding an arbitrary point to the g^1_2, and $W^1_3(C)$ is thus isomorphic to the curve itself. If C is non-hyperelliptic, but possesses a g^1_3, then the residual of this g^1_3 is a g^2_5 which, by Clifford's theorem, cannot have base points. The curve C can then be realized as a plane quintic Γ, and the genus formula tells us that Γ has, as its sole singularity, a node or an ordinary cusp, the g^1_3 being cut out on C by the pencil of lines through this double point. Conversely, any plane quintic Γ having as its only singularity a node or a cusp is a genus 5 curve possessing a g^1_3. We also claim that, in this case, $W^1_3(C)$ consists of a single point. In fact a g^1_3 on C must be cut out, on Γ, by adjoint curves of degree 2 (i.e., conics through the singular point) passing through five residual points. But six points belong to the base locus of a pencil of conics only if five of them are on a line l, the pencil of conics being the sum of the pencil of lines through the remaining point plus l. On the other hand, the only way for a pencil of lines to cut a g^1_3 on Γ is to be centered at the singular point of Γ, proving our claim.

Let us now examine the case $d = 4, r = 1$. Here the Brill–Noether number ρ is equal to 1. Let us first assume that C is hyperelliptic. In this case any complete g^1_4 is obtained from the g^1_2 by adding two base points, so that $W^1_4(C)$ is isomorphic to C_2.

Suppose now that C is non-hyperelliptic. We shall prove that, in this case, $W^1_4(C)$ is one-dimensional. By Martens' theorem (Theorem (5.20) in Chapter IV) we already know that it is at most one-dimensional. Consider then a canonical model Γ of C. As we have seen, Γ lies on three independent quadrics in $\mathbb{P}^4$. Let us consider a complete g^1_4, $|D|$, on C. Its residual $|K(-D)|$ is also a g^1_4, and we denote by F and F' the fixed divisors of $|D|$ and $|K(-D)|$, respectively. We will distinguish two cases according to whether $|D|$ coincides with its residual or not. Suppose first that $|D| \neq |K(-D)|$; choose hyperplanes H_{ij} in $\mathbb{P}^4$ such that

$$H_{ij} \cdot \Gamma = D_i + F + F' + D'_j, \qquad 1 \leq i, j \leq 2,$$

where $D_i + F \in |D|$, $i = 1, 2$, $D'_j + F' \in |K(-D)|$, $j = 1, 2$, and where D_1 and D_2 (resp. D'_1, D'_2) have disjoint supports. If $L_{ij} = 0$ is the defining equation of H_{ij}, then

$$L_{11}L_{22}/L_{12}L_{21}$$

defines a non-zero meromorphic function on C which is, in fact, holomorphic. This means that, for some non-zero constant k, the canonical curve Γ is contained in the singular quadric

$$L_{11}L_{22} - kL_{12}L_{21} = 0.$$

This quadric is ruled by two systems of two-planes. One of the rulings cuts out the series $|D|$, the other one cuts out the series $|K(-D)|$. In the case where $|D| = |K(-D)|$ we just consider hyperplanes H_{ij} such that

$$H_{ij} \cdot \Gamma = D_i + 2F + D_j, \qquad 1 \leq i \leq j \leq 2,$$

where D_1 and D_2 have disjoint supports and $D_i + F \in |D|$, $i = 1, 2$. Using the same notation as before we see that the canonical curve Γ is contained in the rank 3 quadric

$$L_{11}L_{22} - kL_{22}^2 = 0.$$

This quadric is simply ruled by a system of two-planes cutting out, on Γ, the given g_4^1. Also observe that in both cases the g_4^1, being complete, uniquely determines the quadric. In fact, by the geometric version of the Riemann–Roch theorem, any divisor D belonging to the g_4^1 spans a $\mathbb{P}^2$ so that, as D varies in the g_4^1, its linear span sweeps out the quadric. If C is not trigonal the above construction can be reversed. Namely, the two rulings of a quadric of rank 4 containing Γ cut out, on Γ, two mutually residual g_4^1's; while the single ruling of a rank 3 quadric containing Γ cuts out an autoresidual g_4^1. It is also clear that autoresidual g_4^1's are in finite number, any two of them differing by a two-torsion point in the Jacobian variety of C. To describe $W_4^1(C)$ in the non-trigonal case we now proceed as follows. Consider three linearly independent quadrics containing the canonical curve Γ, and denote by Q_0, Q_1, Q_2 the symmetric matrices defining them. The locus of singular quadrics containing C is then described, in the homogeneous coordinates $\lambda_0, \lambda_1, \lambda_2$, by the equation

$$\det(\lambda_0 Q_0 + \lambda_1 Q_1 + \lambda_2 Q_2) = 0.$$

This equation cannot be identically satisfied; otherwise, there would be a two-plane of singular quadrics containing Γ and hence at least ∞^2 g_4^1's on C. But this possibility, as we already pointed out, is ruled out by Martens' Theorem. It follows that the locus of singular quadrics containing Γ is a quintic curve in the plane with homogeneous coordinates $\lambda_0, \lambda_1, \lambda_2$. We can then conclude that $W_4^1(C)$ is a two-sheeted ramified cover of a plane quintic, the involution of this covering being given by residuation. An interesting situation occurs when C is trigonal and non-hyperelliptic. As we already observed when studying $W_3^1(C)$, the residual of the g_3^1 on C is then a g_5^2 which exhibits C as a plane quintic $\bar{C}$ having, as unique singularity, a node or a cusp. It is a useful exercise to check that, in this case, $W_4^1(C)$ consists of two copies of the curve C meeting in two points, or doubly at a point if $\bar{C}$ has a cusp, the two components being interchanged by residuation. One copy of C is the locus of g_4^1's of type $g_3^1 + p$, $p \in C$. The other is the locus of g_4^1's which are cut by lines through a fixed point of $\bar{C}$.

Let us now examine the last case: $g = 6$. Here the values of d and r to be considered are

$$\begin{aligned} d &= 3, & r &= 1 & (\rho &< 0), \\ d &= 4, & r &= 1 & (\rho &= 0), \\ d &= 5, & r &= 1 & (\rho &= 2), \\ d &= 5, & r &= 2 & (\rho &< 0). \end{aligned}$$

The case $d = 3$, $r = 1$ is dealt with exactly as before. Namely, if C is hyperelliptic, then any g_3^1 is obtained from the g_2^1 by adding an arbitrary base point so that $W_3^1(C)$ is isomorphic to the curve itself. If C is non-hyperelliptic but, nevertheless, contains a g_3^1, then $W_3^1(C)$ reduces to a single point. Suppose in fact that C possesses two distinct g_3^1's. By Clifford's theorem the sum of the two g_3^1's is a complete g_6^2. This g_6^2 defines a morphism ϕ of C into $\mathbb{P}^2$, which cannot be a triple covering of a conic, otherwise the two g_3^1's would coincide. On the other hand ϕ cannot be a two-sheeted covering of an elliptic cubic (the g_6^2 would not contain a g_3^1), nor can it be birational, otherwise the image curve would be a sextic with two triple points, in contrast with the genus formula. This is absurd and our claim is proved.

Let us now consider the other case in which the Brill–Noether number ρ is negative, namely, when $d = 5$, $r = 2$. Again, when ρ is negative, the existence theorem doesn't say anything, but, as before, we shall describe what $W_5^2(C)$ looks like, when non-empty. First of all, when C is hyperelliptic, any g_5^2 on C can be obtained from the double of the g_2^1 by adding an arbitrary base point. This shows that, in this case, $W_5^2(C)$ is isomorphic to C itself. Suppose now that C is not hyperelliptic. A g_5^2 on C must then define a birational morphism of C onto a plane quintic, which, by the genus formula, must be smooth. We claim that the g_5^2 is unique. Suppose in fact that another g_5^2 exists. Then it must be cut out, on the plane quintic, by a two-dimensional linear system of conics passing through five points. The only way this can happen is for this system of conics to be the union of the system of lines in $\mathbb{P}^2$ and a fixed line. This exactly means that the given g_5^2 is the one cut out by the lines in $\mathbb{P}^2$. In conclusion, for a smooth plane quintic, $W_5^2(C)$ consists of exactly one point.

We now come to the most interesting case; the one for which $d = 4$, $r = 1$. Here the Brill–Noether number is equal to zero. Let us dispose of two special cases. First of all, if C is hyperelliptic then any complete g_4^1 on C is obtained by adding two base points to the g_2^1, so that $W_4^1(C)$ is birationally equivalent to the second symmetric product of the curve. If C is isomorphic to a smooth plane quintic then, as in the case of g_5^1's, one shows that any g_4^1 is obtained from the g_5^2 by subtraction of a point, so that $W_4^1(C)$ is isomorphic to the plane quintic itself. Suppose now that C is neither hyperelliptic nor isomorphic to a plane quintic. The existence theorem assures us that there are g_4^1's on C: a direct argument is as follows. We can obviously assume that C is not trigonal. Let $p. q, r$ be any three points of C. Since C is not trigonal, $H^1(C, K(-p-q-r))$ maps isomorphically to $H^1(C, K(-p))$. Thus the evaluation map

$$H^0(C, K(-p)) \to \mathbb{C}_q \oplus \mathbb{C}_r$$

is onto, and $|K(-p)|$ embeds C in $\mathbb{P}^4$. On the other hand

$$h^0(\mathbb{P}^4, \mathcal{O}(2)) - h^0(C, K^2(-2p)) = 2,$$

so there is a pencil of quadrics in $\mathbb{P}^4$ containing C. Among these there is at least one (five, in general) quadric of rank at most four. One ruling of this quadric cuts a g_4^1 on C.

We shall now give an idea of what the various possibilities for $W_4^1(C)$ are. The residual of a g_4^1 is a g_6^2 which defines a morphism ϕ onto a plane curve Γ. Excluding the cases where C is hyperelliptic or else isomorphic to a smooth plane quintic, we see that only few cases can occur: Γ is a conic and ϕ a three-sheeted covering; Γ is a smooth cubic and ϕ a two-sheeted covering; Γ is a plane sextic and has a triple point; Γ is a sextic having only double points as singularities. In the first three cases $W_4^1(C)$ contains a one-dimensional component. In fact, in the first and third case, C is trigonal and therefore infinitely many g_4^1's can be obtained from the g_3^1 by adding an arbitrary base point. In the second case C is bi-elliptic, i.e., a ramified double covering of an elliptic curve, and infinitely many g_4^1's on C can be obtained by varying the g_2^1 on the elliptic curve Γ and composing this with ϕ. As far as the fourth case is concerned, we shall limit ourselves to the general case of a plane sextic with four distinct double points p_1, p_2, p_3, p_4, no three of them collinear. To see that this does indeed occur, given four points $p_1, \dots, p_4$ in the plane, no three of them collinear, set

$$F = Q_1 + Q_2 + L_1 + L_2,$$
$$G = Q_3 + Q_4 + L_3 + L_4,$$

where $Q_1, \dots, Q_4$ are distinct smooth conics through $p_1, \dots, p_4$ and $L_1, \dots, L_4$ are distinct lines with no points in common, and not passing through any one of the points $p_1, \dots, p_4$. By Bertini's theorem, a general member of the pencil spanned by F and G is a sextic Γ with the required properties. We shall prove that, in this case, $W_4^1(C)$ consists of exactly five points. These correspond to the four g_4^1's cut out on Γ by the pencils of lines though $p_1, \dots, p_4$ and to the one cut out by the conics through $p_1, \dots, p_4$. To see that indeed these are the only g_4^1's on C, just observe that a g_4^1 must be cut out on Γ by a one-dimensional system of cubics through $p_1, \dots, p_4$ and through six residual points $p_5, \dots, p_{10}$. By Bezout's theorem this is possible only if this system possesses a base curve, which must then be a line or a conic. From this it is immediate to deduce that the given g_4^1 must coincide with one of the five we previously described.

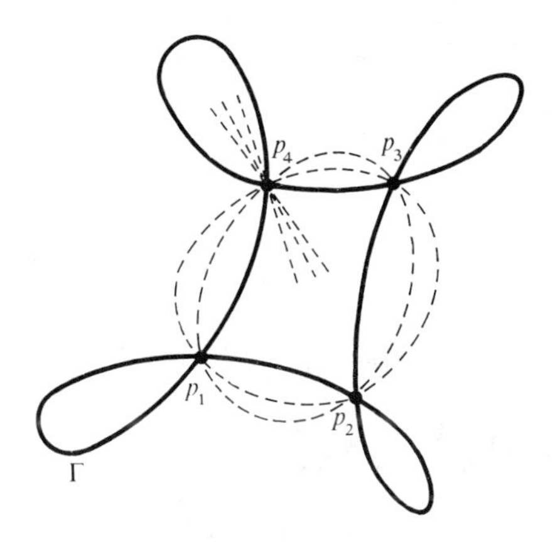

The final case is $d = 5$, $r = 1$. Obviously, so far as existence is concerned, we have nothing left to prove. If C is hyperelliptic, any complete g_5^1 is obtained from the g_2^1 by adding three base points, so that $W_5^1(C)$ is birationally equivalent to the third symmetric product of C. If C is not hyperelliptic, then Martens' theorem says that the dimension of $W_5^1(C)$ does not exceed two. Since $\rho = 2$, $W_5^1(C)$ is two-dimensional. What $W_5^1(C)$ actually looks like depends on the particular nature of C, as may be seen from the following two examples. Suppose first that C is a smooth plane quintic. Then, as the reader may easily check, any complete g_5^1 on C is of the form $|\mathcal{O}_C(1)(-p + q)|$, where p and q are distinct points. Thus $W_5^1(C) - W_5^2(C)$ is isomorphic to the product of C with itself minus the diagonal. On the other hand, if C is trigonal, $W_5^1(C)$ has two components. One component is made up of the series $g_3^1 + p + q$, where p and q are points of C; the other component is made up of the residuals of the series in the first component.

From the preceding discussion it appears that, whenever $\rho \geq 0$, and with the exception of cases that can be geometrically described, and directly dealt with, the variety $W_d^r(C)$ has dimension equal to ρ. Moreover, whenever $\rho = 0$, we were able to count the points of $W_d^r(C)$. The results of these direct computations are particular cases of a formula, discovered by Castelnuovo, stating that, when $\rho = 0$, the number of g_d^r's is

$$g! \prod_{i=0}^{r} \frac{i!}{(g - d + r + i)!}. \tag{1.2}$$

Both Kempf and Kleiman–Laksov generalized this formula and obtained, along with the existence theorem, an explicit expression for the class of $W_d^r(C)$ in the cohomology ring of the Jacobian. As we shall see in Chapter VII, their proofs are based on the determinantal nature of $W_d^r(C)$ and on Porteous' formula.

(1.3) Theorem. *Let C be a smooth curve of genus g. Let θ be the class of the theta divisor in the cohomology ring of the Jacobian $J(C)$. Let w^r_d be the class of $W^r_d(C)$. Then, if $W^r_d(C)$ is of the expected dimension ρ,*

$$w^r_d = \left(\prod_{i=0}^{r} \frac{i!}{(g-d+r+i)!}\right) \cdot \theta^{g-\rho}.$$ [1]

Recalling that the theta divisor provides a principal polarization of the Jacobian, and hence its g-fold intersection number is $g!$, we get, as a special case, Castelnuovo's count for the number of g^r_d's when $\rho = 0$.

It was remarked by Fulton and Lazarsfeld that the existence theorem is, in a sense, a reflection of an ampleness property of the explicit complex we used to define $W^r_d(C)$ itself. From this remark they went on to prove the following fundamental Lefschetz (or Bertini) type of result.

(1.4) Connectedness Theorem. *Let C be a smooth curve of genus g. Let d, r be integers such that $d \geq 1$, $r \geq 0$. Assume that*

$$\rho = g - (r+1)(g-d+r) \geq 1.$$

Then $G^r_d(C)$, and hence $W^r_d(C)$, are connected.

The reader is invited to go back to the examples we treated and directly show that, whenever $\rho \geq 1$, the variety $W^r_d(C)$ is connected.

The existence theorem gives a lower bound for the dimension of $G^r_d(C)$. We may now ask whether this bound is sharp and, as a special case, whether $W^r_d(C)$ is empty for negative ρ. In this generality the answer to this question is obviously no. Hyperelliptic curves in genus $g \geq 3$, trigonal curves in genus $g \geq 5$, and smooth plane quintics in case $g = 6$, give examples for which $\rho < 0$ but $W^r_d(C)$ is non-empty. To get some feeling for which kind of geometrical implications the condition $\rho < 0$ gives, we shall offer two examples. The first is the one of g^1_d's. For these the Brill–Noether number is given by

$$\rho = 2d - g - 2.$$

A base-point-free g^1_d on C defines a d-sheeted ramified covering of $\mathbb{P}^1$ which, by Hurwitz' formula, has $2d + 2g - 2$ ramification points. Since, up to projectivities, we can always assume that the d-sheeted covering in question is ramified over the points $0, 1, \infty$, we conclude that a curve of genus g which

[1] It is useful to observe that the enumerative formula for the class of $W^r_d(C)$ also holds in the Chow ring of $\mathrm{Pic}^d(C)$ with minor formal modifications. Namely, one has to substitute $W_{g-1}(C)$ for the theta divisor, and also one has to be careful in choosing the right identification between $\mathrm{Pic}^d(C)$, where $W^r_d(C)$ lives, and $\mathrm{Pic}^{g-1}(C)$, which is the natural ambient space for $W_{g-1}(C)$.

can be expressed as a d-sheeted covering of $\mathbb{P}^1$ depends on no more that $2d + 2g - 5$ parameters. Now the condition $\rho < 0$ can be rewritten in this way:

$$2d + 2g - 5 < 3g - 3.$$

We can therefore assert that, if C is a general curve of genus g, and if $\rho < 0$, then $W^1_d(C)$ is empty. Another example of this phenomenon is provided by smooth plane curves of degree d. Let C be one such; then the genus of C is given by

$$g = \frac{(d-1)(d-2)}{2}.$$

On the other hand, the Brill–Noether number for g^2_d's is

$$\begin{aligned}\rho &= 3d - 2g - 6\\ &= d(6 - d) - 8.\end{aligned}$$

For g and ρ given as above, the condition $\rho < 0$ can also be written as

$$\frac{d(d+3)}{2} - 8 < 3g - 3.$$

Now the number on the left-hand side is clearly an upper bound for the number of moduli of curves having smooth, degree d plane models. Here again we see that, if $\rho < 0$, then a general curve of genus g cannot be realized as a smooth plane curve of degree d. Now let us go back for one moment to the example following the statement of the Existence Theorem. Let us restrict our attention to the cases for which

$$\dim W^r_d(C) > \max(\rho, -1).$$

One then observes that these are

$$\begin{aligned}&g = 4, 5, 6 \quad && C \text{ hyperelliptic},\\ &g = 5, 6 && C \text{ trigonal},\\ &g = 6, && C \text{ a smooth quintic; } C \text{ bi-elliptic}.\end{aligned}$$

In each of these cases a direct parameter count shows that C cannot be a general curve of genus g. All these considerations point in the direction of the following dimension theorem, which was first stated by Brill and Noether, and proved by Griffiths–Harris.

(1.5) Dimension Theorem. *Let C be a general curve of genus g. Let d and r be integers such that $d \geq 1$, $r \geq 0$. Then if*

$$\rho = g - (r + 1)(g - d + r) < 0,$$

$G^r_d(C)$ is empty. If $\rho \geq 0$, then G^r_d is reduced and of pure dimension ρ.

The study of Zariski tangent spaces that we carried out in Section 4 of Chapter IV makes it natural to ask what can be said about the infinitesimal structure of $G^r_d(C)$ and $W^r_d(C)$ when C is a general curve. This question is answered by the following result, whose proof is due to D. Gieseker.

(1.6) Smoothness Theorem. *Let C be a general curve of genus g. Let d, r be integers such that $d \geq 1$ and $r \geq 0$. Then $G^r_d(C)$ is smooth of dimension ρ.*

Corollary (Fulton–Lazarsfeld). *If $\rho \geq 1$, $G^r_d(C)$, and hence $W^r_d(C)$, are irreducible.*

The corollary follows at once from Gieseker's theorem by Fulton and Lazarsfeld's connectedness theorem. It is noteworthy that, for special curves C, $W^r_d(C)$ may well be reducible even if it has the expected dimension ρ and $\rho > 0$. The first instance of this phenomenon occurs for g^1_4's on curves of genus 5. As we already observed, if C is a non-hyperelliptic, trigonal curve of genus five, $W^1_4(C)$ is the union of two components, each of them isomorphic to C, meeting at two points. By contrast, for a general genus 5 curve, $W^1_4(C)$ is an irreducible smooth double cover of a smooth plane quintic.

The Smoothness Theorem can also be seen as a singularity theorem for $W^r_d(C)$ on a general curve C. In fact, by Proposition (4.4) of Chapter IV, the smoothness result for $G^r_d(C)$, when C is general, makes it possible to conclude that, when $r > d - g$, then for a general curve C

$$W^r_d(C)_{\text{sing}} = W^{r+1}_d(C).$$

To better understand this result we recall the expression of the Zariski tangent spaces to $G^r_d(C)$ given in Proposition (4.1) of Chapter IV. Let w be a point of $G^r_d(C)$; assume that w corresponds to an $(r + 1)$-dimensional subspace W of $H^0(C, \mathcal{O}(D))$, for some effective degree d divisor D on C; then

$$\dim T_w(G^r_d(C)) = \rho + \dim(\ker \mu_{0,W}),$$

where

$$\mu_{0,W}: W \otimes H^0(C, K(-D)) \to H^0(C, K)$$

is the cup-product mapping. It follows that an equivalent version of the preceding theorem is

(1.7) Smoothness Theorem (Second Version). *Let C be a general curve of genus g. Let D be an effective divisor on C. Then the cup-product homomorphism*

$$\mu_0 : H^0(C, \mathcal{O}(D)) \otimes H^0(C, K(-D)) \to H^0(C, K)$$

is injective.

This is the version that Gieseker proved and this is the one that K. Petri stated, in a very parenthetical way, in a long forgotten paper [2].[1]

We shall now illustrate in a couple of examples the content of the above theorem, or more precisely the role and geometrical meaning of the map μ_0. The first example is the one of g_3^1's on curves of genus 4. Here we know that, for a non-hyperelliptic curve, $W_3^1(C)$ consists of either two points or of a single point, and that the second case occurs exactly when the given g_3^1 is autoresidual and cut out by the ruling of a quadric cone Q containing a canonical model of C. Let $|D|$ be a g_3^1 on C. By the base-point-free pencil trick, the map

$$\mu_0 : H^0(C, \mathcal{O}(D)) \otimes H^0(C, K(-D)) \to H^0(C, K)$$

is injective if $|D|$ is not autoresidual and has a one-dimensional kernel otherwise. In the latter case, since the range and the domain of μ_0 are both four-dimensional, the cokernel of μ_0 is also one-dimensional, and the reader will observe that its projectivized dual can be canonically identified with the vertex of Q.

The second example is the one of g_4^1's on a curve C of genus 6. Let us put ourselves in the case when C can be realized as a plane sextic with at most double points as singularities, and more precisely with four nodes (or cusps) $p_1, \ldots, p_4$, no three of them collinear. For such a sextic we already noticed that the only g_4^1's are the ones cut out by the pencils of lines centered at $p_1, \ldots, p_4$, respectively, and by the pencil of conics through $p_1, \ldots, p_4$. All of this is in accordance with Castelnuovo's count (1.2) which says that $W_4^1(C)$ consists of five points. To completely check Castelnuovo's count it would of course be necessary to verify that the five g_4^1's on C are all simple

[1] "Haben $\phi_0^{(i)}, \phi_0^{(i)}, \ldots, \phi_r^{(i)}$ eine zu $G_{n'}$ korresiduale Gruppe gemein, während sie in der gleichen G_n treffen wie $\phi_0, \phi_1, \ldots, \phi_r$, so liefert der Riemann–Rochsche Satz weiter die Relationen für $r' > 0$:

$$\phi_0 : \phi_1 : \cdots : \phi_r = \phi_0^{(1)} : \phi_1^{(1)} : \cdots : \phi_r^{(1)} = \cdots = \phi_0^{(r')} : \phi_1^{(r')} : \cdots : \phi_r^{(r')} \tag{2}$$

wenn $g_{n'}^{r'}$ die zu g_n^r residuale Schar bedeutet.

Im allgemeinen Körper sind diese $(r+1)(r'+1)$ Formen ϕ bekanntlich linear unabhängig, bei speziellen Körpern braucht dies nicht der Fall zu sein.

Jede solche Relation legt den Parametern des Körpers eine Bedingung auf und die Zahl der Parameter wird im allgemeinen gerade un die Zahl λ_1 dieser Relationen verkleinert." (*loc. cit.*, p. 184)

points of $W_4^1(C)$. This is what we are now going to do. Consider one of the five g_4^1's, for example the one corresponding to lines through the double point p. Let us examine μ_0 in this case. An element of $\ker\mu_0$ would correspond to a relation

$$l_1Q_2 - l_2Q_1 = 0,$$

where l_1, l_2 are distinct lines through p, and Q_1, Q_2 are conics through the remaining double points. If such a relation held there would exist a line l such that $Q_i = ll_i$, $i = 1, 2$. This is absurd since l would then have to contain the remaining three double points. This argument breaks down precisely when the three remaining double points are collinear. Suppose, for example, that our plane sextic has four nodes and that three of them are collinear; let p be the remaining one. In this case $W_4^1(C)$ is no longer reduced. In fact it has exactly one double point, corresponding to the coming together of two of the previously described g_4^1's, namely the one cut out by conics and the one cut out by lines through p. The above argument also shows that, for this g_4^1, the kernel of μ_0 is one-dimensional. This situation may degenerate further and the most degenerate case occurs when all the g_4^1's come together. In this case $W_4^1(C)$ consists of a single five-fold point. An example of this is given by the curve whose affine equation is

$$(y - x^3)^2 + y^3 + y^5 = 0.$$

It can be easily checked however, that, in all these cases the kernel of μ_0 is at most one-dimensional.

The theorems we have illustrated in this chapter, while constituting the cornerstone of the study of the varieties $G_d^r(C)$ and $W_d^r(C)$, give only a faint idea of the richness and diversity of the results that have been obtained in this direction. Some of these, like the Martens and Mumford theorems of Chapter IV, or like the singularity theorems of Riemann and Kempf, to be proved in the next chapter, deal mainly with the structure of the varieties $W_d^r(C)$ themselves. Others are concerned with the geometry of the linear series these varieties parametrize. As an illustration, we mention, among these, the following result.

(1.8) Theorem. *Let C be a general curve of genus g, and $\mathscr{D} \in G_d^r(C)$ a general point. If $r \geq 3$ then the map $\phi_{\mathscr{D}}\colon C \to \mathbb{P}^r$ is an embedding.*

Observe that by this theorem for $r \geq 3$ a general curve of genus g may be represented as a smooth non-degenerate curve of degree d in $\mathbb{P}^r$ if and only if $\rho = g - (r + 1)(g - d + r) \geq 0$.

Bibliographical Notes

The first instance of the theory of special divisors can be found in Riemann's famous work on abelian functions [1], but the first extensive study of this theory appears in the fundamental paper of Brill and Noether [1]. It is there that the Brill–Noether matrix makes its first appearance and so does the Brill–Noether number. The influence of this paper has been vast. Enriques and Chisini devote a large part of the third volume of their treatise [1, Ch. I, 1h; Ch. III:30, 31, 32, 33] to the problem of special linear series and so does Severi in his Lectures [2]. The same attention to the problem is testified to by Castelnuovo's paper [2], which contains the count of g^r_d's when $\rho = 0$.

In the early sixties, A. Mayer [1], D. Mumford, and J. Lewittes [1] revisited Riemann's singularity theorem, bringing back attention to the theory of special divisors.

The existence theorem for g^1_d's was first proved by Meis [1]. Later, at a time when the techniques of enumerative geometry were better understood, the first fundamental theorem of the theory was established with a completely modern approach. In fact (partly under the influence of unpublished work of Mumford) simultaneously Kempf and Kleiman–Laksov gave the first rigorous proof of the Existence Theorem, and of Theorem (1.3). (See Kempf [1], Kleiman–Laksov [1, 2].)

The Connectedness Theorem is totally new; it has no classical origin and is completely due to the joint efforts of Fulton and Lazarsfeld [1].

The Dimension Theorem, on the contrary, has a long history, one which cannot be recounted in these short notes. Brill, Noether, Enriques, and Severi had a clear idea of the statement and of the reasons why it holds true, but were never able to produce an acceptable proof. Maybe the only one who somehow had in mind a convincing strategy was Severi [1].

The first results in this direction (dealing essentially with the case of g^1_d's) were obtained by Farkas [1] and Martens [2], [3]. It was then Kleiman who, in [1], going back to a degeneration argument of Castelnuovo [2], was able to reduce the Dimension Theorem to an enumerative problem on a rational curve.

The first proof of the Dimension Theorem comes a few years later in Griffiths–Harris [4], where a more direct degeneration argument is given and where the enumerative problem is solved. A substantially simpler proof of the Dimension Theorem is presented in Eisenbud–Harris [1].

As we pointed out in the text, the Smoothness Theorem was stated, in a very indirect way, by Petri in [2]. When Petri's study of the ideal of the canonical curve was revisited, this statement was brought to light. The first proof of the Smoothness Theorem is due to Gieseker [1] and a more direct and simpler proof appears in Eisenbud–Harris [2]. It should also be mentioned that the Smoothness Theorem for G^1_d first appears in

Lax [1], while the one for G_d^2 is to be found first in Arbarello–Cornalba [1]. Theorem (1.8) is proved in Eisenbud–Harris [1].

The bibliographical notes to Chapters VII, VIII, and XI will be a natural complement to these.

Exercises

A. $W_4^1(C)$ on a Curve C of Genus 6

We have seen that every smooth curve C of genus 6 has a g_6^2; in this sequence of exercises we will let $\mathscr{D}$ be one such.

A-1. Show that, if $\mathscr{D}$ has two base points, then $\phi_{\mathscr{D}}$ is a 2-to-1 map onto a conic.

A-2. Show that, if $\mathscr{D}$ has exactly one base point, then $\phi_{\mathscr{D}}$ embeds C as a smooth plane quintic.

A-3. Assume that $\mathscr{D}$ has no base points and show that one of the following occurs:

(a) $\phi_{\mathscr{D}}$ maps C in a 3-to-1 manner onto a conic (thus C is trigonal);
(b) $\phi_{\mathscr{D}}$ maps C in a 2-to-1 manner onto a smooth plane cubic (thus C is bi-elliptic);
(c) $\phi_{\mathscr{D}}$ maps C in a 2-to-1 manner onto a singular plane cubic (again, C is hyper-elliptic);
(d) $\phi_{\mathscr{D}}$ maps C birationally to a plane sextic curve C_0. In this case C_0 cannot have a point of multiplicity ≥ 4, and if C_0 has a triple point, then C is trigonal.

A-4. Show that, conversely, if C is hyperelliptic then either $\phi_{\mathscr{D}}$ has two base points or else condition (c) in the preceding exercise holds.

A-5. Show that if C is trigonal, then either condition (a) or condition (d) in Exercise A-3 holds, and if the latter, then C_0 has a triple point.

A-6. Show that if C is a smooth plane quintic, then condition (a) in Exercise A-3 holds.

A-7. Show that if C is bi-elliptic, then condition (b) in Exercise A-3 holds.

A-8. Suppose now that C is hyperelliptic and let $|D|$ be the g_2^1 on C. Show that set-theoretically

$$W_4^1(C) = u(D) + W_2(C),$$

and that $W_4^1(C)$ is singular at the point $u(2D)$. In addition, show that $W_4^1(C)$ is reduced away from $u(2D)$. (We would guess that it is reduced at $u(2D)$ as well, but we haven't checked this.)

A-9. Suppose that C is trigonal and let $|D|$ be the g_3^1 on C. Show that set-theoretically

$$W_4^1(C) = (u(D) + W_1(C)) \cup \{K(-2D)\},$$

where $K(-2D)$ may or may not be in $u(D) + W_1(C)$ (thus, the above decomposition may or may not be redundant). Show that $\phi_{\mathscr{D}}$ is of type (a) in Exercise A-3 if $\mathscr{D} = |2D|$, while it is of type (d) if $\mathscr{D} = |K(-D)(-p)| \neq |2D|$ for some $p \in C$.

(*Hint*: If $|E|, |F|$ are a g_3^1 and a g_4^1, respectively, show that either $|F| = |E| + p$ or the map

$$\phi = (\phi_E, \phi_F): C \to \mathbb{P}^1 \times \mathbb{P}^1$$

is an embedding and use adjunction to show that $K_C \cong \mathcal{O}_C(2E + F)$.)

A-10. Show that if $K(-2D) \notin u(D) + W_1(C)$, then $W_4^1(C)$ is reduced; while if

$$K(-2D) \in u(D) + W_1(C),$$

then $K(-2D)$ is an embedded point of $W_4^1(C)$. In the latter case, identify the tangent space to $W_4^1(C)$ at $K(-2D)$.

A-11. Suppose C is bi-elliptic with two-sheeted map $\pi: C \to E$ onto an elliptic curve E. Show that $W_4^1(C)$ is reduced and

$$W_4^1(C) = \pi^* W_2^1(E) \cong E,$$

where

$$\pi^*: J(E) \to J(C)$$

is the induced map.

A-12. Suppose that $C \subset \mathbb{P}^2$ is a smooth plane quintic with $\mathcal{O}_C(1) = \mathcal{O}_C(D)$. Show that set-theoretically

$$W_4^1(C) = u(D) - W_1(C).$$

In addition, show that $W_4^1(C)$ is reduced.

Combining the preceding four exercises, we have established the following:

Suppose that C is a smooth curve of genus 6. Then the following are equivalent:

(i) *C is not hyperelliptic, trigonal, bi-elliptic, or a smooth plane quintic;*

(ii) *C is birationally equivalent to a plane sextic curve C_0 having double points as singularities;*

(iii) $\dim W_4^1(C) = 0$

For the remainder of this sequence of exercises we will restrict our attention to curves satisfying the conditions of the theorem and we will denote by $\phi: C \to C_0$ a birational map of C to a plane sextic C_0 having only double points.

A-13. Show that, if $p_1, \ldots, p_7 \in \mathbb{P}^2$ are seven distinct points that fail to impose independent conditions on the linear system $|\mathcal{O}_{\mathbb{P}^2}(3)|$ of plane cubics, then five of the p_i are collinear, and conversely.

A-14. Using the preceding exercise, show that if eight distinct points $p_1, \ldots, p_8 \in \mathbb{P}^2$ fail to impose independent conditions on $|\mathcal{O}_{\mathbb{P}^2}(3)|$, then either five are collinear or all eight lie on a conic, and conversely.

A-15. Using the preceding exercise, show that the only g_4^1's on C are those cut out on C by pencils of lines through double points of C_0, and the one cut out on C by conics passing through all four nodes of C_0 (more generally, satisfying the adjoint conditions on C, cf. Appendix A).

A-16. Compute $\ker \mu_0$ for the g_4^1 cut out on C by lines through a double point p of C_0 in the following cases:

(i) p is a node and the remaining double points are not collinear;
(ii) p is a node and the remaining double points are collinear (we assume that the double points are distinct):
(iii) p is a tacnode:
(iv) p is the intersection of two smooth branches having contact of order 3 (we call this an *oscnode*).

A-17. Show that a plane sextic with four distinct nodes, of which no three are collinear, can never be birational to a plane sextic with a tacnode.

A-18. Show that a plane sextic C_0 with four distinct nodes, three of which are collinear, is birational to a plane sextic C_0' with one tacnode and two ordinary nodes. Show that the tangent line to C_0' at the tacnode cannot pass through either of the remaining nodes.

A-19. Let C_0 be a plane sextic having two tacnodes such that the tangent line to C_0 at each tacnode does not pass through the other. Show that C_0 cannot be birational to a plane sextic with an oscnode (cf. A-16) and a node.

A-20. Show that the support of $W_4^1(C)$ consists of one point if and only if C_0 has a singularity consisting of two smooth arcs meeting with contact of order 4 at a point which is a flex of each arc.

B. Embeddings of Small Degree

In this batch of exercises we will not assume Theorem (1.8) of Chapter V.

B-1. Let C be any curve of genus g and D a *general* effective divisor of degree $g + 3$. Show that:

$$r(D) = 3;$$

$$|D| \text{ is base-point-free;}$$

$$\phi_D: C \to \mathbb{P}^3 \text{ is an embedding.}$$

B-2. Show that a hyperelliptic curve C of genus $g \geq 2$ can never be embedded in $\mathbb{P}^r$ with degree at most $g + 2$.

B-3. Suppose that C is trigonal and D is a general effective divisor of degree d. Show that $\phi_{K(-D)}$ is not an embedding unless $D = 0$. (Thus we cannot biregularly project $\phi_K(C)$ from a linear space spanned by general points on the canonical curve.)

B-4. Similarly, suppose that C is a smooth bi-elliptic curve of genus $g \geq 6$ and D a general effective divisor of degree d on C. Show that $\phi_{K(-D)}$ is not an embedding unless $d \leq 1$.

Conversely, we will now see that, if C is not hyperelliptic, trigonal, or bi-elliptic, then $\phi_K(C)$ may be biregularly projected to a curve of degree $g + 2$.

B-5. Let C be a curve of genus $g \geq 4$ not hyperelliptic, trigonal, or bi-elliptic, and D a general effective divisor of degree $g - 4$. Show, using Mumford's theorem, that

$$\phi_{K(-D)}: C \to \mathbb{P}^3$$

embeds C as a curve of degree $g + 2$.

B-6. Let C be a *general* curve of genus g with $4 \leq g \leq 7$. Show that $g + 2$ is the smallest degree of an embedding of C in $\mathbb{P}^r$. On the other hand, for $g = 8$ show that

$$W_9^3(C) \neq \varnothing, \qquad W_7^2(C) = \varnothing$$

and conclude that C may be embedded as a curve of degree 9 in $\mathbb{P}^3$.

C. Projective Normality (II)

This series of exercises will investigate the question: What is the smallest r such that a general curve C of genus g can be embedded in $\mathbb{P}^r$ as a projectively normal curve? By Max Noether's theorem on the projective normality of the canonical curve we know that $r \leq g - 1$.

C-1. Show that a general curve of genus $g \geq 6$ cannot be realized as a projectively normal curve in $\mathbb{P}^3$.

(*Hint*: Use (1.5) to estimate the degree of C, and then study $H^0(C, \mathcal{O}_C(2))$).

C-2. More generally, by a similar argument show that, if $C \subset \mathbb{P}^r$ is a projectively normal general curve of genus g, then

$$g \leq r(r + 1)/2.$$

C-3. Using properties of scrolls show that if $C \subset \mathbb{P}^r$ is linearly normal and hyperelliptic, then the linear series cut out on C by quadrics has vector space dimension $\leq 3r$, and hence that if $C \subset \mathbb{P}^r$ is projectively normal and hyperelliptic then

$$\begin{cases} d \geq 2g + 1, \\ r \geq g + 1. \end{cases}$$

C-4. Suppose that C is a smooth curve of genus g and D is a non-special effective divisor of degree $d \geq g + 1$ with $|D|$ base-point-free. Show that the map

$$H^0(C, \mathcal{O}(D)) \otimes H^0(C, \mathcal{O}(kD)) \to H^0(C, \mathcal{O}((k+1)D))$$

is surjective for $k \geq 2$.

(*Hint*: Let $V \subset H^0(C, \mathcal{O}(D))$ correspond to a base-point-free pencil in $|D|$ and use the base-point-free pencil trick to deduce the exact sequence

$$0 \to \mathcal{O}((k-1)D) \to V \otimes \mathcal{O}(kD) \to \mathcal{O}(kD) \to 0.$$

Now use the Riemann-Roch theorem together with the non-speciality of D.)

A consequence of this exercise is the assertion:

(*) *If $C \subset \mathbb{P}^r$ is any smooth curve for which $\mathcal{O}_C(1)$ is non-special, then C is projectively normal if, and only if, it is linearly and quadratically normal.*

In the remaining series of exercises we will prove the following:

(**) **Theorem.** *If C is a general curve of genus $g \geq 3$ and D a general divisor of degree $d = [3g/2] + 2$, then $\phi_D(C)$ is projectively normal.*

Corollary. *A general curve of genus g may be realized as projectively normal curve in $\mathbb{P}^r$ for*

$$r = \left[\frac{g}{2}\right] + 2.$$

C-5. Let C be a general curve of genus g and D a general divisor of degree d.

(i) Show that, if $d < 3g/2 + 2$, then D is not linearly equivalent to a sum of divisors D_1, D_2 with $r(D_1), r(D_2) \geq 1$.

(ii) Show that, if $d = 3g/2 + 2$, then D is the sum of at most finitely many such pairs.

C-6. Let C be a smooth curve and D a divisor such that there exist linearly independent sections

$$s_0, s_1, s_2, s_3 \in H^0(C, \mathcal{O}(D))$$

satisfying

$$s_0 s_1 - s_2 s_3 = 0.$$

Show that D is linearly equivalent to a sum of divisors that move in pencils. Conclude, using the preceding exercise, that if C is a general curve of genus g and D a general divisor on C of degree $d < 3g/2 + 2$ (resp., $d = 3g/2 + 2$), then the image $\phi_D(C) \subset \mathbb{P}^{d-g}$ lies on no (resp., at most finitely many) quadrics of rank ≤ 4.

C-7. Using Exercise A-2 of Chapter II, show that if a curve $C \subset \mathbb{P}^r$ lies on no quadrics of rank ≤ 4, then it lies on at most $\binom{r-2}{2}$ linearly independent quadrics. Similarly, if C lies on finitely many quadrics of rank ≤ 4, then it lies on at most $\binom{r-2}{2} + 1$ linearly independent quadrics. The theorem now follows.

Referring to the question posed at the beginning of this series of exercises, we have

$$\frac{-1 + \sqrt{8g+1}}{2} \leq r \leq \frac{3g}{2} + 2.$$

So far as we know, it is an open question just what the best estimate for r is.

D. The Difference Map ϕ_d: $C_d \times C_d \to J(C)$ (I)

In this sequence of exercises we shall study the map

$$\phi_d : C_d \times C_d \to J(C)$$

defined by

$$\phi_d(D, E) = u(D - E).$$

The image will be denoted by $V_d(C)$, or simply by V_d if no confusion is possible. It is an analytic variety whose fundamental class we denote by v_d.

D-1. Show that the differential of ϕ_d has maximal rank $\min(2d, g)$ at a general point, and hence that $\dim(V_d) = \min(2d, g)$.

D-2. Show that

$$\phi_1 : C \times C \to V_1(C)$$

is birational if C is non-hyperelliptic and has degree 2 if C is hyperelliptic. More generally, show that if C is non-hyperelliptic then ϕ_d is birational for $d < g/2$.

D-3. Show that, if C is non-hyperelliptic and $d < g/2$, then

$$v_d = \frac{\theta^{g-2d}}{(d!)^2}.$$

Similarly, show that for any smooth curve of genus $g = 2d$, the mapping ϕ_d has degree $\binom{2d}{d}$.

(**Remark.** In proving this by a computation similar to the proof of Poincaré's formula you need not assume anything about the rank of $(\phi_{g/2})_*$. Having proved that the degree is non-zero, you may conclude that every divisor of degree 0 is the difference of two effective divisor classes of degree $g/2$.)

D-4. Suppose that C is a non-hyperelliptic curve of genus 4 and suppose that D is a divisor of degree 0 on C. Show that

$$\dim \phi_2^{-1}(u(D)) = \begin{cases} 2 & \text{if } u(D) = 0, \\ 1 & \text{if } u(D) \in V_1 \cup (K - 2W_3^1(C)), \\ 0 & \text{otherwise,} \end{cases}$$

(the middle equation means that either $D \sim p - q$ or else $D \sim E - E'$ where E, E' are divisors from the two g_3^1's on C).

D-5. Keeping the notations of the preceding problem, assume that $[E] \neq [E']$ and show that

$$W_{g-1}(C) \cap (W_{g-1}(C) + u(E - E')) = V_2(C) + u(E - E').$$

Chapter VI

The Geometric Theory of Riemann's Theta Function

The results of the Brill–Noether theory stated in the preceding chapter have been proved in full generality only fairly recently. However, special but important cases of them were classically known and, in a sense, provided a motivation for the entire theory. What we have in mind here are the classical theorems concerning the geometry of $W_{g-1}(C)$, that is, the geometry of Riemann's theta function. Of course, these results are more than mere exemplifications of the general ones of Chapter V. Rather, they are to be viewed as illustrations of how those general results can be used in the study of concrete geometrical problems. Our analysis will be carried out partly by means of classical methods, and partly in the language of Chapter IV.

In the first section we shall be mostly concerned with the proof of the beautiful Riemann singularity theorem, which gives a precise geometrical interpretation of the multiplicities of the singular points of the theta divisor.

In the second, following Kempf, we shall show that Riemann's singularity theorem, as well as a generalization of it, due to Kempf himself, are indeed formal consequences of the general theory exposed in Chapters II and IV.

In the third section we shall give Andreotti's proof of Torelli's theorem.

In the fourth, and final, section we shall present the theory of Andreotti and Mayer, whose goal is to try and geometrically solve the Schottky problem of giving explicit equations for the locus of Jacobians inside the generalized Siegel upper half-plane. Our presentation will make use of a theorem, whose proof is due to Mark Green, to the effect that the vector space of quadrics through a canonical curve is spanned by the tangent cones to the theta divisor at double points.

§1. The Riemann Singularity Theorem

Throughout this section, C will denote a fixed smooth curve of genus g. We fix a basis $\omega_1, \ldots, \omega_g$ for the space $H^0(C, K)$ of abelian differentials on C. As usual, we shall denote by

$$\phi_K: C \to \mathbb{P}^{g-1}$$

the canonical map, and we recall the abelian sum map

$$u\colon C_d \to J(C)$$

given, for $D = \sum_{i=1}^{d} p_i$, by

$$u(D) = \left(\sum_i \int_{p_0}^{p_i} \omega_1, \ldots, \sum_i \int_{p_0}^{p_i} \omega_g\right).$$

One of the most beautiful results concerning special divisors relates the dimension $r(D)$ of the linear system in which D moves to the multiplicity of the point $u(D)$ on $W_d(C)$, where d is the degree of the divisor in question. This theorem dates from Riemann, and, in its modern form, is due to Mayer, Lewittes, and Mumford in case the degree is $g - 1$, and to Kempf in general. We shall begin by proving the special case when the degree is $g - 1$, and then, in the next section, shall prove the general result.

We choose on C a base point that will be kept fixed throughout the discussion. This allows us to identify the Jacobian $J(C)$ with $\text{Pic}^d(C)$. Recall that, by Riemann's theorem, the divisors Θ and $W_{g-1}(C)$ are translates of one another. More precisely

$$W_{g-1}(C) = \Theta - \kappa,$$

where κ is Riemann's constant. The result we wish to prove is the following.

Riemann Singularity Theorem. *Let C be a smooth curve of genus g. Then for every effective divisor D of degree $g - 1$,*

$$\text{mult}_{u(D)+\kappa}(\Theta) = h^0(C, \mathcal{O}(D)).$$

Before proving the theorem we would like to make a couple of remarks. The first one concerns notation. It will be convenient to use, instead of u, the map

$$\pi\colon C_{g-1} \to J(C),$$

which we defined at the end of Section 5 of Chapter I by letting $\pi(D)$ be equal to $u(D) + \kappa$. As we explained there, one nice feature of π is that

$$-\pi(D) = \pi(K - D).$$

The Riemann singularity theorem can now be restated as saying that

$$\text{mult}_{\pi(D)}(\Theta) = h^0(C, \mathcal{O}(D)).$$

If $h^0(C, \mathcal{O}(D)) = r + 1$ and θ is Riemann's theta function, this amounts to saying that

$$\frac{\partial^s \theta}{\partial u_{\alpha_1} \cdots \partial u_{\alpha_s}}(\pi(D)) = 0 \quad \text{for} \quad s \leq r, \tag{1.1}$$

$$\frac{\partial^{r+1} \theta}{\partial u_{\alpha_0} \cdots \partial u_{\alpha_r}}(\pi(D)) \neq 0 \quad \text{for some multi-index } (\alpha_0, \ldots, \alpha_r). \tag{1.2}$$

We shall begin by proving the first relation.

The geometric idea is that the whole projective space $|D|$ is contracted to the point $\pi(D)$ on Θ, and we shall explore the consequences obtained by "differentiating" this fact. Choose any divisor $D_0 = \sum p_i \in |D|$ and let $p_i(t)$, $i = 1, \ldots, g - 1$, be holomorphic arcs in C such that $p_i(0) = p_i$. Set

$$D_t = \sum p_i(t).$$

Then, by Riemann's theorem,

$$\theta(\pi(D_t)) \equiv 0.$$

We shall successively differentiate this equation with respect to t and then evaluate at $t = 0$. The first step gives

$$\sum_\alpha \frac{\partial \theta}{\partial u_\alpha}(\pi(D)) \frac{\partial u_\alpha}{\partial t}(0) = 0.$$

In a neighborhood of p_i we choose a local coordinate z_i and write

$$\omega_\alpha = g_{\alpha i}(z_i)\, dz_i,$$
$$p_i(t) = z_i(t).$$

We then have

$$\sum_i \sum_\alpha \frac{\partial \theta}{\partial u_\alpha}(\pi(D)) g_{\alpha i}(p_i) z_i'(0) = 0,$$

and since the $z_i'(0)$ are arbitrary we infer that

$$\sum_\alpha \frac{\partial \theta}{\partial u_\alpha}(\pi(D)) g_{\alpha i}(p_i) = 0 \quad \text{for} \quad i = 1, \ldots, g - 1. \tag{1.3}$$

This says that, in $\mathbb{P}^{g-1}$ with homogeneous coordinates $[\omega_1, \ldots, \omega_g]$, the equation

$$\sum_{\alpha} \frac{\partial \theta}{\partial u_\alpha}(\pi(D))\omega_\alpha = 0$$

is satisfied by all points

$$\phi_K(p_i) = [\ldots, g_{\alpha i}(p_i), \ldots].$$

Equivalently, the hyperplane above contains all the points $\phi_K(p_i)$ on the canonical curve.

Thus, if $r(D) \geq 1$, so that there is a divisor $D_0 \in |D|$ containing any given point of C, the above linear relation holds on the entire canonical curve. Since this curve is non-degenerate we conclude that $\pi(D)$ is a singular point of Θ. Of course this we already knew, even in greater generality (cf. Corollary (4.5) in Chapter IV). All we have done here is to present the classical approach.

The remainder of the proof of (1.1) proceeds similarly. Assuming that $r(D) \geq 1$ we take the second derivative of $\theta(\pi(D_t)) \equiv 0$ and evaluate at $t = 0$, using the vanishing of the first derivatives, to obtain

$$\sum_{\alpha, \beta, i, j} \frac{\partial^2 \theta}{\partial u_\alpha \, \partial u_\beta}(\pi(D)) g_{\alpha i}(p_i) g_{\beta j}(p_j) z_i'(0) z_j'(0) = 0.$$

Since the $z_i'(0)$'s are arbitrary this implies that

$$\sum_{\alpha, \beta} \frac{\partial^2 \theta}{\partial u_\alpha \, \partial u_\beta}(\pi(D)) g_{\alpha i}(p_i) g_{\beta j}(p_j) = 0, \tag{1.4}$$

for any i and j. If $r(D) \geq 2$, we may find a divisor $D_0 \in |D|$ passing through any given pair of points on the curve. Then, because the canonical image of C is non-degenerate, the bilinear form

$$Q(\omega, \omega') = \sum_{\alpha, \beta} \frac{\partial^2 \theta}{\partial u_\alpha \, \partial u_\beta}(\pi(D)) \omega_\alpha \omega_\beta'$$

must vanish identically. More generally, if $r(D) \geq k$, the same kind of argument shows that the k-linear form

$$Q(\omega^1, \ldots, \omega^k) = \sum \frac{\partial^k \theta}{\partial u_{\alpha_1} \cdots \partial u_{\alpha_k}}(\pi(D)) \omega_{\alpha_1}^1 \cdots \omega_{\alpha_k}^k$$

vanishes identically. This concludes the proof of (1.1).

We shall now prove (1.2). The case when $r(D) = 0$ is a special case of (4.5) in Chapter IV, but it is immediate to give an elementary proof. Notice that, if $h^0(C, \mathcal{O}(D)) = 1$, then the rank of the Brill–Noether matrix at D is exactly $g - 1$, so that the map π has maximal rank at D. Since, by Abel's theorem, π is one-to-one at D, the only thing that might prevent Θ from being smooth at $\pi(D)$ would be if Θ were non-reduced. But this is forbidden by Riemann's theorem.

The proof of (1.2) uses the fact that, if D belongs to C_d, then

$$\mathbb{P}u_* T_D(C_d) = \overline{\phi_K(D)}. \tag{1.5}$$

This is easy. In local coordinates, u_* is represented by the Brill–Noether matrix, whose columns therefore span $u_* T_D(C_d)$. Recalling the explicit calculation of the Brill–Noether matrix in Section 1 of Chapter IV, the formula follows.

We will now prove (1.2), and we begin by establishing some notations. Consider a complete linear series $|D|$ of degree $g - 1$ and dimension r. Given a general effective divisor $E = p_1 + \cdots + p_{r+1}$ of degree $r + 1$, we set

$$E_i = p_1 + \cdots + \hat{p}_i + \cdots + p_{r+1};$$

we then have

(i) $|D + E| = |D| + E$ (i.e., $r(D + E) = r$),
(ii) $|D - E| = \varnothing$,
(iii) $|D - E_i| \neq \varnothing$ for $i = 1, \ldots, r + 1$

(the third condition is automatic; we state it only for reference). Consider the divisor

$$X_{p_i} = \{E' \in C_{r+1} : E' - p_i \geq 0\}$$

in C_{r+1}. This divisor is clearly isomorphic to C_r and hence smooth. We now consider the holomorphic mapping

$$v: C_{r+1} \to J(C)$$

defined by

$$v(E') = u(D + E - E') + \kappa.$$

The proof of (1.2) will consist in examining the divisor $v^*\Theta$ on C_{r+1}. We first claim that the support of this divisor is given by

$$\operatorname{supp}(v^*\Theta) = \left(\bigcup_i X_{p_i}\right) \cup \{E' : |D - E'| \neq \varnothing\}.$$

Indeed, by Riemann's theorem and condition (i) above, the support of $v^*\Theta$ consists of all $E' \in C_{r+1}$ such that, for some $D_0 \in |D|$,

$$D_0 + E - E' \geq 0.$$

Our claim follows from conditions (i) and (iii) above. In a moment we will show that X_{p_i} occurs with multiplicity one in $v^*\Theta$. Assuming this, (1.2) follows by observing that, by condition (ii) above, E does not belong to $\{E' : |D - E'| \neq \emptyset\}$ and consequently, in a neighborhood of $E \in C_{r+1}$, the divisor $v^*\Theta$ is equal to $X_{p_1} + \cdots + X_{p_{r+1}}$. Thus

$$\begin{aligned} r + 1 &= \operatorname{mult}_E(v^*\Theta) \\ &\geq \operatorname{mult}_{v(E)}(\Theta) \\ &= \operatorname{mult}_{\pi(D)}(\Theta), \end{aligned}$$

which proves (1.2).

It remains to prove that X_{p_i} occurs with multiplicity one in $v^*\Theta$. For this it suffices to show that, for a general point

$$E_i' = p_i + q_1 + \cdots + q_r$$

of X_{p_i}, the tangent spaces to Θ and $v(C_{r+1})$ intersect transversely at $v(E_i') = u(D + E - E_i') + \kappa$, i.e., that

$$v_* T_{E_i'}(C_{r+1}) \not\subset T_{v(E_i')}(\Theta).$$

To show this we will use (1.5). We first choose general points $q_1, \ldots, q_r$ so that $r(D - q_1 - \cdots - q_r) = 0$, and then for a unique $D_0 \in |D|$ we will have

$$D_0 = D_0' + q_1 + \cdots + q_r, \qquad D_0' \geq 0.$$

Having chosen the q_α's, we then choose the p_i's to satisfy

$$\phi_K(p_i) \notin \overline{\phi_K(D_0' + E_i)}$$

as well as the conditions (i), (ii), (iii) above. It is clear that all these conditions are satisfied in a non-empty Zariski open subset of C_{r+1}. By (1.5) we have that

$$\mathbb{P}T_{v(E_i')}(\Theta) = \overline{\phi_K(D_0' + E_i)}.$$

On the other hand,

$$\begin{aligned} \mathbb{P}v_* T_{E_i'}(C_{r+1}) &= \mathbb{P}u_* T_{E_i'}(C_{r+1}) \\ &= \overline{\phi_K(E_i')}. \end{aligned}$$

But then

$$\phi_K(p_i) \notin \mathbb{P}T_{v(E_i')}(\Theta),$$
$$\phi_K(p_i) \in \mathbb{P}v_* T_{E_i'}(C_{r+1}),$$

proving that $v_* T_{E_i'}(C_{r+1})$ is not contained in $T_{v(E_i')}(\Theta)$. This concludes the proof of the Riemann singularity theorem.

The Riemann singularity theorem gives us information about the first non-vanishing term in the power series expansion of $\theta(u)$ around a point $\pi(D) \in \Theta$. Specifically, if $D \in C_{g-1}$ and $r(D) = r$, then $\theta(u + \pi(D)) = V_{r+1}(u) + V_{r+2}(u) + \cdots$ where $V_m(u)$ is a homogeneous polynomial of degree m in $u \in \mathbb{C}^g$ and $V_{r+1} \neq 0$. We will now give a result concerning the terms

$$V_{r+1}, \ldots, V_{2r+1}.$$

To explain the theorem we write

$$\frac{1}{m!} \frac{\partial^m \theta}{\partial u_{\alpha_1} \cdots \partial u_{\alpha_m}}(\pi(D)) = c_{m; \alpha_1, \ldots, \alpha_m}$$

and define a symmetric, m-linear form $\tilde{V}_m$ on $\mathbb{C}^g$ by

$$\tilde{V}_m(u^1, \ldots, u^m) = \sum_{\alpha_1, \ldots, \alpha_m} c_{m; \alpha_1, \ldots, \alpha_m} u^1_{\alpha_1} \cdots u^m_{\alpha_m}.$$

Then $\tilde{V}_m$ is just the *polarization* of V_m; i.e.,

$$V_m(u) = \tilde{V}_m(u, \ldots, u).$$

We observe that V_{r+1}, viewed as a polynomial on $T_{\pi(D)}(J(C))$, is well defined up to scalar multiples (here, well defined means allowing arbitrary local holomorphic changes of coordinates). In general, V_m is well defined, up to scalar multiples, modulo the ideal in $\mathbb{C}[u_1, \ldots, u_g]$ generated by $V_{r+1}, \ldots, V_{m-1}$. By abuse of language we will refer to "the hypersurface V_m in $\mathbb{P}^{g-1}$," by which we mean the hypersurface in $\mathbb{P}H^0(C, K_C)^*$ defined by the form V_m. By our remarks above, the intersections

$$V_{r+1} \cap \cdots \cap V_m$$

are all intrinsically defined subvarieties of canonical space $\mathbb{P}^{g-1}$.

For our last piece of terminology we recall that the *k-secant variety* to a curve $\Gamma \subset \mathbb{P}^n$ is the closure of the union of the linear subspaces in $\mathbb{P}^n$ spanned by a k-tuple of distinct points of Γ. Thus, the 1-secant variety is the curve itself, and in general the k-secant variety contains the locus of osculating $(k-1)$-planes to the curve Γ.

(1.6) Theorem. *Assume that $|D|$ is a base-point-free linear system of degree $g-1$ and dimension r. Then:*

(i) *For every divisor $\Delta \in |D|$, V_{r+1} contains the $(g-r-2)$-plane $\overline{\phi_K(\Delta)}$. In particular, V_{r+1} contains the r-secant variety of $\phi_K(C)$.*
(ii) *For $s \geq 0$, V_{r+2+s} contains the $(r-s)$-secant variety of $\phi_K(C)$. In particular, $V_{r+1}, \ldots, V_{2r+1}$ all contain $\phi_K(C)$.*
(iii) *V_{r+2} is zero modulo V_{r+1} if and only if D is semi-canonical.*

Informally, (ii) states that "$\phi_K(C)$ is *very much* contained in Θ."

For an application we consider a non-hyperelliptic curve C of genus four. As we know, the canonical curve

$$\phi_K(C) = Q \cap S$$

is the complete intersection of a quadric Q and cubic S in $\mathbb{P}^3$. In general, the quadric is smooth and its two rulings cut out the two distinct g_3^1's on C; Q is singular exactly when C has a semi-canonical g_3^1.

Let $|D|$ be a g_3^1 on C. By the Riemann singularity theorem and part (i) of the theorem, the tangent cone to Θ at $\pi(D)$ is the quadric Q. From (iii) we then have the

Corollary. *Let C be a genus four canonical curve having no semi-canonical pencil, $\theta(u)$ its theta function, and $w \in \mathbb{C}^4$ a point at which $\theta(w) = \partial\theta/\partial u_\alpha(w) = 0$ $(\alpha = 1, \ldots, 4)$. Writing*

$$\theta(u + w) = V_2(w) + V_3(w) + \cdots,$$

the curve C is the locus in $\mathbb{P}^3$ defined by $V_2(u) = V_3(u) = 0$.

We observe that this corollary gives, in this case, an explicit reconstruction of the curve from its polarized Jacobian.

The proof of theorem (1.6) will occupy the remainder of this section. Being a generalization of the Riemann singularity theorem, the proof will be an extension of the argument used there.

If $\Delta = p_1 + \cdots + p_{g-1} \in C_{g-1}$, with the points p_i being distinct, if t_i is a local coordinate centered at p_i, and $p_i(t_i)$ stands for the point corresponding to the value t_i of the coordinate, we set

$$\pi(t_1, \ldots, t_{g-1}) = \pi(p_1(t_1) + \cdots + p_{g-1}(t_{g-1})),$$

$$\omega(p_i) = \frac{\partial \pi}{\partial t_i}(0),$$

$$\omega^{(k)}(p_i) = \frac{\partial^{k+1}\pi}{\partial t_i^{k+1}}(0).$$

It has already been observed that $\omega(p_i) \in \mathbb{C}^g - \{0\}$ projects onto $\phi_K(p_i) \in \mathbb{P}^{g-1}$. More generally, $\omega(p_i), \omega'(p_i), \ldots, \omega^{(k)}(p_i)$ span in $\mathbb{C}^g$ a linear space that projects onto the osculating space $\overline{\phi_K((k+1)p_i)}$ to $\phi_K(C)$ at $\phi_K(p_i)$.

We now prove part (i) of the theorem. By Riemann's theorem, $\theta(\pi(t)) \equiv 0$. Taking derivatives, for every collection $k_1, \ldots, k_{g-1}$ with $\sum_i k_i = r + 1$ we have

$$\begin{aligned} 0 &= \frac{\partial^{r+1}\theta(\pi(t))}{\partial t_1^{k_1} \cdots \partial t_{g-1}^{k_{g-1}}} \\ &= \tilde{V}_{r+1}(\underbrace{\omega(p_1), \ldots, \omega(p_1)}_{k_1}, \ldots, \underbrace{\omega(p_{g-1}), \ldots, \omega(p_{g-1})}_{k_{g-1}}), \end{aligned}$$

since all lower-order derivatives of θ are zero, by (1.1). The restriction of the multilinear form $\tilde{V}_{r+1}$ to the $(g - r - 2)$-plane $\overline{\phi_K(\Delta)}$ is thus identically zero, i.e.,

$$\overline{\phi_K(\Delta)} \subset V_{r+1}.$$

Since every r-secant linear space to $\phi_K(C)$ is contained in $\overline{\phi_K(\Delta)}$ for some $\Delta \in |D|$, this proves (i).

Part (ii) will be established by a similar argument together with the following

Lemma. *Let $x(t) = (x_1(t), \ldots, x_n(t)) \in \mathbb{C}^n - \{0\}$ define an analytic arc in $\mathbb{P}^{n-1}$, and let $\tilde{Q}(v_1, \ldots, v_l)$ be a symmetric l-linear form with $Q(v) = \tilde{Q}(v, \ldots, v)$ the corresponding polynomial. Suppose that Q vanishes on the r-secant plane locus of the arc, i.e.,*

$$\tilde{Q}(x(t_{j_1}), \ldots, x(t_{j_l})) = 0 \tag{1.7}$$

for all $t_1, \ldots, t_r$ and all l-tuples $j_1, \ldots, j_l$ in $\{1, \ldots, r\}$. Then for any $t_1, \ldots, t_{r-s}$, $k_1, \ldots, k_l$ with $\sum k_i \leq s + 1$, and $i_1, \ldots, i_l \in \{1, \ldots, r - s\}$, we have

$$\tilde{Q}(x^{(k_1)}(t_{i_1}), \ldots, x^{(k_l)}(t_{i_l})) = 0.$$

Informally, if the number N of derivatives taken plus the number P of points of the arc involved in the argument of $\tilde{Q}$ satisfy $N + P \leq r + 1$, then $\tilde{Q}$ is zero. Observe that this is one better than the obvious: if $\sum k_i \leq s$ then the argument of $\tilde{Q}$ would lie in a r-secant plane to the arc, and by assumption $\tilde{Q}$ would vanish.

Proof of the Lemma. We use induction on s, the chain rule

$$\frac{\partial \tilde{Q}}{\partial t}(v_1(t), \ldots, v_l(t)) = \sum_i \tilde{Q}(v_1(t), \ldots, v_i'(t), \ldots, v_l(t)),$$

and the symmetry of $\tilde{Q}$. Differentiating (1.7) with respect to t_{i_1} gives

$$\tilde{Q}(x'(t_{j_1}), x(t_{j_2}), \ldots, x(t_{j_l})) = 0$$

for all $t_1, \ldots, t_r$ and $j_1, \ldots, j_l \in \{1, \ldots, r\}$, which is the lemma for $s = 0$.

If $t_1, \ldots, t_{r-s}$, $k_1, \ldots, k_l$, and $i_1, \ldots, i_l \in \{1, \ldots, r - s\}$ are as in the statement of the lemma, and if $k_l > 0$, then by induction

$$\tilde{Q}(x^{(k_1)}(t_{i_1}), \ldots, x^{(k_{l-1})}(t_{i_{l-1}}), x^{(k_l - 1)}(t)) \equiv 0$$

for any t. Differentiating this equation with respect to t and then setting $t = t_{i_l}$ yields the result. Q.E.D.

We now turn to the proof of (ii). In this and in the following computations use will be made of the chain rule, which as we now explain simplifies considerably since Θ is a variety of translation type. Namely, recalling that

$$\begin{aligned}\pi(t_1, \ldots, t_{g-1}) &= \pi(p_1(t_1) + \cdots + p_{g-1}(t_{g-1}))\\ &= u(p_1(t_1)) + \cdots + u(p_{g-1}(t_{g-1})) + \kappa,\end{aligned}$$

it follows that

$$\frac{\partial^k \pi(t_1, \ldots, t_{g-1})}{\partial t_{i_1} \cdots \partial t_{i_k}} = 0 \quad \text{unless} \quad i_1 = \cdots = i_k,$$

so that in applying the chain rule all crossed mixed partials are zero.

Thus, for integers $l_1, \ldots, l_r$ such that $\sum l_i = r + 2$, differentiation of $\theta(\pi(t)) \equiv 0$ gives

$$\begin{aligned}0 = \frac{\partial^{r+2}\theta(\pi(t))}{\partial t_1^{l_1} \cdots \partial t_r^{l_r}}(0) = \sum_{i=1}^{r} \binom{l_i}{2} \tilde{V}_{r+1}(&\underbrace{\omega(p_1), \ldots, \omega(p_1)}_{l_1}, \ldots,\\ &\underbrace{\omega(p_i), \ldots, \omega(p_i)}_{l_i - 1}, \ldots, \omega'(p_i), \ldots, \underbrace{\omega(p_r), \ldots, \omega(p_r)}_{l_r})\\ &+ \tilde{V}_{r+2}(\underbrace{\omega(p_1), \ldots, \omega(p_1)}_{l_1}, \ldots, \underbrace{\omega(p_r), \ldots, \omega(p_r)}_{l_r}),\end{aligned}$$

again since the lower-order derivatives of $\theta(u)$ are zero at $u = \pi(\Delta)$. By the case $s = 0$ of the lemma, each term in the first sum is zero. Since the points $p_1, \ldots, p_r$ may be chosen arbitrarily we conclude that V_{r+2} contains the r-secant variety to $\phi_K(C)$, which is the case $s = 0$ of (ii).

We shall now do the case $s = 1$; the computation will make the general pattern clear. For integers $l_1, \ldots, l_{r-1}$ with $\sum l_i = r + 3$, differentation of $\theta(\pi(t)) \equiv 0$ gives

$$0 = \frac{\partial^{r+3}\theta(\pi(t))}{\partial t_1^{l_1} \cdots \partial t_{r-1}^{l_{r-1}}} = A_1 + A_2 + A_3 + A_4 + \tilde{V}_{r+3}(\underbrace{\omega(p_1), \ldots, \omega(p_1)}_{l_1}, \ldots, \underbrace{\omega(p_{r-1}), \ldots, \omega(p_{r-1})}_{l_{r-1}}),$$

where

$$A_1 = \sum_{i \neq j} \binom{l_i}{2}\binom{l_j}{2} \tilde{V}_{r+1}(\underbrace{\omega(p_1), \ldots, \omega(p_1)}_{l_1}, \ldots, \underbrace{\omega(p_i), \ldots, \omega(p_i), \omega'(p_i)}_{l_i - 1}, \ldots, \underbrace{\omega(p_j), \ldots, \omega(p_j), \omega'(p_j)}_{l_j - 1}, \ldots)$$

$$A_2 = \sum_i 3\binom{l_i}{4} \tilde{V}_{r+1}(\underbrace{\omega(p_1), \ldots, \omega(p_1)}_{l_1}, \ldots, \underbrace{\omega(p_i), \ldots, \omega'(p_i), \omega'(p_i)}_{l_i - 2}, \ldots, \underbrace{\omega(p_{r-1}), \ldots, \omega(p_{r-1})}_{l_{r-1}})$$

$$A_3 = \sum_i \binom{l_i}{3} \tilde{V}_{r+1}(\underbrace{\omega(p_1), \ldots, \omega(p_1)}_{l_1}, \ldots, \underbrace{\omega(p_i), \ldots, \omega(p_i), \omega''(p_i)}_{l_i - 2}, \ldots, \underbrace{\omega(p_{r-1}), \ldots, \omega(p_{r-1})}_{l_{r-1}})$$

$$A_4 = \sum_i \binom{l_i}{2} \tilde{V}_{r+2}(\underbrace{\omega(p_1), \ldots, \omega(p_1)}_{l_1}, \ldots, \underbrace{\omega(p_i), \ldots, \omega(p_i), \omega'(p_i)}_{l_i - 1}, \ldots, \underbrace{\omega(p_{r-1}), \ldots, \omega(p_{r-1})}_{l_{r-1}}).$$

By the lemma in the case $s = 1$, each term in the sums A_1, A_2, and A_3 is zero. Moreover, each term in the sum A_4 is zero because of the case $s = 0$ of (ii) and the fact that $\omega(p_1), \ldots, \omega(p_{r-1}), \omega'(p_i)$ lie in the r-secant space

$$\overline{\phi_K(p_1 + \cdots + 2p_i + \cdots + p_{r-1})}$$

to $\phi_K(C)$. This shows that V_{r+3} contains $\overline{\phi_K(p_1 + \cdots + p_{r-1})}$, and since the p_i were arbitrary this gives the case $s = 1$ of (i) in the theorem.

We now turn to part (iii). One direction is elementary from symmetry considerations: by Exercise 44 of Appendix B, if D is semi-canonical then the function

$$e^{l(u)}\theta(u + \pi(D)),$$

where $l(u)$ is a suitable linear form on $\mathbb{C}^g$, is either odd or even. In either case, if $V_0 = \cdots = V_r = 0$ and $V_{r+1} \neq 0$, then $V_{r+2} = 0$. As in the proof of (1.2), the other half of (iii) involves new considerations, but the idea is simple. In parts (i) and (ii) we proved that both V_{r+1} and V_{r+2} contain the r-secant variety but, as we proved in (i), the distinguishing feature of V_{r+1} is that

$$V_{r+1} \supset \bigcup_{\Delta \in |D|} \overline{\phi_K(\Delta)}.$$

By Clifford's theorem, except in the case when C is hyperelliptic and $|D|$ is multiple of the g_2^1, this is a stronger statement than just saying that V_{r+1} contains the r-secant variety. Assuming that $2D \notin |K|$ (which in particular rules out the hyperelliptic case), to prove that $V_{r+2} \not\equiv 0$ modulo V_{r+1}, it will be sufficient to show that

$$V_{r+2} \not\supset \overline{\phi_K(D)}. \tag{1.8}$$

Since D is base-point-free, we may assume that $D = p_1 + \cdots + p_{g-1}$, where the p_i are distinct. Since $2D \notin |K|$ we have

$$\overline{\phi_K(2D)} = \mathbb{P}^{g-1};$$

i.e., the tangent lines to the canonical curve at the points $\phi_K(p_i)$ span $\mathbb{P}^{g-1}$. Renumbering if necessary, we may assume that

$$\overline{\phi_K(D + p_1 + \cdots + p_{r+1})} = \mathbb{P}^{g-1}.$$

By the geometric form of the Riemann–Roch theorem this is equivalent to

$$|D + p_1 + \cdots + p_{r+1}| = |D| + p_1 + \cdots + p_{r+1}. \tag{1.9}$$

We define

$$v: C_{r+1} \to J(C)$$

by

$$v(E) = u(D + p_1 + \cdots + p_{r+1}) - u(E).$$

As in the proof of (1.2), the argument for (iii) will consist in examining the divisor $v^*(\Theta)$.

We set

$$X_{p_i} = C_r + p_i \subset C_{r+1},$$
$$\Sigma = \{E \in C_{r+1} : r(D - E) \geq 0\},$$

and will prove the following

Lemma. (i) *Σ is smooth at $p_1 + \cdots + p_{r+1}$.*
(ii) *in a neighborhood of $p_1 + \cdots + p_{r+1}$, we have*

$$v^*(\Theta) = \sum_{i=1}^{r+1} X_{p_i} + \Sigma;$$

moreover, Σ and X_{p_i} intersect transversely.

Assuming the lemma we may complete the proof of (iii) as follows. Using the notations from the proof of (ii), we set

$$v(t) = u(D + p_1 + \cdots + p_{r+1} - p_1(t_1) - \cdots - p_{r+1}(t_{r+1})).$$

Then the function $\theta(v(t))$ is defined in a neighborhood of the origin in $\mathbb{C}^{r+1}$, and

$$\mu_0 \theta(v(t)) = r + 2.$$

More precisely, the lemma tells us that, up to a change of coordinates, the tangent cone to the divisor of $\theta(v(t))$ consists of the union of the coordinate hyperplanes in $\mathbb{C}^{r+1}$ together with a hyperplane not containing any of the coordinate axes. In particular,

$$\frac{\partial^{r+2}\theta(v(t))}{\partial t_1^2\, \partial t_2 \cdots \partial t_{r+1}}(0) \neq 0.$$

By the computation in the proof of (ii),

$$\frac{\partial^{r+2}\theta(v(t))}{\partial t_1^2\, \partial t_2 \cdots \partial t_{r+1}}(0) = \tilde{V}_{r+1}(-\omega'(p_1), -\omega(p_2), \ldots, -\omega(p_{r+1}))$$
$$+ \tilde{V}_{r+2}(-\omega(p_1), -\omega(p_1), -\omega(p_2), \ldots, -\omega(p_{r+1})).$$

Thus

$$\tilde{V}_{r+1}(\omega'(p_1), \omega(p_2), \ldots, \omega(p_{r+1}))$$
$$- \tilde{V}_{r+2}(\omega(p_1), \omega(p_1), \omega(p_2), \ldots, \omega(p_{r+1})) \neq 0.$$

On the other hand, again by the computation in the proof of (ii), differentiation of $\theta(\pi(t)) \equiv 0$ gives

$$\begin{aligned} 0 &= \frac{\partial^{r+2}\theta(\pi(t))}{\partial t_1^2\, \partial t_2 \cdots \partial t_{r+1}}(0) \\ &= \tilde{V}_{r+1}(\omega'(p_1), \omega(p_2), \ldots, \omega(p_{r+1})) \\ &\quad + \tilde{V}_{r+2}(\omega(p_1), \omega(p_1), \omega(p_2), \ldots, \omega(p_{r+1})). \end{aligned}$$

Subtracting the last two relations we obtain

$$\tilde{V}_{r+2}(\omega(p_1), \omega(p_1), \omega(p_2), \ldots, \omega(p_{r+1})) \neq 0,$$

i.e., the hypersurface V_{r+2} does not contain the r-secant $(g - r - 2)$-plane $\overline{\phi_K(D)}$, so that (1.8) holds. The proof of (iii) will be complete once we establish the lemma.

Proof of Part (i) *of the Lemma.* Let $s_1, \ldots, s_{r+1}$ be a basis for $H^0(C, \mathcal{O}(D))$ and, by choosing local trivializations of $\mathcal{O}(D)$, represent s_j by a holomorphic function $f_j(t_i)$ in a neighborhood of p_i. Setting $f(t_i) = (f_1(t_i), \ldots, f_{r+1}(t_i)) \in \mathbb{C}^{r+1}$, Σ has local defining function

$$\begin{aligned} \sigma(t_1, \ldots, t_{r+1}) &= \det(f_j(t_i)) \\ &= f(t_1) \wedge \cdots \wedge f(t_{r+1}). \end{aligned}$$

Thus, for example,

$$\frac{\partial \sigma}{\partial t_1}(t_1, \ldots, t_{r+1}) = f'(t_1) \wedge f(t_2) \wedge \cdots \wedge f(t_{r+1}).$$

Since the unique section of $\mathcal{O}(D)$ vanishing at $p_2, \ldots, p_{r+1}$ vanishes to order exactly one at p_1, this is non-zero at the origin. Therefore, Σ is smooth at $p_1 + \cdots + p_{r+1}$.

Proof of Part (ii) *of the Lemma.* This is very much like the proof of (1.2). First, by (1.9) we have

$$\operatorname{supp} v^*(\Theta) = \left(\bigcup_i X_{p_i}\right) \cup \Sigma.$$

If $E + p_i$ is a general point of X_{p_i},

$$r(D + p_1 + \cdots + \hat{p}_i + \cdots + p_{r+1} - E) = 0,$$

so that Θ is smooth at $v(E + p_i)$. Moreover, if $D_0 + E$ is the unique divisor in $|D|$ containing E, the tangent plane to Θ at $v(E + p_i)$ is given by

$$\mathbb{P}T_{v(E+p_i)}(\Theta) = \overline{\phi_K(D_0 + p_1 + \cdots + \hat{p}_i + \cdots + p_{r+1})}.$$

Since, by hypothesis, this does not contain $\phi_K(p_i)$, whereas

$$v_* \mathbb{P}T_{E+p_i}(C_{r+1}) = \overline{\phi_K(E + p_i)}$$

does, we conclude that Θ meet $v(C_{r+1})$ transversely at $v(E + p_i)$; in particular

$$\text{mult}_{X_{p_i}} v^*(\Theta) = 1.$$

Similarly, if $E \in \Sigma$ is near $p_1 + \cdots + p_{r+1}$ we have

$$r(D - E) = 0.$$

If $D_0 + E$ is the unique divisor in $|D|$ containing E, then Θ is smooth at $v(E)$ with tangent plane

$$\mathbb{P}T_{v(E)}(\Theta) = \overline{\phi_K(D_0 + p_1 + \cdots + p_{r+1})}.$$

On the other hand,

$$v_* \mathbb{P}T_E(C_{r+1}) = \overline{\phi_K(E)}.$$

Since, by (1.9). D_0, E, and $p_1, \ldots, p_{r+1}$ do not all lie in a hyperplane, near $p_1 + \cdots + p_{r+1}$

$$\text{mult}_\Sigma v^*(\Theta) = 1.$$

This completes the proof of the theorem.

§2. Kempf's Generalization of the Riemann Singularity Theorem

In this section we momentarily interrupt our exposition of the classical geometric theory of the Riemann theta function to briefly talk about an important generalization of the Riemann singularity theorem, due to Kempf. The result is an exact analogue for $W_d(C)$ of what Riemann did for the theta divisor, or, what is the same, for $W_{g-1}(C)$. One of the main difficulties that Kempf had to overcome is that $W_d(C)$ is no longer a divisor, so that the expression of its tangent cones cannot be given by taking derivatives of a single function. The proof will be a fairly easy application of the machinery developed in Chapters II and IV. Actually, we will find it convenient to prove a singularity theorem for $W^r_d(C)$ which reduces to Kempf's when $r = 0$.

We begin by recalling the content of Proposition (4.4) in Chapter IV. Suppose C is a smooth genus g curve and let d, r be non-negative integers such that $d < g + r$. Assume that $G^r_d(C)$ is smooth of the "correct" dimension

$$\rho = g - (r + 1)(g - d + r).$$

Then $W^r_d(C)$ is a Cohen–Macaulay, reduced, normal variety and its singular locus is precisely $W^{r+1}_d(C)$.

The problem now is to describe the tangent cone, $\mathcal{T}_L(W^r_d(C))$, to $W^r_d(C)$ at a point L, and to compute its degree in terms of $h^0(C, L)$. Of course, this tangent cone will be viewed as a subvariety of $H^0(C, K)^*$, which is the Zariski tangent space to $\mathrm{Pic}^d(C)$ at L. In order to state the result, we have to choose bases $x_1, \ldots, x_l$, $y_1, \ldots, y_i$ for $H^0(C, L)$ and $H^0(C, KL^{-1})$, respectively. For simplicity, we shall write $x_s y_t$ to denote the image of $x_s \otimes y_t$ under the cup-product mapping

$$\mu_0 \colon H^0(C, L) \otimes H^0(C, KL^{-1}) \to H^0(C, K).$$

As usual, we shall denote by

$$c \colon G^r_d(C) \to \mathrm{Pic}^d(C)$$

the natural mapping. The result we wish to prove is the following.

(2.1) Theorem. *Let L be a point of $W^r_d(C)$. Assume that $G^r_d(C)$ is smooth of the "correct" dimension*

$$\rho = g - (r + 1)(g - d + r)$$

in a neighborhood of $c^{-1}(L)$, and that $d < g + r$. Then:

(i) *$\mathcal{T}_L(W^r_d(C))$ is Cohen–Macaulay, reduced, normal, and, set-theoretically*

$$\mathcal{T}_L(W^r_d(C)) = \bigcup_{W \in G(r+1, H^0(C, L))} \mu_0(W \otimes H^0(C, KL^{-1}))^{\perp} \subset H^0(C, K)^*.$$

(ii) *The ideal of $\mathcal{T}_L(W^r_d(C))$, as a subvariety of $H^0(C, K)^*$, is generated by the $(l - r) \times (l - r)$ minors of the matrix*

$$(x_s y_t)_{\substack{s=1,\ldots,l \\ t=1,\ldots,i}}.$$

(iii) *The degree of* $\mathcal{T}_L(W^r_d(C))$ *is*

$$\prod_{h=0}^{r} \frac{(i+h)!\,h!}{(l-1-r+h)!\,(i-l+r+1+h)!}.$$

Since $G^0_d(C) = C_d$ is smooth of dimension $d = \rho(g, 0, d)$, when $r = 0$ we get, as a special case,

Kempf's Singularity Theorem. *Let L be a point of $W_d(C)$ with $d < g$. Then $\mathcal{T}_L(W_d(C))$ is a Cohen–Macaulay, reduced, normal subvariety of $H^0(C, K)^*$ of degree*

$$\binom{i}{l-1} = \binom{g-d+l-1}{l-1},$$

whose ideal is generated by the maximal minors of the matrix $(x_s y_t)_{\substack{s=1,\dots,l,\\ t=1,\dots,i}}$ *and is supported on*

$$\bigcup_{W \in \mathbb{P}H^0(C,L)} \mu_0(W \otimes H^0(C, KL^{-1}))^{\perp}.$$

From a geometrical point of view it is interesting to look at the projectivization of the tangent cones to $W^r_d(C)$. These sit naturally in the space of the canonical curve. Under the hypotheses of Theorem (2.1), we get

$$\mathbb{P}\mathcal{T}_L(W^r_d(C)) = \bigcup_{\mathcal{D} \text{ is a } g^r_d \text{ contained in } |L|} \left(\bigcap_{D \in \mathcal{D}} \overline{\phi_K(D)} \right).$$

In Kempf's case, we find:

$$\mathbb{P}\mathcal{T}_L(W_d(C)) = \bigcup_{D \in |L|} \overline{\phi_K(D)}.$$

We shall now prove Theorem (2.1). To describe the tangent cone to $W^r_d(C)$ at L we shall use Lemma (1.3) in Chapter II. This requires computing the normal bundle to $c^{-1}(L)$, which we shall denote by N. The answer is provided by Proposition (4.1) in Chapter IV. By the smoothness assumption on $G^r_d(C)$, we find that, for any $(r+1)$-dimensional subspace W of $H^0(C, L)$, the restriction of μ_0 to $W \otimes H^0(C, KL^{-1})$ is injective. Since the normal space to

$$c^{-1}(L) = G(r+1, H^0(C, L))$$

at W is nothing but the image of

$$c_*\colon T_W(G^r_d(C)) \to T_L(\operatorname{Pic}^d(C)),$$

we also find that

$$\begin{aligned} N &= \{(W, v) \colon v \perp \mu_0(W \otimes H^0(C, KL^{-1}))\} \\ &\subset G(r+1, H^0(C, L)) \times H^0(C, K)^*, \end{aligned}$$

and that the differential

$$c_* \colon N \to T_L(\mathrm{Pic}^d(C))$$

is just projection onto the second factor. The rest of the proof is reduced to a linear algebra result. In order to state it, we let A, B, E be complex vector spaces of dimensions a, b, e, and let

$$\phi \colon A \otimes B \to E$$

be a linear map. We choose bases $x_1, \ldots, x_a$ and $y_1, \ldots, y_b$ of A and B, respectively, and write, for simplicity

$$x_i y_j = \phi(x_i \otimes y_j).$$

Lemma. *Let w be an integer such that $1 \leq w \leq a$. Assume that $a < b + w$ and that, for each W in $G(w, A)$, the restriction of ϕ to $W \otimes B$ is injective. Set*

$$\begin{aligned} I &= \{(W, \gamma) \colon \gamma \perp \phi(W \otimes B)\} \\ &\subset G(w, A) \times E^*. \end{aligned}$$

Let

$$p \colon G(w, A) \times E^* \to E^*$$

be the projection. Then:

(i) *$p(I)$ is Cohen–Macaulay, reduced and normal.*
(ii) *The ideal of $p(I)$ is generated by the $(a - w + 1) \times (a - w + 1)$ minors of the matrix $(x_i y_j)_{\substack{i=1,\ldots,a \\ j=1,\ldots,b}}$.*
(iii) *The degree of $p(I)$ is*

$$\prod_{h=0}^{w-1} \frac{(b+h)!\, h!}{(a-w+h)!\, (b-a+w+h)!}$$

(iv) *p maps I birationally onto $p(I)$.*

Before proving the lemma, we deduce from it the theorem. In our situation, A is $H^0(C, L)$, B is $H^0(C, KL^{-1})$, E is $H^0(C, K)$, ϕ is μ_0, and w is $r + 1$. The hypothesis that $a < b + w$ is an exact translation of $d < g + r$, by the Riemann–Roch theorem. Also, I is just N, and the restriction of p to I is c_*.

To apply Lemma (1.3) of Chapter II, we first need to know that N maps birationally to $c_*(N)$, which is part (iv) of the lemma, and secondly that $c_*(N)$ is normal, which follows by part (i). Thus we conclude that $\mathscr{T}_L(W_d^r(C))$ is reduced and supported on

$$c_*(N) = \bigcup_{W \in G(r+1, H^0(C, L))} \mu_0(W \otimes H^0(C, KL^{-1}))^{\perp}.$$

That it is Cohen–Macaulay and normal follows from part (i) of the lemma, while parts (ii) and (iii) of the theorem are just translations of the corresponding parts of the lemma.

Proof of the Lemma. We begin by noticing that the injectivity hypothesis on ϕ implies that I is a rank $(e - bw)$ vector bundle over $G(w, A)$, and hence is smooth of dimension $w(a - w) + e - bw$. We let J be the subvariety of E^* whose ideal is generated by the $(a - w + 1) \times (a - w + 1)$ minors of $(x_i y_j)$. To say that γ belongs to J means that the matrix $(x_i y_j(\gamma))$ has rank at most $a - w$, that is, that the space of vectors $(\xi_1, \ldots, \xi_a)$ such that $\sum \xi_i x_i y_j(\gamma) = 0$ has dimension at least w. This amounts to saying that there is a w-dimensional vector subspace W of A such that γ annihilates $\phi(W \otimes B)$, that is, that γ belongs to $p(I)$. Thus the support of J is $p(I)$. Now the dimension of $p(I)$, and hence of J, does not exceed

$$\dim I = e - w(b + w - a).$$

On the other hand, since J is determinantal and its "expected" dimension is also equal to

$$e - (b - (a - w))(a - (a - w)) = \dim I,$$

it turns out that J is a determinantal variety of the "correct" dimension, and hence, in particular, is Cohen–Macaulay. Thus, to prove parts (i) and (ii) of the lemma it will be enough to show that J is generically reduced and normal. By Serre's criterion (cf. Proposition (5.4) in Chapter II) this will follow if we can show that J is smooth outside a codimension two subset. We define a map

$$\alpha: E^* \to M$$

into the variety of $b \times a$ matrices by letting $\alpha(\gamma)$ be the transpose of the matrix $(x_i y_j(\gamma))$. Thus J is the pull-back, via α, of the variety M_{a-w} of all $b \times a$ matrices of rank at most $a - w$. We also let $\tilde{M}_{a-w}$ be the canonical desingularization of M_{a-w} defined in Chapter II, Section 2. In other words, $\tilde{M}_{a-w}$ is the variety of couples $((m_{ij}), V)$, where m_{ij} is a $b \times a$ matrix and V is a w-dimensional vector subspace of $\mathbb{C}^a$ satisfying the "equations"

$$\sum_j m_{ij} \xi_j = 0$$

for every i and every $(\xi_1, \ldots, \xi_a)$ in V. It is clear from the definitions of I and α that I is just the fiber product of $\tilde{M}_{a-w}$ and E^* over M. Since I is smooth and since M_{a-w} and $\tilde{M}_{a-w}$ are isomorphic off M_{a-w-1}, J must be smooth off $\alpha^{-1}(M_{a-w-1})$. But now, to say that γ belongs to $\alpha^{-1}(M_{a-w-1})$ means that there is a vector subspace W' of A of dimension strictly greater than w such that γ annihilates $\phi(W' \otimes B)$. Thus $p^{-1}(\gamma)$ has dimension at least w, so, if the codimension of $\alpha^{-1}(M_{a-w-1})$ in J is not two or more,

$$\begin{aligned} \dim I &\geq \dim \alpha^{-1}(M_{a-w-1}) + w \\ &\geq \dim I. \end{aligned}$$

This implies that $J = \alpha^{-1}(M_{a-w-1})$; but then the above inequalities cannot hold, since $w \geq 1$ in any case. The only way out is for $\alpha^{-1}(M_{a-w-1})$ to have codimension two or more. This finishes the proof of (i) and (ii).

To prove (iv), suppose two points of I project to the same one in $p(I)$. In other words suppose there are γ in E^* and distinct elements W, W' of $G(w, A)$ such that

$$\gamma \perp \phi(W \otimes B),$$

$$\gamma \perp \phi(W' \otimes B).$$

Thus γ annihilates $\phi((W + W') \otimes B)$, and therefore belongs to $\alpha^{-1}(M_{a-w-1})$. Since we just showed that $\alpha^{-1}(M_{a-w-1})$ is strictly smaller than $p(I)$, this proves that I maps birationally onto $p(I)$.

As for part (iii) of the lemma, it follows from formula (5.1) in Chapter II giving the degree of a generic determinantal variety and from the remark that $p(I)$ is the pull-back of M_{a-w} via a linear map. This concludes the proof the lemma and of Theorem (2.1).

The following example is a nice illustration of the Riemann singularity theorem for a g^2_{g-1}. Suppose that $C \subset \mathbb{P}^2$ is a smooth plane quintic curve. Then the genus of C is $g = 6$ and C is non-hyperelliptic. The linear series $|\mathcal{O}_C(1)|$ cut out on C by the lines in $\mathbb{P}^2$ is a g^2_{g-1}; accordingly, we expect to find a triple point on Θ.

In fact we may see that there is a unique triple point, as follows. By the Riemann singularity theorem any triple point on Θ comes from a g^2_5 on the curve. Since, by Clifford's theorem, there is no g^2_4 on the curve, the g^2_5 must be free of base points. The holomorphic map

$$C \to \mathbb{P}^2$$

given by the g^2_5 will map onto a plane curve C' of degree $d' = 5/\delta$ where δ is the degree of $C \to C'$. It follows that $\delta = 1$, C' is biholomorphic to C, and then, as was shown in Chapter V, that the g^2_5 is $|\mathcal{O}_C(1)|$.

To describe the tangent cone to Θ at the triple point we first make a remark concerning the Veronese surface Ξ, defined to be the image of the standard mapping

$$\phi_{\mathcal{O}(2)}: \mathbb{P}^2 \to \mathbb{P}^5.$$

It is a smooth surface of degree 4, whose chordal and tangential varieties we shall denote by $\sigma(\Xi)$ and $\tau(\Xi)$, respectively. By an obvious dimension count, $\sigma(\Xi)$ is supposed to fill out $\mathbb{P}^5$. The Veronese surface has the remarkable property of being the only smooth and non-degenerate surface for which $\sigma(\Xi) \neq \mathbb{P}^5$. For the Veronese surface the varieties $\sigma(\Xi)$ and $\tau(\Xi)$ coincide, and the tangential variety $\tau(\Xi)$ is a cubic hypersurface in $\mathbb{P}^5$ that is singular along Ξ.

The canonical bundle of C is $\mathcal{O}_C(2)$, and the canonical mapping on the curve extends to $\mathbb{P}^2$ to give a diagram

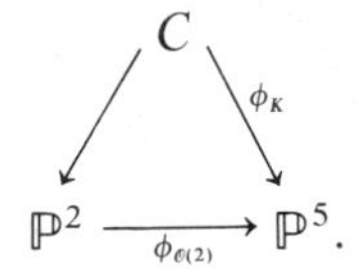

Thus the canonical curve lies on the Veronese surface. For each $\lambda \in \mathbb{P}^{2*}$ we denote by l_λ the corresponding line in $\mathbb{P}^2$, and by D_λ the intersection $l_\lambda \cdot C$. Then the divisor D_λ maps to five points on the conic $\phi_{\mathcal{O}(2)}(l_\lambda)$, and hence

$$\dim \overline{\phi_K(D_\lambda)} = 2$$

(as must be the case, by the geometric Riemann–Roch theorem). In particular, each $\overline{\phi_K(D_\lambda)}$ lies in the chordal variety to the Veronese surface, and since both the tangent cone to Θ at $u(g_5^2)$ and the chordal-tangential variety are cubic hypersurfaces in $\mathbb{P}^5$ we conclude that the *projectivized tangent cone to the theta divisor at the g_5^2 on a smooth plane quintic is the tangential variety to the Veronese surface.*

In Exercise B-2 we will return to a discussion of this example.

§3. The Torelli Theorem

To each smooth curve C of genus g there is attached the abelian variety $J(C)$ together with the principal polarization given by the theta divisor Θ. It is a fundamental result, due to Torelli, that the curve C can actually be reconstructed from the abstract principally polarized abelian variety $(J(C), \Theta)$. This can be formally stated as follows.

Torelli's Theorem. *Let C, C' be smooth curves. Let $(J(C), \Theta)$ and $(J(C'), \Theta')$ be the corresponding Jacobians. Assume there is an isomorphism of principally* *polarized abelian varieties between the Jacobians of C and C'. Then C and C' are isomorphic.*

The proof that we shall present below is due to Andreotti. Of course, when the genus of C and C' is zero or one there is nothing to be proved. Accordingly, we shall assume C and C' to have genus at least 2 throughout. We denote by Θ_{reg} the set of smooth points of Θ. Consider the Gauss map

$$\gamma: \Theta_{\text{reg}} \to (\mathbb{P}^{g-1})^*$$

that assigns to each point $u \in \Theta_{\text{reg}}$ the translate to the origin of the projectivized tangent hyperplane $\mathbb{P}T_u(\Theta)$. Let Γ be the closure of the graph of γ and denote by $\hat{\Gamma}$ its normalization. Let

$$\alpha: \hat{\Gamma} \to (\mathbb{P}^{g-1})^*$$

be the induced morphism. Clearly α is completely determined by $(J(C), \Theta)$. We are going to show, first of all, that α is finite. Next, denoting by B_α the branch locus of α, we shall prove that:

(i) *If C is not hyperelliptic,*

$$B_\alpha = C^*,$$

where C^* stands for the dual hypersurface to the canonical curve $C \subset \mathbb{P}^{g-1}$.

(ii) *If C is hyperelliptic,*

$$B_\alpha = C_0^* \cup p_1^* \cup \cdots \cup p_{2g+2}^*,$$

where C_0 is the rational normal curve which is the image of C under the canonical mapping, C_0^* is its dual, and $p_1^*, \ldots, p_{2g+2}^*$ are the stars of hyperplanes in $\mathbb{P}^{g-1}$ through the branch points $p_1, \ldots, p_{2g+2}$ of the canonical map.

We shall then complete the proof of Torelli's theorem by indicating how C can be reconstructed from C^* or $C_0^* \cup p_1^* \cup \cdots \cup p_{2g+2}^*$.

Since the Gauss map remains unchanged if the theta divisor is replaced by a translate, we are allowed to identify Θ with $W_{g-1}(C)$. Accordingly, we shall consider the theta divisor as consisting of all abelian sums $u(D)$, where D varies over C_{g-1}. The next remark to be made is the following. Let u be a point in Θ_{reg}. Then u is of the form $u(D)$ for a unique D in C_{g-1}, and the projectivized tangent space to Θ at u is the linear span of $\phi_K(D)$, so that we have

$$\gamma(u(D)) = \overline{\phi_K(D)}.$$

Let us now assume that C is not hyperelliptic. Then

$$\Gamma = \{(u(D), H): \bar{D} \subset H\} \subset \Theta \times (\mathbb{P}^{g-1})^*$$

(here, as usual, we are writing $\bar{D}$ for $\overline{\phi_K(D)}$). To see this it suffices to notice that, by the general position theorem, any point in Γ is the limit of points $(u(D), \gamma(u(D)))$, where $u(D)$ is a smooth point of Θ. It is clear from the explicit description of Γ that the projection from Γ to $(\mathbb{P}^{g-1})^*$ is a finite map; therefore α is also finite. Now set

$$\Gamma' = \{(D, H): \bar{D} \subset H\} \subset C_{g-1} \times (\mathbb{P}^{g-1})^*,$$

and let α' be the projection from Γ' to $(\mathbb{P}^{g-1})^*$. The variety Γ' is irreducible and the map $(u, 1): \Gamma' \to \Gamma$ is finite and generically one-to-one. By the very definition of normalization, we thus get a commutative diagram

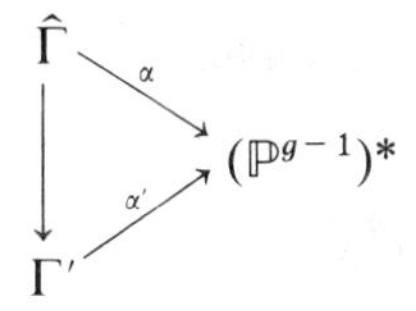

The map α' is clearly a finite map of degree $\binom{2g-2}{g-1}$, and its branch locus is the locus of tangent hyperplanes to C, that is, the dual hypersurface C^*; thus $B_\alpha \subset C^*$. On the other hand, the ramification locus of α', which we shall denote by R, is the set of couples (D, H), where $\bar{D} \subset H$ and D contains multiple points. Thus, if π is the projection from Γ' onto C_{g-1}, $\pi(R)$ is the irreducible codimension one subvariety of C_{g-1} consisting of all points of the form $2p + p_1 + \cdots + p_{g-3}$. If D is a point of C_{g-1}, $\bar{D}$ is a hyperplane if and only if D does not belong to C^1_{g-1}, hence π is biregular off $\pi^{-1}(C^1_{g-1})$, while it has positive-dimensional fibers over C^1_{g-1}. Hence C^1_{g-1} has codimension at least two in C_{g-1} and a general point of R is a smooth point of Γ'. In particular, at such a point, Γ' and $\hat{\Gamma}$ are isomorphic. Thus B_α contains an open subset of C^*. As we already observed that C^* contains B_α, and C^* is irreducible, we conclude that B_α is equal to C^*.

Let us now deal with the case when C is hyperelliptic. We first claim that the rational map $\gamma \circ u$ from C_{g-1} to $(\mathbb{P}^{g-1})^*$ extends to a regular map β. We may in fact set

$$\beta\left(\sum_{i=1}^{g-1} p_i\right) = \overline{\sum_i \phi_K(p_i)}.$$

We thus get a commutative diagram

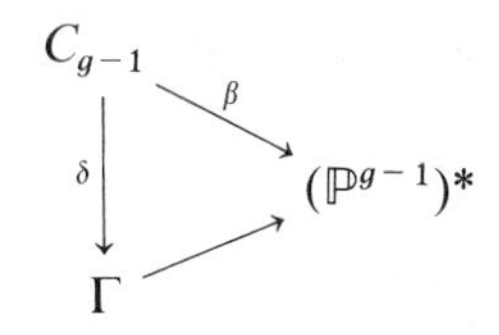

where $\delta(D) = (u(D), \beta(D))$. Since it is obvious from its definition that δ is finite and generically one-to-one, this allows us to identify C_{g-1} with $\hat{\Gamma}$ and β with α. But then β is a finite map of degree 2^{g-1} whose branch locus is the locus of tangent hyperplanes to C_0, plus the locus of hyperplanes passing through some branch point of the canonical map.

Properties (i) and (ii), whose proof we just finished, allow us to reconstruct C from the knowledge of $(J(C), \Theta)$. It is, in fact, an elementary general local result that the double duality

$$(C^*)^* = C$$

is valid (cf. Griffiths–Harris [1, page 397]), thereby establishing the Torelli theorem in the non-hyperelliptic case. In the hyperelliptic case one need only notice that C is entirely determined by the position of the branch points of ϕ_K on C_0.

A direct way of reconstructing C (or C_0) from its dual is the following: for each point $p \in C$ the locus $T_p(C)^*$ of hyperplanes containing the tangent line $T_p(C)$ is a $(g-3)$-plane in $(\mathbb{P}^{g-1})^*$. Conversely, if $\Lambda \subset (\mathbb{P}^{g-1})^*$ is any $(g-3)$-dimensional linear system of hyperplanes in $\mathbb{P}^{g-1}$ all of which are tangent to C, then by Bertini's theorem the general one can only be tangent at a base point p of Λ, so that Λ must be contained in—and hence equal to—$T_p(C)^*$. Of course, it may happen that $T_p(C)^* = T_q(C)^*$ for some pair $p \neq q$ of points on C, but this can only happen finitely many times. In this way we have now identified the curve C as being the normalization of the variety of $(g-3)$-planes contained in C^*.

If one wishes to "picture" Andreotti's proof we recommend the case of a non-hyperelliptic curve of genus 3. The canonical model is then a smooth plane quartic curve, and there are two kinds of hyperplane sections—those that are general such as l, and those that are tangent such as l_0:

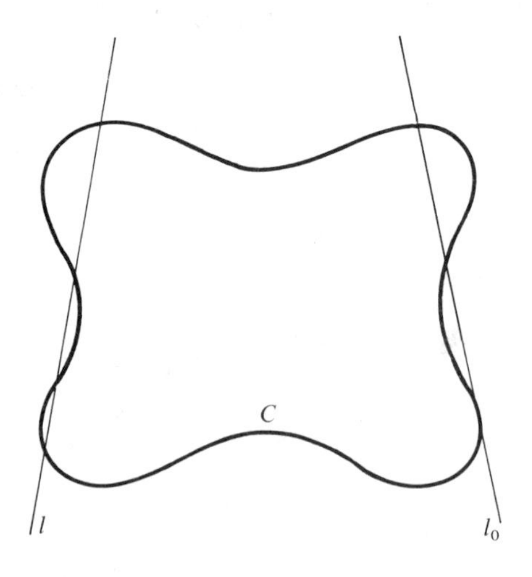

Of course, there are degenerate forms of l_0, such as bi-tangents and flex-tangents, but these need not concern us since they vary in a lower-dimensional family. The point is that, thinking of Θ as all sums $u(p + q)$ where $p, q \in C$, the Gauss mapping

$$\gamma: \Theta \to (\mathbb{P}^2)^*$$

is everywhere defined by

$$\gamma(u(p + q)) = \overline{pq},$$

where it is understood that $\overline{2p}$ is the tangent line at p. The Gauss mapping ramifies exactly over the dual curve $C^* \subset (\mathbb{P}^2)^*$ of tangent lines to C.

In the hyperelliptic case the picture is

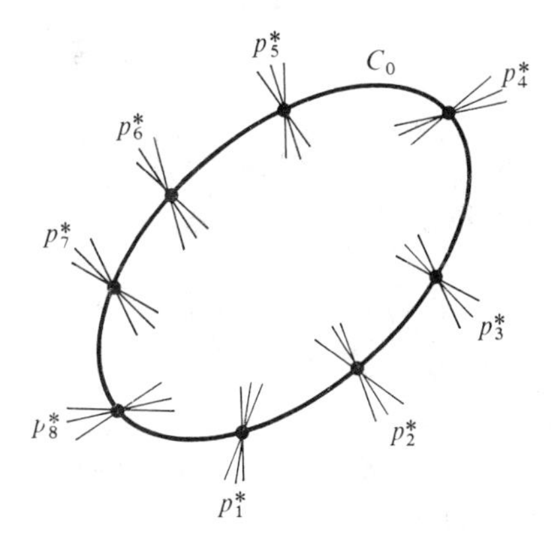

Here C_0 is the canonical curve and B_α is the reducible curve whose components are the curve C_0^* consisting of the lines l_0 tangent to C_0, and the pencils p_j^* of lines l'_j through p_j, where $p_1, \ldots, p_8$ are the branch points of the hyperelliptic involution.

§4. The Theory of Andreotti and Mayer

The classical Schottky problem is the one of finding explicit equations for the locus $\mathscr{J}_g$ of Jacobians in the Siegel upper-half plane $\mathscr{H}_g$ in terms of the theta function. The problem has been solved by means of techniques that bear no relationship to the theory of special divisors.

Among the classical approaches to the Schottky problem, which, although so far unsuccessful, have been quite fruitful in a number of ways, there is however one which is close to the ideas we have been developing. This is due to Andreotti and Mayer.

The essential tool used by Andreotti and Mayer in their beautiful paper [1] is the heat equation

$$\frac{\partial^2 \theta}{\partial u_\alpha \, \partial u_\beta}(u, Z) = 2\pi i(1 + \delta_{\alpha\beta}) \frac{\partial \theta}{\partial Z_{\alpha\beta}}(u, Z)$$

satisfied by the theta function. This will come into play later.

The starting point of their analysis is the following classical remark. Let Θ be the theta divisor in the Jacobian $J(C)$ of a smooth genus g curve. Then the dimension of the singular locus of Θ is given by

$$\dim(\Theta_{\text{sing}}) = g - 4 \quad \text{if } C \text{ is not hyperelliptic,}$$
$$\dim(\Theta_{\text{sing}}) = g - 3 \quad \text{if } C \text{ is hyperelliptic.}$$

The proof of this fact is as follows. By Riemann's theorem, Θ is a translate of $W_{g-1}(C)$ and by Riemann's singularity theorem Θ_{sing} is a translate of $W^1_{g-1}(C)$. If C is hyperelliptic, then the assertion that Θ_{sing} is $(g - 3)$-dimensional is a special case of the last part of Martens' theorem (Theorem (5.1) in Chapter IV). Again by Martens' theorem, the dimension of Θ_{sing} cannot exceed $g - 4$ when C is not hyperelliptic. On the other hand, for $W^1_{g-1}(C)$, the Brill–Noether number ρ equals $g - 4$. Hence, by Lemma (3.3) in Chapter IV, the result follows, provided we can show that, for $g \geq 4$, $W^1_{g-1}(C)$ is not empty. Of course, this is a special case of the Existence Theorem (cf. Chapter V), but one which can be proved by an "ad hoc" argument. We shall actually show that any effective divisor $p_1 + \cdots + p_{g-3}$ of degree $g - 3$ on C is contained in a degree $g - 1$ divisor D that moves in a pencil. In proving this, we shall view C as being canonically embedded in $\mathbb{P}^{g-1}$. If the secant plane $\overline{p_1 + \cdots + p_{g-3}}$ meets C at a further point or is tangent to C at one of the p_i, then we can find a further point q such that $p_1 + \cdots + p_{g-3} + q$ moves in a pencil and we are done. Consequently, we may assume that we have $p_1, \ldots, p_{g-3}$ such that the secant plane

$$\overline{p_1 + \cdots + p_{g-3}} = \mathbb{P}^{g-4}$$

does not meet C away from the points p_i and is not tangent to C at any of the p_i. We claim that the linear projection

$$\pi: C \to \mathbb{P}^2$$

from this secant plane cannot be an embedding. For, if π were an embedding of C as a smooth plane curve then we would have

$$\deg \pi(C) = 2g - 2 - (g - 3) = g + 1,$$

and hence g would equal $g(g - 1)/2$, which is a contradiction if $g \geq 4$. In this case, then, π cannot be an embedding and it follows that either there must be distinct points p, q such that $\pi(p) = \pi(q)$, or a point $p = q$ where π has vanishing differential. Then

$$D = p_1 + \cdots + p_{g-3} + p + q$$

moves in a pencil, since $\bar{D}$ is a $\mathbb{P}^{g-3}$.

We have thus seen that the theta divisor of any Jacobian is singular in dimension $g - 4$ or more. Andreotti and Mayer's idea is to attempt to characterize Jacobians by this condition. This is motivated by the fact that, as they observe, one can write down explicit equations defining the locus $N_{g-4} \subset \mathscr{H}_g$ corresponding to principally polarized abelian varieties whose theta divisor is singular in dimension $g - 4$ or more.

The formal set-up is as follows. One may construct a universal family of principally polarized abelian varieties $\pi: \mathscr{X} \to \mathscr{H}_g$ by setting

$$\mathscr{X} = \mathbb{C}^g \times \mathscr{H}_g/\rho(\mathbb{Z}^{2g}),$$

where ρ is the representation of $\mathbb{Z}^{2g}$ in the automorphisms group of $\mathbb{C}^g \times \mathscr{H}_g$ defined by

$$\rho(\gamma)(u, Z) = (u + (I, Z)\gamma, Z).$$

We then have a commutative diagram

$$\begin{array}{ccc} & \mathbb{C}^g \times \mathscr{H}_g & \\ \swarrow^{\mu} & & \searrow^{p} \\ \mathscr{X} & \xrightarrow{\pi} & \mathscr{H}_g, \end{array}$$

where μ is the quotient map (a local homeomorphism), p is the projection and π is a proper map with the property that

$$\pi^{-1}(Z) = \mathbb{C}^g/\Lambda_Z,$$

where Λ_Z is the lattice generated by the columns of (I, Z). We now define, in terms of Θ, a number of analytic subvarieties of $\mathbb{C}^g \times \mathscr{H}_g$. We set

$$A = \{(u, Z) \in \mathbb{C}^g \times \mathscr{H}_g : \theta(u, Z) = 0\},$$

$$B = \left\{(u, Z) \in A : \frac{\partial \theta}{\partial u_\alpha}(u, Z) = 0, \text{ for all } \alpha\right\}.$$

We then set $\tau = p|_B$ and

$$B_k = \{x \in B : \dim \tau^{-1}\tau(x) \geq k\}.$$

Since π is proper and μ is a local homeomorphism, $p(B_k)$ is an analytic subvariety of $\mathscr{H}_g$. Following Andreotti and Mayer we set

$$N_k = p(B_k) \subset \mathscr{H}_g, \qquad k \geq 0,$$

so that N_k is the sublocus of $\mathscr{H}_g$ consisting of the points that correspond to principally polarized abelian varieties whose theta divisor has a singular locus of dimension at least k. Clearly,

$$\cdots \subset N_k \subset N_{k-1} \subset \cdots \subset N_0.$$

We first want to show that N_0, and therefore N_k, is a *proper* subvariety of $\mathscr{H}_g$.

This is a nice and elementary consequence of the properties of the solutions of the Cauchy problem for the heat equation. Suppose, in fact, that B_0 surjects onto $\mathscr{H}_g$. Then there is an open subset $U \times V$ contained in $\mathbb{C}^g \times \mathscr{H}_g$ and a section

$$s: V \to U \times V$$

such that

$$\begin{cases} \theta(s(Z), Z) = 0, \\ \dfrac{\partial \theta}{\partial u_\alpha}(s(Z), Z) = 0, \qquad Z \in V, \quad \alpha = 1, \ldots, g. \end{cases}$$

By shrinking V and changing coordinates, if necessary, the theta function transforms into a function $v(u, Z)$ satisfying an equation

$$\frac{\partial^2 v}{\partial u_i \, \partial u_j} = \sum a_{ij}^{\alpha\beta}(u, Z) \frac{\partial v}{\partial Z_{\alpha\beta}} + \sum b_{ij}^l(u, Z) \frac{\partial v}{\partial u_l} + c_{ij}(u, Z) v.$$

We may also assume that our change of coordinates makes $s(Z)$ equal to zero, so that the preceding conditions read as follows

$$\begin{cases} v(0, Z) = 0, \\ \dfrac{\partial v}{\partial u_\alpha}(0, Z) = 0. \end{cases}$$

But now the differential equations above give, by iteration, that all partials of v vanish at $(0, Z)$, which is absurd.

We shall denote by $\overline{\mathscr{J}}_g$ the analytic Zariski closure of the Jacobian locus $\mathscr{J}_g$ in $\mathscr{H}_g$. We have already observed that

$$\overline{\mathscr{J}}_g \subset N_{g-4} \subset \mathscr{H}_g.$$

The main result of Andreotti and Mayer is the following.

Andreotti–Mayer Theorem. *$\bar{\mathscr{J}}_g$ is the unique irreducible component of N_{g-4} that contains $\mathscr{J}_g$.*

It is known that, for $g \geq 4$, the locus N_{g-4} has other components besides $\bar{\mathscr{J}}_g$. However the analysis of the "ramification loci" N_k introduced by Andreotti and Mayer has proved to be an essential tool in the study of the moduli space of principally polarized abelian varieties.

The proof of Andreotti and Mayer's theorem uses the fact that $\bar{\mathscr{J}}_g$ is irreducible of dimension at least $3g - 3$. This follows, for example, from the fact that the Teichmüller space $\mathscr{T}_g$ of all marked Riemann surfaces of genus g is a bounded domain in $\mathbb{C}^{3g-3}$ and that the map from $\mathscr{T}_g$ to $\mathscr{H}_g$ associating to each curve its Jacobian has discrete fibers. We refer to Bers [1] for these statements. An overview of the moduli theory of curves will also be found in the second volume.

To prove the theorem all we have to show is that the dimension of any component of N_{g-4} containing $\bar{\mathscr{J}}_g$ is at most $3g - 3$. We shall need the following.

Lemma. *Let M be an irreducible component of N_k. Let Δ be a component of B_k such that $p(\Delta) = M$. Let Z_0 be a simple point of M. Let $\sum q_{\alpha\beta}(\partial/\partial Z_{\alpha\beta}) \in T_{Z_0}(M)$, Then for every $(u_0, Z_0) \in p^{-1}(Z_0) \cap \Delta$ we have*

$$\sum \frac{\partial\theta}{\partial Z_{\alpha\beta}}(u_0, Z_0) q_{\alpha\beta} = 0.$$

Proof. Set

$$\Delta' = \Delta - p^{-1}(M_{\text{sing}}) - \Delta_{\text{sing}} - \{x \in \Delta: \text{rank}_x(p) \text{ is not maximum}\}.$$

Clearly Δ' is open in Δ and $p(\Delta')$ is dense in M. Therefore it suffices to prove the lemma with $(u_0, Z_0) \in \Delta'$. Locally Δ' is parametrized by

$$u = u(y, t), \qquad Z = Z(t),$$

$$t \in \mathbb{C}^s, \quad s = \dim M, \qquad y \in \mathbb{C}^r, \quad r = \dim \Delta' - \text{rank } p|_{\Delta'}.$$

Since $\Delta' \subset B_k \subset B_0$ we have

$$\theta(u(y, t), Z(t)) \equiv 0,$$

$$\frac{\partial\theta}{\partial u_\alpha}(u(y, t), Z(t)) \equiv 0.$$

Differentiating the first of these equations we find

$$\sum_{\alpha,\beta} \frac{\partial\theta}{\partial Z_{\alpha\beta}}(u(y,t), Z(t))\frac{\partial Z_{\alpha\beta}}{\partial t_i} = 0.$$

Since $\sum q_{\alpha\beta}(\partial/\partial Z_{\alpha\beta}) \in T_{Z_0}(M)$, there exist constants λ_i such that $q_{\alpha\beta} = \sum \lambda_i(\partial Z_{\alpha\beta}/\partial t_i)$. Q.E.D.

Returning to the proof of the theorem, we let M be an irreducible component of N_{g-4} containing $\bar{\mathcal{J}}_g$. In order to show that the dimension of M is at most $3g - 3$ it suffices, by the lemma, to find a point $Z_0 \in M$ such that:

(1) Z_0 is a simple point of M;
(2) in $p^{-1}(Z_0) \cap \Delta$ there are $N = (g-2)(g-3)/2$ points

$$(u^{(1)}, Z_0), \ldots, (u^{(N)}, Z_0)$$

such that the vectors

$$v_i = \left(\frac{\partial\theta}{\partial Z_{\alpha\beta}}(u^{(i)}, Z_0)\right) \in \mathbb{C}^{g(g+1)/2}, \qquad i = 1, \ldots, N,$$

are linearly independent.

Andreotti and Mayer's beautiful idea is the following. Choose as Z_0 a point corresponding to a Jacobian $J(C)$, where C is a non-hyperelliptic smooth curve, and take $u^{(1)}, \ldots, u^{(N)}$ to be general points in Θ_{sing} (these have to be double points). To show that the vectors v_i above are linearly independent is the same as showing that the quadrics

$$Q_i = \left\{\sum_{\alpha\beta} \frac{\partial\theta}{\partial Z_{\alpha\beta}}(u^{(i)}, Z_0)\omega_\alpha\omega_\beta = 0\right\} \subset \mathbb{P}^{g-1}$$

are linearly independent. Now use the heat equation to get

$$Q_i = \left\{\sum_{\alpha\beta} \frac{\partial\theta}{\partial u_\alpha\,\partial u_\beta}(u^{(i)}, Z_0)\omega_\alpha\omega_\beta = 0\right\}.$$

These are exactly the tangent cones to Θ at the points $u^{(i)}$, which, as we saw in the proof of Riemann's singularity theorem, all contain the canonical image of C (cf. (1.4)).

Denote then by $I_2(\Theta)$ the linear span of the tangent cones to double points of Θ inside the vector space I_2 of all quadrics through the canonical curve. Since, by Noether's theorem, the dimension of I_2 is $N = (g-2)(g-3)/2$, the fact that the vectors $v_1, \ldots, v_N$ are linearly independent, and hence the

Andreotti–Mayer theorem, are a consequence of the following result, whose proof is due to Mark Green.

(4.1) Theorem. *For any non-hyperelliptic smooth curve C of genus $g \geq 4$, $I_2(\Theta) = I_2$.*

The problem of deciding whether the above is true or not originated with Andreotti and Mayer's paper: in it the result is proved for general C, and that, of course, is all they needed to conclude the proof of their theorem.

In proving (4.1) we shall need the following general remark. Let C be any smooth genus g curve, let

$$D = p_1 + \cdots + p_{d-h}, \qquad h \geq 1$$

be an effective divisor on C, and consider the morphism α from C to C_d defined by

$$\alpha(p) = hp + D.$$

Then α induces isomorphisms

$$(4.2) \qquad \begin{aligned} H^0(C, K_C) &\cong H^0(C_d, \Omega^1_{C_d}), \\ H^1(C, \mathcal{O}_C) &\cong H^1(C_d, \mathcal{O}_{C_d}). \end{aligned}$$

We may argue as follows. The map α factors

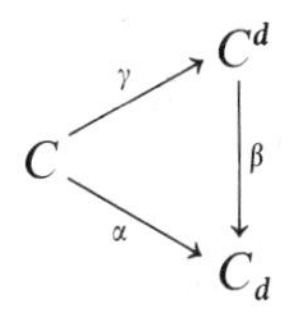

where β is the quotient morphism, and

$$\gamma(p) = (p_1, \ldots, p_{d-h}, p, \ldots, p).$$

To prove the first part of (4.2) we have to show that the map

$$\alpha^*: H^0(C_d, \Omega^1_{C_d}) \to H^0(C^d, \Omega^1_{C^d})^{S_d} \cong H^0(C, K_C)$$

is an isomorphism (here, and in what follows, $(\ )^{S_d}$ stands for the invariants under the action of the symmetric group S_d, and the isomorphism on the right follows from Künneth's formula). Injectivity is obvious. As for surjectivity, we must show that any invariant holomorphic 1-form ω on C^d descends to C_d. This is clear away from the diagonals: by Hartogs' theorem it suffices then to show that ω descends to C_d near a general point of the diagonals. At such a point, relative to suitable local coordinates on C^d, the morphism β is

$$(z_1, z_2, \ldots, z_d) \to (u = z_1 + z_2, v = z_1 z_2, z_3, \ldots).$$

If we write

$$\omega = \sum a_i \, dz_i,$$

we have that, for any $i \geq 3$, interchanging z_1 and z_2 leaves a_i unchanged. Therefore a_i is a function of $u, v, z_3, \ldots, z_d$. On the other hand, the invariance of ω implies that

$$a_1(z_1, z_2, z_3, \ldots) = a_2(z_2, z_1, z_3, \ldots),$$

hence

$$\omega = a \, du + b \, dv + a_3 \, dz_3 + \cdots,$$

where

$$a = (a_1 + a_2 - (z_1 + z_2)b)/2; \qquad b = \frac{a_1 - a_2}{z_2 - z_1}.$$

Since a, b are holomorphic and invariant under the interchange of z_1, z_2, this shows that ω is holomorphic on C_d.

The second part of (4.2) follows from the first by Hodge theory. Alternatively, we may proceed as follows. As before, it is clear that

$$H^1(C^d, \mathcal{O}_{C^d})^{S_d} \cong H^1(C, \mathcal{O}_C).$$

On the other hand, since β is finite,

$$H^1(C^d, \mathcal{O}_{C^d}) = H^1(C_d, \beta_*(\mathcal{O}_{C^d})).$$

The projection onto the invariants

$$f(a) = \frac{1}{d!} \sum_{\sigma \in S_d} \sigma(a)$$

operates in a compatible way both on $H^1(C^d, \mathcal{O}_{C^d})$ and on $\beta_*(\mathcal{O}_{C^d})$, hence

$$\begin{aligned} H^1(C^d, \mathcal{O}_{C^d})^{S_d} &= f(H^1(C^d, \mathcal{O}_{C^d})) \\ &= H^1(C_d, f(\beta_*(\mathcal{O}_{C^d}))) \\ &= H^1(C_d, \mathcal{O}_{C_d}). \end{aligned}$$

Another general remark that we shall need is the following. Let

$$\phi: E \to F$$

be an injective homomorphism of locally free sheaves of ranks n and $n + 1$ on an irreducible variety X. One may identify F with $\Lambda^{n+1}F \otimes \Lambda^n F^*$, hence one gets a homomorphism

$$\psi = 1_{\Lambda^{n+1}F} \otimes \Lambda^n \phi^*: F = \Lambda^{n+1}F \otimes \Lambda^n F^* \to \Lambda^{n+1}F \otimes \Lambda^n E^*.$$

We claim that $\psi \circ \phi = 0$. This is best seen by taking duals; in other words we have to show that the composite homomorphism

$$\Lambda^{n+1}F^* \otimes \Lambda^n E \xrightarrow{\psi^*} F^* \xrightarrow{\phi^*} E^*$$

vanishes. But this is clear, since

$$\psi^*(f_1^* \wedge \cdots \wedge f_{n+1}^* \otimes e_1 \wedge \cdots \wedge e_n)$$

acts on F by

$$f \mapsto f_1^* \wedge \cdots \wedge f_{n+1}^*(f \wedge \phi(e_1) \wedge \cdots \wedge \phi(e_n)),$$

which is zero whenever f comes from E. The above formula also makes it clear that, if $e_1, \ldots, e_n$ and $f_1, \ldots, f_{n+1}$ are local frames for E and F and we write

$$\phi(e_h) = \sum m_{hk} f_k,$$

then

$$\psi(f_i) = (-1)^{i-1} M_i \, f_1 \wedge \cdots \wedge f_{n+1} \otimes e_1^* \wedge \cdots \wedge e_n^*,$$

where $\{e_i^*\}$ is the dual frame to $\{e_i\}$ and M_i stands for the minor obtained by deleting the ith column from the matrix (m_{hk}). In particular, by Cramer's rule, the sequence

$$E \xrightarrow{\phi} F \xrightarrow{\psi} \Lambda^{n+1}F \otimes \Lambda^n E^* \tag{4.3}$$

is exact off the locus where (m_{hk}) is not of maximal rank. In case this locus is of codimension two or more, and X is normal, we get exactness everywhere, by Hartog's theorem. This applies in particular to the Brill–Noether homomorphism

$$\Theta_{C_{g-1}} \to u^*\Theta_{J(C)}$$

for a non-hyperelliptic curve C. In fact, since the singular locus of the theta divisor has dimension $g - 4$, C^1_{g-1}, which is the degeneracy locus of the Brill–Noether matrix, has codimension two in C_{g-1}. From now on, for simplicity, we write D instead of C^1_{g-1}; we also identify $u(C_{g-1})$ with (a suitable translate of) the theta divisor Θ in $J(C)$, and write L for the pull-back of $\mathcal{O}(\Theta)$ to C_{g-1}. By adjunction

$$K_{C_{g-1}} \cong L$$

on $C_{g-1} - D = \Theta - \Theta_{\text{sing}}$; since D has codimension two, this isomorphism holds in fact on all of C_{g-1}, again by Hartogs' theorem. With these conventions, (4.3) yields an exact sequence

$$0 \to \Theta_{C_{g-1}} \xrightarrow{\phi} \operatorname{Hom}(H^0(C, K_C), \mathcal{O}_{C_{g-1}}) \xrightarrow{\psi} \mathcal{I}_D \otimes L \to 0,$$

where $\mathcal{I}_D$ stands for the ideal sheaf of D in C_{g-1}: in fact $\mathcal{I}_D$ is locally generated by the maximal minors M_i of the Brill–Noether matrix.

We shall next construct a diagram

(4.4)

$$\begin{array}{ccccc} H^1(C_{g-1}, \Theta_{C_{g-1}}) & \xrightarrow{\phi} & \operatorname{Hom}(H^0(C_{g-1}, \Omega^1_{C_{g-1}}), H^1(C_{g-1}, \mathcal{O}_{C_{g-1}})) & \xrightarrow{\psi} & H^1(C_{g-1}, \mathcal{I}_D \otimes L) \\ \uparrow \gamma & & \uparrow \sigma & & \uparrow \delta \\ H^1(C, \Theta_C) & \xrightarrow{h} & \operatorname{Sym}^2 H^1(C, \mathcal{O}_C) & \xrightarrow{f} & H^0(C_{g-1}, L \otimes \mathcal{O}_D) \end{array}$$

commutative up to constants, where ϕ, ψ are as defined above, δ is a coboundary map, σ is defined via the identification (4.2), and h is the transpose of the cup-product map

$$\operatorname{Sym}^2 H^0(C, K_C) \to H^0(C, K_C^2).$$

A moment of reflection will convince the reader that ϕ and

$$h: H^1(C, \Theta_C) \to \operatorname{Sym}^2 H^1(C, \mathcal{O}_C) \to \operatorname{Hom}(H^0(C, K_C), H^1(C, \mathcal{O}_C))$$

are induced by the contraction homomorphisms

$$\Theta_{C_{g-1}} \otimes \Omega^1_{C_{g-1}} \to \mathcal{O}_{C_{g-1}}; \qquad \Theta_C \otimes K_C \to \mathcal{O}_C.$$

The only maps in (4.4) that have not been defined so far are f and γ. We shall postpone the definition of γ, as well as the proof that γ is an *isomorphism*, until the second volume. Here we shall only define f and show that the rightmost rectangle in (4.4) commutes. In doing so we shall use the same symbols to denote the theta function or its derivatives and their pull-backs to C_{g-1}. It will be clear from the context which ones we are talking about.

We begin by choosing coordinates $u_1, \ldots, u_g$ in $J(C)$ in such a way that the Chern class of $\mathcal{O}(\Theta)$ is represented by

$$\frac{\sqrt{-1}}{2} \sum du_i \wedge d\bar{u}_i,$$

and denote by ω_i the pull-back of du_i to C (or to C_d, when needed), and by ω_i^* the pull-back of $d\bar{u}_i$, viewed as a class in $H^1(C, \mathcal{O}_C)$ or $H^1(C_{g-1}, \mathcal{O}_{C_{g-1}})$. We then set

$$f(\omega_i^* \otimes \omega_j^* + \omega_j^* \otimes \omega_i^*) = \left.\frac{\partial^2 \theta}{\partial u_i\, \partial u_j}\right|_D,$$

where θ is the Riemann theta function and $|_D$ stands for pull-back to D. The meaning of the right-hand side is the following. If we view θ as a section of $\mathcal{O}(\Theta)$, that is as a collection $\{\theta_\alpha\}$ of local holomorphic functions subject to the relations

$$\theta_\beta = \xi_{\beta\alpha} \theta_\alpha,$$

where the $\xi_{\beta\alpha}$ are transition functions for $\mathcal{O}(\Theta)$, then, on the theta divisor, and hence, by pull-back, on C_{g-1}, one has

$$\xi_{\alpha\beta} \frac{\partial^2 \theta_\beta}{\partial u_i\, \partial u_j} = \frac{\partial^2 \theta_\alpha}{\partial u_i\, \partial u_j} + \frac{\partial}{\partial u_i} (\log \xi_{\beta\alpha}) \frac{\partial \theta_\alpha}{\partial u_j} + \frac{\partial}{\partial u_j} (\log \xi_{\beta\alpha}) \frac{\partial \theta_\alpha}{\partial u_i}.$$

Since the pull-back of θ vanishes to second order on D, this shows that $\{\partial^2 \theta_\alpha / \partial u_i\, \partial u_j|_D\}$ represents a section of $L \otimes \mathcal{O}_D$. The formula also shows that $\delta f(\omega_i^* \otimes \omega_j^* + \omega_j^* \otimes \omega_i^*)$ is represented by the cocycle

$$\left\{\frac{\partial}{\partial u_i} (\log \xi_{\beta\alpha}) \frac{\partial \theta_\alpha}{\partial u_j} + \frac{\partial}{\partial u_j} (\log \xi_{\beta\alpha}) \frac{\partial \theta_\alpha}{\partial u_i}\right\}.$$

Up to constants, the Chern class of $\mathcal{O}(\Theta)$ is represented by the cocycle

$$\partial(\log \xi_{\beta\alpha}) = \sum \frac{\partial}{\partial u_i} (\log \xi_{\beta\alpha})\, du_i,$$

and by our choice of coordinates, we conclude that, up to constants, the Čech cocycle representing $d\bar{u}_i$ is the pull-back to C_{g-1} of $\partial/\partial u_i(\log \xi_{\beta\alpha})$. It follows that

$$\delta f(\omega_i^* \otimes \omega_j^* + \omega_j^* \otimes \omega_i^*) = k\left(\frac{\partial \theta}{\partial u_i} \omega_j^* + \frac{\partial \theta}{\partial u_j} \omega_i^*\right),$$

where k is a constant. On the other hand

$$\psi\sigma(\omega_i^* \otimes \omega_j^* + \omega_j^* \otimes \omega_i^*) = (-1)^{i-1}M_i\omega_j^* + (-1)^{j-1}M_j\omega_i^*,$$

where the M_i are the maximal minors of the Brill–Noether matrix (m_{hk}) relative to the frame $\{\omega_i^*\}$ in $H^1(C, \mathcal{O}_C) \otimes \mathcal{O}_{C_{g-1}}$ and any fixed frame in $\Theta_{C_{g-1}}$. To conclude commutativity, up to constants, of the rightmost rectangle in (4.4), we have to show that there is a constant k_1 such that

$$\frac{\partial\theta}{\partial u_i} = k_1(-1)^{i-1}M_i, \qquad i = 1, \ldots, g,$$

where both sides are viewed as sections of L. To do this, notice that

$$\sum \frac{\partial\theta}{\partial u_i} m_{ji} = 0, \qquad j = 1, \ldots, g-1.$$

This is the content of formula (1.3). Then, by Cramer's rule, off D we have

$$\frac{\partial\theta}{\partial u_i} = F(-1)^{i-1}M_i, \qquad i = 1, \ldots, g,$$

where F is a never-vanishing function. By Hartogs' theorem, F must be constant, and the above holds on all of C_{g-1}.

We are now ready to prove theorem (4.1). As we have said, the crucial fact, to be proved in the second volume, is that the map γ in (4.4) is an isomorphism. This immediately implies that

$$\operatorname{Ker}(f) \subset \operatorname{Im}(h).$$

Suppose now that $I_2(\Theta) \neq I_2$. Since $\operatorname{Sym}^2 H^1(C, \mathcal{O}_C)$ is the dual to the vector space of quadrics in canonical space, there must be an element

$$Q = \sum_{i \le j} q_{ij}(\omega_i^* \otimes \omega_j^* + \omega_j^* \otimes \omega_i^*) \in \operatorname{Sym}^2 H^1(C, \mathcal{O}_C)$$

which does not annihilate I_2, but is such that

$$\sum q_{ij} \frac{\partial^2\theta}{\partial u_i\, \partial u_j} \equiv 0 \quad \text{on } D.$$

This means that $f(Q) = 0$. As a consequence, Q belongs to the image of h, and hence annihilates the kernel of h^*, that is I_2, a contradiction. This ends the proof of Theorem (4.1).

We may observe that this theorem gives another constructive proof of the Torelli theorem, at least for curves whose canonical ideal is generated by quadrics, namely for curves which are neither hyperelliptic, non-trigonal, nor isomorphic to a smooth plane quintic.

Let us finish by taking a closer look at the tangent cones to the theta divisor at double points. As we know (cf. Kempf's singularity theorem), the tangent cone to such a point $u(D)$ is a quadric Q_D whose equation can be obtained as follows. Choose bases x_1, x_2 and y_1, y_2 for $H^0(C, \mathcal{O}(D))$ and $H^0(C, K(-D))$ and set $\omega_{ij} = x_i y_j$. Then Q_D is given by

$$\omega_{11}\omega_{22} - \omega_{12}\omega_{21} = 0.$$

It is clear from this that $Q_D = Q_{K-D}$; from a different point of view, this is a reflection of the evenness of the theta function. It is also clear that Q_D has rank at most four (we recall that the rank of a quadric is the rank of the corresponding symmetric matrix). When $|D| = |K - D|$ we may choose $x_1 = y_1$, $x_2 = y_2$ so that $\omega_{12} = \omega_{21}$, and Q_D has rank three. Geometrically we know that

$$Q_D = \bigcup_{\Delta \in |D|} \bar{\Delta} = \bigcup_{\Delta' \in |K-D|} \bar{\Delta}' \subset \mathbb{P}^{g-1};$$

this explicitly exhibits, in this particular case, the rulings in $\mathbb{P}^{g-3}$'s that exist on any quadric of rank three or four. The two rulings are distinct, and hence Q_D has rank four, exactly when $|D| \neq |K - D|$.

Bibliographical Notes

The original version of Riemann's singularity theorem is to be found in Riemann [1]. Modern proofs have been provided by Mayer [1], Lewittes [1], and Mumford (unpublished). Kempf proves his singularity theorem in [2]. The generalization to $W^r_d(C)$, which closely follows Kempf's ideas, is adapted from Arbarello–Cornalba [1].

Proofs of Torelli's theorem abound. There is of course Torelli's original argument [1]. Two proofs are due to Andreotti [1, 2]: his second proof is essentially the one we have given in the text. Other proofs are due to Comessatti [2] (for a modern version see Ciliberto [1]), Weil [1], Matsusaka [1], Martens [1], Saint-Donat [2] and Little [1]. As we observed in the text, theorem (4.1), whose original proof is in Green [1], also leads to a constructive proof of Torelli's theorem. See also exercise batches C and E.

The literature on the Schottky problem is vast. We do not even attempt to give an overview of it, since the problem is only cursorily touched upon in the text. Any reader interested in learning more about the subject would do well to consult Mumford's lectures [3], Mumford's two volumes [5, 6], and Appendix D to Chapter VII in Mumford–Fogarty [1]. Solutions to the Schottky problem have been provided by Arbarello–De Concini [1, 2] (using results of Gunning [2] and Welters [1]), by Mulase [1] and Shiota [1]. Our main concern in this chapter is with Andreotti and Mayer's approach as found in their paper [1]. It is a known fact that the Andreotti–Mayer locus N_{g-4} contains other components beside the Jacobian locus if $g \geq 4$. In his paper [1], Beauville analyzes in detail the cases $g = 4, 5$. His analysis is completed, for $g = 5$, by Donagi [1]. A more direct proof, for the case $g = 4$, has been provided by Smith–Varley [1].

Exercises

A. The Difference Map ϕ_d (II)

In this sequence of exercises we will retain the notations from exercise batch D of Chapter V.

A-1. Let C be a curve of genus g and $D = p_1 + \cdots + p_d$, $E = q_1 + \cdots + q_d$ general divisors on C. Show that the projectivized image of the differential

$$(\phi_d)_*\colon T_{(D,E)}(C_d \times C_d) \to \mathbb{C}^g$$

is the join $\overline{\phi_K(D), \phi_K(E)}$ of the subspaces $\overline{\phi_K(D)}$ and $\overline{\phi_K(E)}$ in the projective space of the canonical curve (as usual, we identify all tangent spaces to $J(C)$ with $\mathbb{C}^g$).

A-2. Now suppose that C is non-hyperelliptic and let $H \cdot \phi_K(C) = p_1 + \cdots + p_{2g-2}$ be a general hyperplane section. Show that the osculating spaces to $\phi_K(C)$ at the points p_i are in general position; i.e., for any integers $a_1, \ldots, a_{2g-2}$ with $\sum_i a_i = g - 1$ and $D = \sum_i a_i p_i$,

$$\dim \overline{\phi_K(D)} = g - 2.$$

Conclude that no two distinct divisors of degree $d \leq g - 1$ supported on $H \cdot \phi_K(C)$ are linearly equivalent to one another.

(*Suggestion*: See the uniform position theorem.)

A-3. Suppose that C is non-hyperelliptic and $d < g/2$. Using the preceding two exercises show that

$$\phi_d\colon C_d \times C_d \to J(C)$$

is birational onto its image V_d.

A-4. In case C is hyperelliptic and $d < g/2$, show that the mapping ϕ_d has degree 2^d onto its image.

A-5. In the non-hyperelliptic case show that the projectivized tangent cone to V_1 at the origin $0 \in J(C)$ is

$$\mathbb{P}\mathcal{T}_0(V_1) = \phi_K(C) \subset \mathbb{P}^{g-1}.$$

Note. You must show that $\phi_1^{-1}(0)$ is the diagonal $\Delta \subset C \times C$ as a scheme.

A-6. More generally, in case C does not have a g_d^1 show that the projectivized tangent cone to V_d at $0 \in J(C)$ is the d-secant variety to the canonical curve.

A-7. Assuming that C has no g_d^1's, show that

$$\text{mult}_0(V_d) = \sum_{i=0}^{d} \binom{g-d-1-i}{d-i}\binom{g}{i}$$

(cf. Exercise A-3 of Chapter VIII).

A-8. Suppose that C has a single g_d^1 (necessarily complete), say $|D|$. Show that the fiber $\phi_d^{-1}(0)$ consists set-theoretically of the diagonal $\Delta \subset C_d \times C_d$ together with the set of pairs

$$\{(E, E') \colon E, E' \in |D|\}.$$

Conclude that the support of the projectivized tangent cone $\mathbb{P}\mathcal{T}_0(V_d)$ contains the d-secant variety to $\phi_K(C)$ together with the union

$$\Xi = \bigcup_{E, E' \in |D|} \overline{\phi_K(E + E')}.$$

A-9. Retaining the notation of the preceding exercise, assume that

$$H^0(C, K(-2D)) = 0$$

and show the support of $\mathbb{P}\mathcal{T}_0(V_d)$ is exactly the d-secant variety to $\phi_K(C)$ together with Ξ.

B. Refined Torelli Theorems

Let C be a smooth curve with a special linear series g_d^r. By a *refined Torelli theorem* we shall mean a geometric construction of C and g_d^r from the polarized Jacobian $(J(C), \Theta)$. In particular, this means that the g_d^r is unique. In the following sequence of exercises we shall give refined Torelli theorems for the following curves with g_d^r's:

(a) smooth plane quintics;
(b) smooth plane sextics;
(c) extremal curves of degree $3r - 1$ in $\mathbb{P}^r$.

Note that case (c) includes case (a).

In Exercises B-1 and B-2, $C \subset \mathbb{P}^2$ will be a smooth quintic with hyperplane bundle $L = \mathcal{O}_C(1) = \mathcal{O}_C(D)$.

B-1. Using Exercise 18(iii) of Appendix A, show that set-theoretically

$$\begin{aligned} W_5^2(C) &= \{u(D)\}, \\ W_5^1(C) &= \{u(D + p - q): p, q \in C\} \\ &= V_1 + u(D), \end{aligned}$$

where V_1 = image ϕ_1, with ϕ_1 as in exercise batches C of Chapter V and A of Chapter VI.

B-2. Using the Riemann singularity theorem and the preceding exercise show that

$$\begin{aligned} (W_{g-1}(C))_{\text{sing}} &= V_1 + u(D), \\ ((W_{g-1}(C))_{\text{sing}})_{\text{sing}} &= \{u(D)\}, \\ \mathbb{P}\mathcal{T}_{u(D)}((\Theta_{\text{sing}})_{\text{sing}}) &= \phi_K(C). \end{aligned}$$

The last equality implies the refined Torelli theorem for smooth plane quintics.

In Exercises B-3 through B-5, $C \subset \mathbb{P}^2$ will be a smooth plane sextic (genus = 10), $L = \mathcal{O}_C(1) = \mathcal{O}_C(D)$ the hyperplane bundle, and $S = \phi_{\mathcal{O}(3)}(\mathbb{P}^2) \subset \mathbb{P}^9$ the cubic Veronese surface. Note that $K_C = \mathcal{O}_C(3)$ and

$$\phi_K(C) \subset S \subset \mathbb{P}^9.$$

B-3. Show that set theoretically

$$\begin{aligned} W_9^3(C) &= \varnothing, \\ W_9^2(C) &= \{u(D + p_1 + p_2 + p_3): p_i \in C\} \cup \{u(2D - p_1 - p_2 - p_3): p_i \in C\} \\ &= (W_3(C) + u(D)) \cup (-W_3(C) + u(2D)), \\ W_9^2(C)_{\text{sing}} &= \{u(D + p_1 + p_2 + p_3);\ p_i \in C \text{ and } p_1, p_2, p_3 \text{ collinear}\}. \end{aligned}$$

(*Hint*: Show that nine points in $\mathbb{P}^2$ impose seven or fewer conditions on cubics if and only if six lie on a line or all nine lie on a conic. For this prove that the net of cubics through the nine points has a fixed component.)

B-4. Using the preceding exercise show that $\phi_K(C) \subset \mathbb{P}^9$ is characterized by the following construction: Denoting by $\Theta_{(r)}$ the points of multiplicity $\geq r$ on Θ (thus $\Theta_{(2)} = \Theta_{\text{sing}}$), for $v \in \mathbb{P}^9$ set

$$Q(v) = \{u \in \Theta_{(3)} - \Theta_{(3)\text{sing}}: \mathbb{P}T_u(\Theta_{(3)}) \ni v\}.$$

Then

$$\phi_K(C) = \{v \in \mathbb{P}^9: \dim Q(v) \geq 2\}.$$

B-5. Setting

$$\Sigma = \bigcap_{u \in \Theta_{(3)}} \mathbb{P}\mathcal{T}_u(\Theta)$$

establish the set-theoretic equality

$$\Sigma = \bigcup_{\Lambda \in \mathbb{P}^{2*}} \phi_{\mathcal{O}(3)}(\Lambda),$$

where the right-hand side is the union of the $\mathbb{P}^3$'s in $\mathbb{P}^9$ spanned by the images under $\phi_{\mathcal{O}_{\mathbb{P}^2}(3)}$ of the lines Λ in $\mathbb{P}^2$. Show that also

$$\Sigma = \bigcup_{u \in (\Theta_{(3)})_{\mathrm{sing}}} \mathbb{P}T_u(\Theta_{(3)}),$$

where $\mathbb{P}T$ is the projectivized Zariski tangent space.

Combining B-4 and B-5 we may reconstruct both $\phi_K(C)$ and the g^2_6 on C from $(J(C), \Theta)$, thereby establishing the refined Torelli theorem for plane sextics.

B-6. Finally, suppose that $C \subset \mathbb{P}^r$ is an extremal curve of degree $3r - 1$ with hyperplane bundle $L = \mathcal{O}_C(1) = \mathcal{O}_C(D)$.

(i) Show that

$$K_C = \mathcal{O}_C(2).$$

(ii) Show that

$$W^r_{g-1}(C) = \{u(D)\},$$

$$W^{r-1}_{g-1}(C) = \{u(D + p - q) \colon p, q \in C\}$$

$$= V_1 + u(D)$$

(cf. Exercise B-1). As in the case $r = 2$, conclude that

$$\phi_K(C) = \mathbb{P}\mathcal{T}_{u(D)}(\Theta_{(r)}).$$

C. Translates of W_{g-1}, Their Intersections, and the Torelli Theorem

The following series of exercises investigates the relationships among the subvarieties $W^r_d = W^r_d(C)$ (we will use the abbreviated notation throughout) of the Jacobian of a curve under sums and translates.

C-1. Let D be any divisor on a smooth curve C of genus g. Show that, for $p \in C$ a general point,

$$r(D + p) = \begin{cases} r(D) & \text{if } D \text{ is special,} \\ r(D) + 1 & \text{if } D \text{ is non-special,} \end{cases}$$

and

$$r(D - p) = r(D) - 1 \quad \text{if} \quad r(D) \geq 0.$$

C-2. Establish the set-theoretic equality

$$W_d = \bigcap_{E \in C_{e-d}} (W_e - u(E)), \qquad d \le e \le g - 1.$$

(*Hint*: Use the preceding exercise to show that, if D is a divisor of degree d with $d \le e \le g - 1$, then the following are equivalent:

$$\begin{cases} r(D) \ge 0, \\ r(D + E) \ge 0 \quad \text{for all} \quad E \in C_{e-d}. \end{cases}$$)

Show also that at a smooth point of W_d equality holds scheme-theoretically, i.e., that the intersection of the tangent spaces on the right is the tangent space to W_d.

C-3. Generalize the previous exercise to the following set-theoretic equalities:

$$W_d^r = \bigcap_{\substack{E \in C_r \\ F \in C_{e-d+r}}} (W_e + u(E) - u(F)), \qquad d \le e \le g - 1,$$

$$W_d^r = \bigcap_{\substack{E \in C_{r-s} \\ F \in C_{e-d+r-s}}} (W_e^s + u(E) - u(F)), \qquad r \le s \le e - g + 1.$$

C-4. Let E be a divisor of degree zero. If

$$W_{g-1} = W_{g-1} - u(E)$$

show that $E \sim 0$.

(*Suggestion*: Use C-1 to show that

$$h^0(C, K(E - p_1 - \cdots - p_{g-1})) \ge 1 \quad \text{for all} \quad p_1, \ldots, p_{g-1},$$

if and only if $E \sim 0$.)

C-5. Use the preceding exercise and the symmetry of Θ to show that Riemann's constant κ satisfies

$$2\kappa = u(K)$$

independently of the choice of base points.

C-6. Let $p, q \in C$, and establish the set-theoretic equalities

$$\begin{aligned} W_{g-1} \cap (W_{g-1} + u(p - q)) &= (W_{g-2} + u(p)) \cup (W_g^1 - u(q)) \\ &= \bigcap_{r, s \in C} ((W_{g-1} + u(p - r)) \cup (W_{g-1} + u(s - q))). \end{aligned}$$

The next few exercises will establish the following converse to the preceding exercise: if C is non-hyperelliptic, then for any point $u \in J(C)$ there exist $w, w' \in J(C)$ with $w, w' \neq 0, u$ and such that

$$W_{g-1} \cap (W_{g-1} + u) \subset (W_{g-1} + w) \cup (W_{g-1} + w'),$$

if and only if

$$u = u(p - q)$$

for some $p, q \in C$. This in turn will yield the

(*) **Theorem** *Let C be a smooth non-hyperelliptic curve with Jacobian $J(C)$ and theta divisor Θ. Then*:

(i) *there exist points $v \in J(C)$ such that*

$$\Theta \cap (\Theta + v) \subset (\Theta + w) \cup (\Theta + w')$$

for some pair $\{w, w'\} \neq \{0, v\}$ of points in $J(C)$;

(ii) *for any such v, the intersection*

$$X = \Theta \cap (\Theta + v)$$

has two irreducible components X_1 and X_2; and

(iii) *the curve $u(C)$ is (up to $\pm$) a translate of the locus*

$$\{w \in J(C) \colon \Theta + w \supset X_1\}.$$

(As suggested in Mumford [3], this theorem has a corollary the Torelli theorem.)

We adopt the following notation: for $\mathscr{D} = \mathbb{P}V$ a linear system corresponding to a linear subspace $V \subset H^0(C, L)$ and points $p_1, \ldots, p_e \in C$, set

$$V(-p_1 - \cdots - p_e) = V \cap H^0(C, L(-p_1 - \cdots - p_e)),$$

$$\mathscr{D}(-p_1 - \cdots - p_e) = \mathbb{P}V(-p_1 - \cdots - p_e).$$

C-7. Suppose that $\mathscr{D}$ and $\mathscr{E}$ are distinct g^r_d's on a curve C. Show that there exists points $p_1, \ldots, p_{r+1}$ on C such that

$$\mathscr{D}(-p_1 - \cdots - p_{r+1}) \neq \varnothing,$$

$$\mathscr{E}(-p_1 - \cdots - p_{r+1}) = \varnothing.$$

C-8. We keep the above notations and let $\mathscr{D}, \mathscr{E}_1, \ldots, \mathscr{E}_m$ be distinct g^r_d's on C, with $\mathscr{D}$ base-point-free and $\phi_{\mathscr{D}}$ birational. Using the uniform position property, show that there exists points $p_1, \ldots, p_{r+1}$ on C such that

$$\mathscr{D}(-p_1 - \cdots - p_{r+1}) \neq \varnothing,$$

$$\mathscr{E}_i(-p_1 - \cdots - p_{r+1}) = \varnothing \quad \text{for all } i.$$

C-9. Show that the conclusion of the preceding exercise is false if we drop the assumption that $\mathscr{D}$ is base-point-free, and likewise if we drop the assumption that $\phi_{\mathscr{D}}$ is birational.

(*Hint*: For the latter, suppose, for example, that $\mathbb{Z}/2 \times \mathbb{Z}/2$ acts on C with quotient $\mathbb{P}^1$, and that $\phi_{\mathscr{D}}$ is the quotient map $C \to \mathbb{P}^1$.)

C-10. Using C-7, show that, for any divisor D of degree zero on C,

$$X = W_{g-1} \cap (W_{g-1} + u(D))$$

is contained in $W_{g-1} + u(E)$ where $\deg E = 0$ if, and only if,

$$E \sim D \quad \text{or} \quad E \sim 0.$$

C-11. Using C-8 again show that if C is non-hyperelliptic, the variety X of the preceding exercise is not contained in any finite union of translates of W_{g-1} other than W_{g-1} and $W_{g-1} + u(D)$ unless $D \sim p - q$ for points $p, q \in C$.

D. Prill's Problem

The following question was posed by David Prill:

Question *Let $f: C \to B$ be a branched covering of a curve of genus $g(B) \geq 2$, and for $p \in B$ set $D_p = f^{-1}(p)$. Then is it possible that $r(D_p) \geq 1$ for all $p \in B$?*

D-1. Show that not all the divisors D_p are linearly equivalent.

D-2. Show that, for general $p \in B$, there does not exist $q \neq p$ such that $D_p \sim D_q$.

(*Hint*: Use Exercise C-1 of Chapter IV.)

D-3. Let G be a finite group acting on $\mathbb{P}^1$. Show that if G has a fixed point (i.e., a point $p \in \mathbb{P}^1$ such that $gp = p$ for all $g \in G$) then it has two.

D-4. Using the above, conclude that if the mapping $f: C \to B$ in the question is a Galois covering, then $r(D_p) \neq 1$ for general $p \in B$.

E. Another Proof of the Torelli Theorem

In this sequence of exercises we let $C \subset \mathbb{P}^{g-1}$ be a canonical curve of genus g with Jacobian $J(C)$ and theta divisor $\Theta \subset J(C)$. As usual we identify all the projectivized tangent spaces to $J(C)$ with $\mathbb{P}^{g-1} = \mathbb{P}T_0(J(C))$, and for $p \in \mathbb{P}^{g-1}$ we set

$$\Gamma_p = \{u \in \Theta : p \in \mathbb{P}T_u(\Theta)\};$$

the following will be a consequence of the first four exercises below.

Theorem. *If $g \geq 5$, then the set-theoretic equality*

$$\{p \in \mathbb{P}^{g-1}: \Gamma_p \text{ is reducible}\} = C \cup \{\text{finite set}\}$$

is valid.

Corollary. *A non-hyperelliptic curve of genus $g \geq 5$ may be reconstructed from the data* $(J(C), \Theta)$.

A variant of the argument below gives the corollary in the hyperelliptic case, and a closer inspection of the proof will also allow the extension to the cases $g = 3, 4$.

E-1. Suppose that C is bi-elliptic with $\pi: C \to E$ a degree two mapping onto an elliptic curve E. Show that the chords π^*q $(q \in E)$ all pass through a common point $p \in \mathbb{P}^{g-1}$, $p \notin C$.
(*Hint*: Show that for any $q, q' \in E$, $r(\pi^*(q + q')) = 1$ and use the geometric form of the Riemann–Roch theorem.)

E-2. Show that, conversely, if $g \geq 5$ and a point $p \in \mathbb{P}^{g-1} - C$ lies on ∞^1 chords to C, then C is bi-elliptic and the chords to C through p meet C at the fibers of the bi-elliptic pencil. In case $g = 4$ show that, if C is not bi-elliptic then C lies on a quadric cone and p is the vertex of that cone.
(*Hint*: Consider the projection of C from p to $\mathbb{P}^{g-2}$.)

E-3. Let $p \in \mathbb{P}^{g-1}$ be any point and define the incidence correspondence $I_p \subset C_{g-1} \times (\mathbb{P}^{g-1})^*$ by

$$I_p = \{(p_1 + \cdots + p_{g-1}, H): \overline{p_1 + \cdots + p_{g-1}} \subset H, p \in H\}.$$

Show that I_p is reducible if and only if p lies on ∞^1 chords to C.
(*Hint*: Use the uniform position lemma.)

E-4. Complete the proof of the above theorem.

E-5. In this exercise we let C be hyperelliptic with 2-to-1 map $\pi: C \to \mathbb{P}^1$ ramified at $p_1, \ldots, p_{2g+2}$. Let $\phi_K(C) \subset \mathbb{P}^{g-1}$ be the canonical image of C and $q_i = \phi_K(p_i)$. With the above notations, show that

$$\{p \in \mathbb{P}^{g-1}: \Gamma_p \text{ is reducible}\} = \phi_K(C) - \{q_1, \ldots, q_{2g+2}\}.$$

From this you may conclude the Torelli theorem in this case also.

E-6. Show that for any $p \in \mathbb{P}^{g-1}$ and d with $2 < d < g - 1$, p lies on ∞^{d-1} $(d-1)$-planes spanned by their intersection with a canonical curve C if and only if $p \in C$.
(*Hint*: Look at the projection of C from p.)

E-7. Using the preceding exercise, show that a non-hyperelliptic curve C may be recovered from the data $(J(C), W_d(C))$ for $3 \leq d \leq g - 2$ by the set-theoretic relation

$$C = \{p \in \mathbb{P}^{g-1}: \dim \Gamma_p = d - 1\}.$$

where $\Gamma_p = \{u \in W_d(C): p \in \mathbb{P}T_u(W_d(C))\}$

F. Curves of Genus 5

In this sequence of exercises we will use the following notations:

C is a non-hyperelliptic curve of genus 5 identified with its canonical model $C \subset \mathbb{P}^4$;

$|\mathscr{I}_C(2)| \cong \mathbb{P}^2$ is the linear system of quadrics in $\mathbb{P}^4$ containing C, $\Gamma \subset |\mathscr{I}_C(2)|$ is the locus of quadrics of rank ≤ 4, and $\Gamma' \subset \Gamma$ is the locus of quadrics of rank ≤ 3.

F-1. Show that, if C is non-trigonal, then no quadric $Q \in |\mathscr{I}_C(2)|$ may be singular at a point of C. Using Bertini's theorem conclude that the general $Q \in |\mathscr{I}_C(2)|$ is smooth in the non-trigonal case.

F-2. On the other hand, if C is trigonal show that every quadric containing C is singular, and that there is exactly one quadric of rank 3 containing C. (Cf. Section 3 of Chapter III.)

From these exercises we conclude that:

C is non-trigonal if and only if it lies on a smooth quadric.

From now on we will assume that C is non-trigonal.

F-3. Suppose that $\Lambda \subset \mathbb{P}(H^0(\mathbb{P}^4, \mathcal{O}(2)))$ is a general 2-plane. Show that the intersection of the quadrics $Q \in \Lambda$ is a smooth canonical curve of genus 5. Conclude that for a general curve of genus 5, Γ is a smooth plane curve.

F-4.

(i) Show that the singular locus of Γ is exactly Γ', and that the singularities of Γ are ordinary nodes.
(*Suggestion*: Use the description of tangent cones to symmetric determinantal varieties given in Exercise A-4 of Chapter II.)

(ii) Conclude that Γ is a plane quintic curve with no multiple components and, in particular, Γ spans $|\mathscr{I}_C(2)|$.

F-5. Let $D \in C_4$ and suppose that $u(D) \in W_4^1(C)$. Show that the quadric

$$Q_D = \mathbb{P}\mathscr{T}_{u(D)}(W_4^1(C)) \in \Gamma,$$

and conclude that for C a non-hyperelliptic and non-trigonal curve of genus 5

$$\phi_K(C) = \bigcap_{u \in \Theta_{\text{sing}}} \mathbb{P}\mathscr{T}_u(\Theta).$$

This is, of course, a special case of Theorem (4.2) in the text.

F-6. With the preceding notations, show that the following are equivalent:

$$Q_D \in \Gamma',$$
$$\mathcal{O}_C(2D) \cong K_C.$$

From this conclude that a non-hyperelliptic curve of genus 5 can have at most 10 semi-canonical pencils. In the hyperelliptic case there are 56 semi-canonical pencils; while a trigonal curve has at most one.

(*Hint*: Show that a configuration of lines maximizes the number of double points.)

F-7. Show that the map

$$\phi: W_4^1(C) \to \Gamma$$

given by

$$\phi(D) = Q_D$$

has degree two and is branched exactly over Γ'. From this conclude that

$$\begin{aligned}(W_4^1(C))_{\text{sing}} &= \phi^{-1}(\Gamma') \\ &= \{u(D): D \in C_4^1 \text{ and } \mathcal{O}_C(2D) \cong K_C\}.\end{aligned}$$

In the next exercise we will use the following result of Fulton–Hansen [1]:

Theorem. *Let $X \subset \mathbb{P}^r$ be an irreducible subvariety of codimension d, Y an irreducible variety of dimension d', and*

$$f: Y \to \mathbb{P}^n$$

a finite map. If $d' > d$, then $f^{-1}(X)$ is connected.

This theorem will be proved in exercise batch A of Chapter VII.

We will also use the scheme $V_k \subset \mathbb{P}^9$ of quadrics in $\mathbb{P}^4$ of rank $\leq k$. There is a double-covering

$$\tilde{V}_4 \to V_4$$

corresponding to the two rulings of a rank 4 quadric; this covering is ramified over V_3.

F-8. Show that $\tilde{V}_4$ is irreducible.

F-9. Show that, if Γ is smooth (which, by F-3, is the case for C a general curve of genus 5), then $W_4^1(C)$ is irreducible.

(*Hint*: Show that the double covering in F-7 is the pull-back of

$$f: \tilde{V}_4 \to V_4,$$

and now use the Fulton–Hansen theorem).

Remark. This is a special case of the corollary to the Connectedness Theorem (1.4) of Chapter V.

F-10. Using F-7 and an examination of the branching behavior of $W_4^1(C) \to \Gamma$, show that $W_4^1(C)$ is irreducible if and only if Γ is irreducible.

For the next two exercises we recall that C is bi-elliptic if there is a 2-to-1 map $\pi: C \to E$ onto an elliptic curve. Assuming that C has genus 5 and is bi-elliptic, there is an irreducible component $u(\Sigma)$ of $W_4^1(C)$ corresponding to the pull-back of the g_2^1's on E.

F-11. Retaining the notation of F-7 show that $\phi(u(\Sigma))$ is a line component of Γ, and that $\phi(u(\Sigma))$ meets the remainder of Γ transversely in four points corresponding to the four semi-canonical pencils on C pulled back from E (cf. Exercise F-4).

F-12. Conversely, show that, if $L \subset \Gamma$ is a line component and $E \to L$ is the double cover branched at the four points of intersection of L with $\Gamma - L$, then C is a double cover of the elliptic curve E and $L = \phi(u(\Sigma))$.

If we agree to call $\pi: C \to E$ a bi-elliptic pencil, then by the preceding two exercises there is a bijection

$$\{\text{lines in } \Gamma\} \to \{\text{bi-elliptic pencils}\}.$$

In particular, C can have at most five bi-elliptic pencils.

F-13. Show that C has five bi-elliptic pencils if and only if it has 10 semi-canonical pencils.

F-14. Let $(\lambda_{\alpha,i})$ $(1 \le \alpha \le 3, 1 \le i \le 5)$ be a 3×5 matrix with all 3×3 minors non-zero. Show that the intersection of the quadrics

$$Q_\alpha = \sum_i \lambda_{\alpha,i} X_i^2$$

in $\mathbb{P}^4$ is a smooth canonical curve with 10 semi-canonical pencils, and that conversely any such curve can be represented in this way.

(*Hint*: If C has 10 semi-canonical pencils then by the preceding exercise there are five bi-elliptic pencils $\pi_i: C \to E_i$. Show that the corresponding sheet interchanges must all commute, and consider the representation of the group they generate on $\mathbb{P}^4$ (the elements of this group are represented by simultaneously diagonalizable matrices.))

In the following few exercises, we investigate the geometry of C and Γ in case C is a double cover of a curve of genus 3.

Suppose now that B is a curve of genus 3, $C \xrightarrow{\pi} B$ a two-sheeted unramified map, and $\pi^*: J(B) \to J(C)$ the pull-back map. Let $u_0 \in \operatorname{Ker} \pi^* \subset J(B)$ be the divisor class of order 2 on B associated to π (cf. Exercises 13 and 14 in Appendix B).

F-15. Let

$$D = W_2(B) \cap (W_2(B) + u_0).$$

Show that

$$E = \pi^* D \subset W_4^1(C).$$

F-16. Using the adjunction formula on $J(B)$, show that the arithmetic genus of D is

$$p_a(D) = 7.$$

Note. A count of constants indicates that D should be smooth for B outside a codimension 1 subset of moduli.

F-17. Show that the map $\pi^*: D \to E$ is an unramified double cover, and conclude that the arithmetic genus

$$p_a(E) = 4.$$

F-18. Show that if d, w_1 are the classes of D and $W_1(B)$ in $J(B)$, then

$$d = 2w_1;$$

and using Exercises D-10 and D-11 of Chapter VIII, conclude that if e is the class of E in $J(C)$,

$$(e \cdot \theta_{J(C)}) = 6.$$

In particular, conclude that $E \neq W_4^1(C)$.

F-19. Using the description of the theta characteristics on B given in Exercises 47–50 of Appendix B, show that D contains 12 theta characteristics of B, and E correspondingly contains 6 theta characteristics of C.

F-20. Using the preceding two exercises, show that the map $\phi: W_4^1(C) \to \Gamma$ expresses E as a double cover of a cubic curve, branched at six points. Conclude that Γ consists of a cubic curve G and a conic curve H, with E the double cover of G branched over the intersection $G \cap H$.

F-21. Now let $F = \phi^{-1}H$ be the remaining component(s) of $W_4^1(C)$ besides E. Show that F spans a two-dimensional abelian subvariety A of $J(C)$, and hence that $A = J(F)$.

F-22. Conclude from the preceding exercise that the Prym variety of the double cover $C \xrightarrow{\pi} B$ is the Jacobian $J(F)$ of the component F of $W_4^1(C)$ (cf. Appendix C).

F-23. Is the converse of the above analysis correct—i.e., if Γ consists of a cubic curve H and a conic curve G, is C then a double cover of a curve of genus 3? (Show that this is the case if Γ is a sum of five lines.)

We may collect some of the results of the above exercises in the following table:

canonical curve $C \subset \mathbb{P}^4$	locus $\Gamma \subset \mathbb{P}^2 = \mathbb{P}I_C(2)$ of singular quadrics through C.	curve $\Theta_{\text{sing}} \subset J(C)$ with involution -1.
C is trigonal $\Leftrightarrow$	Every quadric through C is singular (i.e., all quadrics containing the scroll S have rank ≤ 4.) In this case $\Gamma = \mathbb{P}^2$ $\Leftrightarrow$	Θ_{sing} is the union of two copies of C conjugate under the involution and meeting at two points
C non-trigonal $\Leftrightarrow$	Γ is a plane quintic having at most nodes (the nodes of Γ are in a 1–1 correspondence with semi-canonical pencils on C). In particular, Γ has no multiple components $\Leftrightarrow$	$\Theta_{\text{sing}}/-1 \cong \Gamma$. In particular, every singular quadric containing C is a tangent cone to Θ_{sing}
C bi-elliptic ($\Rightarrow$ non-trigonal) $\Leftrightarrow$	Γ contains a line (the line components of Γ are in a 1–1 correspondence with the elliptic pencils of degree 2 on C) $\Leftrightarrow$	Θ_{sing} contains elliptic curve components, each mapping 2 to 1 onto the lines in Γ
C non-trigonal and there is a two-sheeted unramified mapping $C \to C'$ where C' has genus 3 $\Rightarrow$	Γ contains a conic (the converse is not known) $\Rightarrow$	$\Theta_{\text{sing}} = \Sigma_1 \cup \Sigma_2$ has two components of genera $p_a(\Sigma_1) = 2$, $p_a(\Sigma_2) = 4$ meeting at six points
C non-trigonal and with no semi-canonical pencils ($\Rightarrow C$ not bi-elliptic) $\Leftrightarrow$	Γ is a smooth plane quintic $\Leftrightarrow$	Θ_{sing} is a smooth irreducible curve of genus 11

A general question of some interest is: For what g can we find a linear system on an algebraic surface that includes the general curve of genus g? This is elementary to do for $g \leq 6$, $g \neq 5$. The following sequence of exercises will sketch a proof of such a theorem when $g = 5$. The essential step is the following

Theorem. *A general curve of genus five may be represented as a plane sextic having two tacnodes and one ordinary node.*

F-24. Suppose that Γ is smooth and let $l \subset \mathbb{P}^2$ be a line cutting out a divisor $D = l \cdot \Gamma$ where

$$D = p_1 + \cdots + p_5 \in |\mathcal{O}_\Gamma(1)|$$

with the p_i distinct. Show that there is a birational plane model C_0 of C where C_0 is a sextic having five nodes r_i with the following property: The g_4^1's on C_0 cut out by the lines through r_i correspond to the points $p_i \in \Gamma$ (recall that each point of Γ corresponds to two g_4^1's on C).

(*Hint*: Look at the surface obtained by intersecting the pencils of quadrics corresponding to $l \subset \mathbb{P}^2 = |\mathcal{I}_C(2)|$. For this it may be helpful to recall the representation of the transverse intersection of two quadrics in $\mathbb{P}^4$ as $\mathbb{P}^2$ blown up at five points (cf. Griffiths-Harris [1])).

F-25. Suppose now that l becomes tangent to Γ with p_i, p_j coming together. Show that the nodes r_i, r_j come together to form a tacnode.

F-26. Prove the theorem above by letting l become an ordinary bitangent to Γ. (Here, you must show that a general plane quintic Γ has at least one ordinary bitangent—in fact, this is true whenever Γ is smooth (why?).)

F-27. Using the above theorem, show that there exists a linear system of plane quintics containing a general curve of genus five.

Note. We don't know if this is possible for any g with $7 \le g \le 22$ (for $g \ge 23$, cf. Chapter XIII).

G. Accola's Theorem

In the following sequence of exercises, we will prove a theorem of Accola's, relating three semi-canonical pencils on a curve of genus 5 and the quadrics $Q \subset \mathbb{P}^4$ that they determine. Specifically, let $C \subset \mathbb{P}^4$ be a canonical curve, Q_1, Q_2, Q_3 distinct quadrics of rank 3 containing C, Λ_i a 2-plane of the ruling of Q_i, and $D_i = \Lambda_i \cdot C$ so that $|D_i|$ is a semi-canonical pencil. We will prove the

Theorem. (a) *if Q_1, Q_2, Q_3 are dependent, C is a two-sheeted cover of an elliptic curve E, the pencils $|D_i|$ are pull-backs from E, and correspondingly*

$$r(2K - D_1 - D_2 - D_3) = 1.$$

(b) *If conversely the Q_i are independent, then $r(2K - D_1 - D_2 - D_3) = 0$.*

The conclusion of (b) means that if $\lambda\colon J_2 \times J_2 \to \mathbb{Z}/2$ is the Weil pairing (cf. Appendix B), $\lambda(D_2 - D_1, D_3 - D_1) = 1$.

G-1. (i) Show that C is not trigonal (cf. Exercise F-2 above).
(ii) Show that, if $Q_1, Q_2, Q_3 \in \mathbb{P}^2 \cong |\mathscr{I}_C(2)|$ are not linearly independent, then C is bi-elliptic (cf. Exercise F-11).

Now deduce part (a) of the theorem.

G-2. Suppose now that Q_1, Q_2, Q_3 are independent. Show that Q_1 is the unique quadric containing C and Λ_1, i.e., no linear combination of Q_2, Q_3 contains Λ_1.
(*Hint*: If Q_2 contained Λ_1, then $Q_1 \cap Q_2 \cap Q_3$ would contain a conic.)

G-3. Show that, for $i \neq j$, Λ_i meets Λ_j in one point p_{ij}.
(*Hint*: $\mathcal{O}(D_i + D_j) \neq K$.)

G-4. Let $D_i = C \cdot \Lambda_i$ be as above. Using G-2, show that the six points of the divisor $D + p_{12} + p_{13}$ do not lie on a conic in Λ_1.
(*Hint*: Show Q_2, Q_3 cut out distinct conics on Λ_1.)

G-5. Show that no quadric in $\mathbb{P}^4$ contains $\Lambda_1 \cup \Lambda_2 \cup \Lambda_3$.
(*Hint*: Show that such a quadric would be singular at $\Lambda_i \cap \Lambda_j$.)

G-6. From the preceding two exercises show that the 12 points $D_1 \cup D_2 \cup D_3$ impose independent conditions on quadrics. Conclude that

$$r(2K - D_1 - D_2 - D_3) = 0,$$

and from this deduce Accola's theorem.

Here is a variant of Accola's theorem: ϕ, Γ are as in exercise batch F.

G-7. Suppose that D_1, D_2, D_3, D_4 are semi-canonical pencils on a canonical curve $C \subset \mathbb{P}^4$, and set $p_i = \phi(u(D_i)) \in \Gamma$. Show that the points p_i are collinear if and only if $\sum_i D_i \in |2K_C|$.

H. The Difference Map ϕ_d (III)

We retain the notations from exercise batches D of Chapter V and A of Chapter VI. This sequence of exercises will give a proof of the following:

Theorem. *Let C be a curve of genus g and $u \in J(C)$. If $d < g - 1$ and*

$$\dim \phi_d^{-1}(u) \geq d - 1,$$

then either:

(i) $u = u(p - q)$ *where* $p, q \in C$;
(ii) C *is bi-elliptic with* $\pi: C \to E$ *the two-sheeted map to an elliptic curve and* $u \in \pi^* J(E)$; *or*
(iii) C *has genus* 4 *and* u *is the difference of the two* g_3^1*'s.*

H-1. Suppose that $\dim \phi_d^{-1}(u) = k$. Show that, for some $e \geq 0$, the fiber $\phi_{d-e}^{-1}(u)$ has dimension $k - e$ and the general point $(D, E) \in \phi_{d-e}^{-1}(u)$ consists of a pair of divisors with disjoint support.

H-2. Suppose now that $\dim \phi_d^{-1}(u) = k$, and that for (D, E) a general point in $\phi_d^{-1}(u)$ the divisors D, E have disjoint support. Show that

$$r(D + E) \geq k.$$

H-3. Combining the preceding exercises conclude that

$$\dim \phi_d^{-1}(u) \leq d, \qquad d \leq g - 1,$$

with equality if and only if $u = 0$.

H-4. Let u_1, u_2 and $w\colon C_d \times C_d \to J(C)$ be defined by

$$u_1(D, E) = u(D), \qquad u_2(D, E) = u(E),$$

$$w(D, E) = u(D + E).$$

Show that for any $u \in J(C)$,

$$\begin{aligned} \dim u_1(\phi_d^{-1}(u)) &= \dim u_2(\phi_d^{-1}(u)) \\ &= \dim w(\phi_d^{-1}(u)); \end{aligned}$$

call this number s.

H-5. Now suppose that for some $u \in J(C)$ and $d \leq g - 2$, we have

$$\dim \phi_d^{-1}(u) = d - 1;$$

let (D, E) be a general point of $\phi_d^{-1}(u)$. Show that

$$r(D) + r(E) + s \geq d - 1.$$

H-6. With the hypotheses of the preceding problem conclude that either:

(i) C is hyperelliptic, with $|D|$ and $|E|$ multiples of the hyperelliptic g_2^1 plus base points and $u = u(p - q)$ for some $p, q \in C$; or
(ii) C is not hyperelliptic, in which case

$$r = r(D) = r(E) = \frac{d - 1 - s}{2}.$$

H-7. In case (ii) above, show, by the Martens–Mumford theorem, that if $r \geq 1$ then conditions (ii) or (iii) of the above theorem hold. Show, on the other hand, that if $r = 0$ then

$$\dim W_{2d}^{d-1}(C) \geq d - 1$$

and conclude that either $d = 1$—i.e., condition (i) of the theorem holds; $d = 2$; or $d = 3$ and $g = 5$. The latter two cases will be dealt with in the following exercises to conclude the proof of the theorem.

H-8. Suppose now that $d = 2$; let $\{(D_\lambda, E_\lambda)\}_{\lambda \in B} = \phi_d^{-1}(u)$. Show that if for general λ, $|D_\lambda + p|$ is a g_3^1 for some p then the same is true of E_λ (for a different g_3^1), and we must have condition (iii) of the theorem.

(*Hint*: Use Exercise D-2 of Chapter IV).

Assuming this is not the case, then, show in case $g \geq 5$ that for general $p \in C$ there is a unique divisor D_λ (resp. $E_{\lambda'}$) containing p, and then that $D_\lambda = E_{\lambda'}$; conclude that the divisors D_λ, E_λ are all fibers of an elliptic pencil.

(*Hint*: Consider the diagram

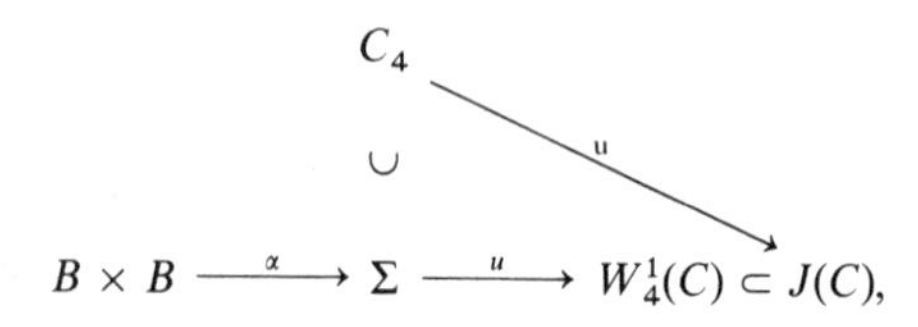

where

$$\begin{cases} \alpha(\mu, \lambda) = D_\mu + E_\lambda \in C_4, \\ \Sigma = \text{image of } \alpha, \\ u \text{ is the Abelian sum mapping.} \end{cases}$$

Show that:

(i) α cannot be birational (i.e., $B \times B$ cannot be fibered by $\mathbb{P}^1$'s);
(ii) $D_\mu = E_{\lambda(\mu)}$ for some map $\lambda: B \to B$;
(iii) a *general* pair of points of B moves in a pencil.

From this conclude the result.)

H-9. To remove the final possibility mentioned in H-7 above, suppose now that $r = 0$, $d = 3$, $g = 5$, and $s = 2$. Conclude that

$$w(\phi_d^{-1}(u)) = u(K_C) - W_2(C),$$

and hence that for general $p, q \in C$ there exists divisors D, E having disjoint support such that

$$K_C = \mathcal{O}(D + E + p + q) \quad \text{and} \quad u = u(D - E).$$

Use the uniform position lemma (monodromy statement) to derive a contradiction.

A similar argument will dispose of the remaining case $g = 4$ in the preceding exercise.

I. Geometry of the Abelian Sum Map u in Low Genera

I-1. Let C be a curve of genus 2. Show that the map

$$u: C_2 \to \mathrm{Pic}^2(C)$$

expresses C_2 as the blow-up of $\mathrm{Pic}^2(C)$ at the point $u(K_C)$.

I-2. Let C be a hyperelliptic curve of genus 3. Show that $W_2(C) \subset \operatorname{Pic}^2(C)$ has an ordinary double point at $u(g_2^1)$.

(*Note*: You could also show that the graph of the hyperelliptic involution in C_2 is a rational curve of self-intersection -2.)

I-3. Let C be any curve of genus 3. Show that the map

$$u \colon C_3 \to \operatorname{Pic}^3(C)$$

expresses C_3 as the blow-up of $\operatorname{Pic}^3(C)$ along the curve $W_3^1(C) = u(K_C) - W_1(C)$.

I-4. Let C be a non-hyperelliptic curve of genus 4. If C has two distinct g_3^1's, show that the map

$$u \colon C_3 \to \operatorname{Pic}^3(C)$$

blows down these g_3^1's to ordinary double points of $W_3(C)$. What happens if C instead has just one, semi-canonical g_3^1?

I-5 (Suggested by A. Collino). Let C be a curve of genus 4 with two g_3^1's, and let $\tilde{W}_3$ be the blow-up of $W_3(C)$ at the nodes of $W_3(C)$ coming from the two g_3^1's. Show that $\tilde{W}_3$ can be realized as the variety

$$\tilde{W}_3 \cong \{(D, E) \colon D + E \in |K|\} \subset C_3 \times C_3$$

and that the projection $\pi \colon \tilde{W}_3 \to W_3(C)$ factors through the projection $\pi_1 \colon \tilde{W}_3 \to C_3$ on the first factor: $\pi = u \circ \pi_1$.

(*Note*: If we blow down opposite rulings on these two quadrics, then we get a smooth non-projective three-fold. This follows from the fact that the two g_3^1's are homologous in C_3 for a general curve C, a fact which in turn follows by computing the intersection numbers of the two g_3^1's with the classes x and θ in Chapter VII below.)

I-6. Let C now be a trigonal curve of genus 5. Show that the projectivized tangent cone $\mathbb{P}\mathcal{T}_{u(g_3^1)}(W_3(C))$ to $W_3(C)$ at its singular point is the rational normal scroll containing the canonical curve $C \subset \mathbb{P}^4$.

I-7. Let C be an arbitrary curve of genus 5, $|D|$ a g_4^1 on C. Show that the projectivized tangent cone $\mathbb{P}\mathcal{T}_{u(D)}(W_4(C))$ is a singular quadric, with vertex the projectivized Zariski tangent space to $W_4^1(C)$.

I-8. Now let C be a general curve of genus 6, $|D_1|, \ldots, |D_5|$ its g_4^1's. Show that the projectivized tangent cone $\mathbb{P}\mathcal{T}_{u(D_i)} W_4(C)$ is a three-dimensional rational normal scroll X_i containing the canonical curve $C \subset \mathbb{P}^5$; and that the intersection $S = \bigcap X_i$

is a smooth del Pezzo surface (surface of degree 5) in $\mathbb{P}^5$. More generally, show that if C has between two and five g_4^1's, the intersection $S = \bigcap X_i$ of the corresponding scrolls is a surface of degree 5 in $\mathbb{P}^5$, singular if and only if C has fewer than five g_4^1's (cf. Arbarello-Harris [1]).

I-9. Let C be a bi-elliptic curve of genus 6 or more. Show that the intersection of the projectivized tangent cones to $W_4(C)$ at points of $W_4^1(C)$ is a cone over an elliptic normal curve. What is the situation in genus 5?

Appendix B

Theta Characteristics

In the following series of exercises we will develop the algebraic theory of theta characteristics on smooth curves (for the case of singular curves, cf. Harris [4]).

§1. Norm Maps

If $\pi: C' \to C$ is any branched cover of Riemann surfaces, f a meromorphic function on C', we may define the meromorphic function $Nm_\pi f$ (or just Nmf) by

$$(Nm_\pi f)(p) = \prod_{q \in \pi^{-1}(p)} f(q)^{\nu(q)},$$

where $\nu(q)$ is the multiplicity with which q appears in the fiber of π over p.

1. Show that $Nm_\pi f$ is indeed a well-defined meromorphic function on C, and that the map

$$Nm_\pi: \mathcal{M}(C')^* \to \mathcal{M}(C)^*$$

is a (multiplicative) group homomorphism.

2. Suppose that $C' \xrightarrow{\pi} C$ is a *Galois extension*, i.e., that the group of deck transformations of C' acts simply transitively on the sheets. Show that the extension of function fields $\mathcal{M}(C) \to \mathcal{M}(C')$ is a Galois extension, and that

$$Nm_\pi = Nm_{\mathcal{M}(C')/\mathcal{M}(C)}: \mathcal{M}(C')^* \to \mathcal{M}(C)^*$$

is the algebraic norm.

3. Letting $\pi: C' \to C$ again be an arbitrary cover, we may define a map

$$Nm_\pi = \pi_*: \mathrm{Div}(C') \to \mathrm{Div}(C),$$

by

$$Nm(\sum q_i) = \sum \pi(q_i).$$

Show that for any function $f \in \mathscr{M}(C')$,

$$Nm((f)) = (Nmf)$$

and conclude that Nm induces a map

$$Nm_\pi: J(C) \to J(C').$$

Show that $Nm_\pi \circ \pi^*: J(C) \to J(C)$ is just multiplication by $n = \deg \pi$.

We now want to prove the following assertion:

$(*)$ *If $C' \xrightarrow{\pi} C$ is any branched cover of Riemann surfaces, the map*

$$Nm_\pi: \mathscr{M}(C')^* \to \mathscr{M}(C)^*$$

is surjective.

To do this, we make one auxiliary definition: We say that a field K is *quasi-algebraically closed* if every polynomial $F(X_1, \ldots, X_n)$ of degree d in n variables over K has a root in K whenever $n > d$.

4. Show that $\mathscr{M}(\mathbb{P}^1) = \mathbb{C}(t)$ is quasi-algebraically closed.
(*Hint*: If $F \in \mathbb{C}[t][X_1 \cdots X_n]$, assume the solution is a collection of polynomials of degree $m \gg 0$, and solve for their coefficients.)

5. Show that a finite extension of a quasi-algebraically closed field is again quasi-algebraically closed.

6. If C is any Riemann surface, show that $\mathscr{M}(C)$ is quasi-algebraically closed.

7. Use the preceding exercise to conclude statement $(*)$.
(*Note*: It suffices to prove this for Galois covers.)

§2. The Weil Pairing

In this sequence of exercises we will use the following notations:

$J_2 = J_2(C)$ is the group of points of order 2 on $J(C)$, sometimes thought of as the line bundles L such that $L \otimes L = \mathcal{O}_C$;

Since $J(C) = H^0(C, K)^*/H_1(C, \mathbb{Z})$ we also make the identification

$$J_2(C) \cong H_1(C, (1/2)\mathbb{Z})/H_1(C, \mathbb{Z}) \cong H_1(C, \mathbb{Z}/2);$$

for f a meromorphic function on C and D a divisor of degree 0 with support disjoint from (f), we define

$$f(D) = \prod_{p \in C} f(p)^{\mathrm{mult}_p(D)}.$$

Given meromorphic functions f, g on C whose divisors D, E have disjoint support, the *Weil reciprocity* law is

$$f(E) = g(D).$$

8. Verify the Weil reciprocity law in case $C = \mathbb{P}^1$ by writing f, g explicitly in terms of a euclidean coordinate on $\mathbb{P}^1$.

9. If $C' \xrightarrow{\pi} C$ is a branched covering of curves, and the norm maps $Nm = Nm_\pi$ are as defined above, show that for any functions f on C', g on C and divisors D on C', E on C,

$$(Nmf)(E) = f(\pi^*E),$$

$$g(NmD) = (\pi^*g)(D).$$

Conclude, in particular, that

$$f((\pi^*g)) = (Nmf)(g).$$

10. Using the previous two exercises, prove the Weil reciprocity law for a pair of functions f, g on any curve C by representing f as the pull-back of a function on $\mathbb{P}^1$.

Note. For a direct proof of the Weil reciprocity law using the residue theorem, see Griffiths–Harris [1].

11. Suppose that $\eta, \omega \in J_2$ and let D, E be divisors with disjoint support representing η, ω, respectively. Write

$$2D = (f), \qquad 2E = (g),$$

and show that

$$\frac{f(E)}{g(D)} = \pm 1.$$

12. Retaining the notation from the preceding exercise define the *Weil pairing*

$$\lambda\colon J_2 \times J_2 \to \mathbb{Z}/2$$

by

$$\lambda(\eta, \omega) = \frac{1}{\pi\sqrt{-1}} \log \frac{f(E)}{g(D)}.$$

Show that this is well defined.

13. Given $\eta \in J_2$ represented by $L \in \mathrm{Pic}^0(C)$ with $L^2 \cong \mathcal{O}_C$, define the double cover

$$\pi\colon \tilde{C}_\eta \to C$$

by

$$\tilde{C}_\eta = \{s \in L\colon s^2 = 1\},$$

where 1 is the constant section of $\mathcal{O}_C$ (alternatively, if D is a divisor representing η with $2D = (f)$, then $\tilde{C}_\eta$ is isomorphic to the normalization of the curve $\{(x, y) \in C \times \mathbb{P}^1\colon y^2 = f(x)\}$). Check that this double cover is smooth and unramified.

14. Show that the kernel of

$$\pi^*\colon J_2(C) \to J_2(\tilde{C}_\eta)$$

is $\{0, \eta\}$.

(*Suggestion*: Consider the mapping induced on the first homology groups by

$$\pi^*\colon J(C) \to J(\tilde{C}_\eta)$$

by making the identification

$$H^1(J(C), \mathbb{Z}) \cong H^1(C, \mathbb{Z}),$$

and similarly for $\tilde{C}_\eta$.)

15. If $\tau\colon \tilde{C} \to \tilde{C}$ denotes the sheet-interchange involution for the double cover constructed above, show that there exist meromorphic functions h on $\tilde{C}$ such that

$$h(q) = -h(\tau q).$$

(*Suggestion*: For any effective divisor D on $\tilde{C}$ invariant under τ, consider the representation of τ on $H^0(\tilde{C}, \mathcal{O}(kD))$ for large k.)

16. If τ is as in the preceding exercise and h is any meromorphic function on $\tilde{C}$ with $h = -h \circ \tau$, show that (Nmh) is divisible by 2; i.e., that we can write

$$(Nmh) = 2D,$$

and that then

$$\tilde{C} \cong C_\eta,$$

where

$$\eta = \mathcal{O}_C(D) \in J_2(C).$$

17. Show that the image of the map

$$1 - \tau: J(\tilde{C}) \to J(\tilde{C}),$$

$$u \mapsto u - \tau u,$$

is an abelian subvariety of dimension $g - 1$ in $J(\tilde{C})$.

18. Show that the kernel of the map

$$Nm: J(\tilde{C}) \to J(C)$$

consists of a finite union of cosets of the image of $1 - \tau$ by points of order 2 on $J_2(\tilde{C})$ (i.e., that $\text{Im}(1 - \tau) = \text{Prym}(\tilde{C}/C)$; cf. Appendix C).

(*Hint*: Use the fact that $\pi^*(NmD) = D + \tau D$.)

19. Using the identifications $J_2(C) = H_1(C, \mathbb{Z}/2)$ and $J_2(\tilde{C}) = H_1(\tilde{C}, \mathbb{Z}/2)$, show that

$$\text{order}((\text{Ker } Nm) \cap J_2(\tilde{C})) = 2^{2g+1},$$

and conclude that $\text{Ker}(Nm)$ has exactly two connected components.

20. Show that every divisor E on $\tilde{C}$ with $NmE \sim 0$ on C is linearly equivalent on $\tilde{C}$ to $D - \tau D$ for some $D \in \text{Div}(\tilde{C})$.

(*Hint*: Write $NmE = (f)$ and (using (*)) $f = Nmg$; and consider the divisor $E + (g)$.)

21. Let

$$\phi_d\colon \mathrm{Div}^d(\tilde{C}) \to J(\tilde{C})$$

be defined by

$$\phi_d(D) = D - \tau D.$$

Show, using the preceding exercise, that the images of ϕ_d and $\phi_{d'}$ are disjoint if and only if $d + d'$ is odd (i.e., that for k large, the images of ϕ_{2k} and ϕ_{2k+1} are the connected components of $\mathrm{Ker}(Nm)$).

22. Prove the preceding exercise directly, arguing that if D is a divisor on $\tilde{C}$ such that

$$D - \tau D = (f),$$

then for any $g \in \mathcal{M}(\tilde{C})$ such that $g = -g \circ \tau$ we have,

$$1 = \frac{f((g))}{g((f))} = \frac{g(\tau D)}{g(D)}$$

so that

$$\deg D \equiv 0 \mod(2).$$

23. Let $\pi\colon \tilde{C}_\eta \to C$ be the double cover associated to $\eta \in J_2(C)$, and for $\omega \in J_2(C)$ write

$$\pi^*E \sim D - \tau D,$$

where E represents ω (this uses Exercise 20 above). Show that the Weil pairing is

$$\lambda(\eta, \omega) \equiv \deg D \mod(2)$$

24. Consider the identifications

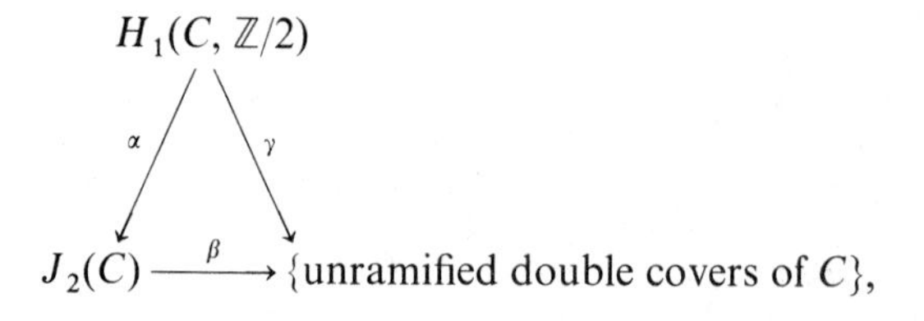

where α is the obvious map, β is the map in Exercise 13, and γ associates to any class $\eta \in H_1(C, \mathbb{Z}/2)$ the double cover corresponding to the subgroup $\eta^\perp$ in $H_1(C, \mathbb{Z}/2)$ with respect to the intersection pairing. Show that these identifications commute.

25. Using the preceding exercise, show that via the identification α the Weil pairing on J_2 corresponds to intersection of cycles on $H_1(C, \mathbb{Z}/2)$.

26. Suppose that $f: C \to \mathbb{P}^1$ is a smooth hyperelliptic curve with ramification divisor $R = p_1 + \cdots + p_{2g+2}$ and branch divisor $B = x_1 + \cdots + x_{2g+2}$ where $x_i = f(p_i)$. Show that every $\eta \in J_2(C)$ can be represented by a divisor

$$E = p_{i_1} + \cdots + p_{i_k} - p_{j_1} - \cdots - p_{j_k},$$

where the i_α and j_β are distinct.

(*Hint*: C is the Riemann surface of $y^2 = \prod_{i=1}^{2g+2}(x - x_i)$.)

27. Keep the notations of the preceding problem and let $|D|$ be the hyperelliptic g_2^1. Then,

$$E \sim kD - p_{l_1} - \cdots - p_{l_{2k}}.$$

Show that any two points $\eta, \omega \in J_2(C)$ may be represented by divisors of the form in the preceding exercise with disjoint support, and use this representation to compute $\lambda(\eta, \omega)$.

28. Let λ_{ij} be an arc in $\mathbb{P}^1$ joining x_i to x_j (and missing the other branch points), γ_{ij} its inverse image in C, and η_{ij} the class of γ_{ij} in $H_1(C, \mathbb{Z}/2)$. Show that the intersection number mod 2 of η_{ij} and η_{kl} is

$$(\eta_{ij} \cdot \eta_{kl}) = \begin{cases} 0 & \text{if } \{i, j\} \cap \{k, l\} = \varnothing \text{ or } \{i, j\}, \\ 1 & \text{otherwise.} \end{cases}$$

29. Show that η_{ij} is represented by the divisor

$$\begin{aligned} E_{ij} &= p_i - p_j \\ &\sim D - p_i - p_j. \end{aligned}$$

30. Comparing the above with Exercise 27, verify directly that the Weil pairing corresponds to the intersection pairing on cycles mod(2).

Remark. By continuity plus irreducibility of the moduli space of curves of genus g this implies the same assertion for any curve of genus g.

§3. Theta Characteristics

Definition. A *theta characteristic* on a smooth curve C is a line bundle L with $L \otimes L \cong K_C$.

31. If C has genus g, show that there are 2^{2g} theta characteristics (cf. the definition of $J_2(C)$ given above).

32. Suppose that C is hyperelliptic with two-sheeted covering $\pi\colon C \to \mathbb{P}^1$ given by the hyperelliptic series $|D|$ and with ramification divisor $p_1 + \cdots + p_{2g+2}$.

(i) Show that every theta characteristic on C is of the form

$$L = \mathcal{O}_C(E),$$

where

$$E = mD + p_{i_1} + \cdots + p_{i_{g-1-2m}},$$

with $-1 \leq m \leq (g-1)/2$, and the p_{i_α} are distinct.

(ii) Show that the above representation is unique if $m \geq 0$, while if $m = -1$ there is a single relation

$$-D + p_{i_1} + \cdots + p_{i_{g+1}} \sim -D + p_{j_1} + \cdots + p_{j_{g+1}}$$

if $\{i_1, \ldots, i_{g+1}, j_1, \ldots, j_{g+1}\} = \{1, \ldots, 2g+2\}$.

(This shows again that the number of theta characteristics is 2^{2g}.)

33. For the theta characteristic E in Exercise 32 show that

$$h^0(C, L) = m + 1.$$

In the next three exercises we will prove the following result, whose formulation requires the concept of a "family of line bundles over a family of curves" that we will make precise in the second volume.

Intuitively, a *family of curves* $\{C_t\}$ is a map $\pi\colon \mathscr{C} \to B$ of analytic spaces whose fibers $C_t = \pi^{-1}(t)$ are all smooth curves; a family of line bundles $\{L_t\}$ on $\{C_t\}$ are the restrictions $L_t = \mathscr{L}|_{C_t}$ of a line bundle $\mathscr{L}$ on $\mathscr{C}$. Finally, a family of theta characteristics is a family $\{L_t\}$ of line bundles such that $L_t^{\otimes 2} \cong K_{C_t}$ for all t.

With this said, our goal is to prove the

(**) **Theorem.** *Let $L_t \to C_t$ be a holomorphic family of theta characteristics. Then $h^0(C_t, L_t)$ is constant modulo 2.*

34. Let L be a theta characteristic on C, and $D = p_1 + \cdots + p_d$ any divisor of degree $d > g - 1$; let V be the $2d$-dimensional vector space

$$V = H^0(C, L(D)/L(-D)).$$

Via the identifications $L^2 \cong K_C$, $L(D)^2 \cong K_C(2D)$ we may define a quadratic form Q on V by

$$Q(\sigma, \tau) = \sum_i \mathrm{Res}_{p_i}(\sigma\tau).$$

Show that Q is well defined and non-degenerate.

35. Let $\Lambda_1 \subset V$ be the subspace $H^0(C, L/L(-D))$. Show that Λ_1 is a d-dimensional isotropic subspace for Q.

36. Let $\Lambda_2 \subset V$ be the image, under the restriction map, of $H^0(C, L(D))$. Show that $\dim \Lambda_2 = d$ and that Λ_2 is an isotropic subspace for Q (use the Riemann–Roch and residue theorems).

37. Show that

$$h^0(C, L) = \dim(\Lambda_1 \cap \Lambda_2).$$

From Exercises 34 through 37 together with statement (*) of Exercise B-5 in Chapter II you may complete the proof of Theorem (**).

The following exercise also requires concepts that will be formulated precisely in the second volume.

38. Let $\mathcal{T}^r_g \subset \mathcal{M}_g$ denote the family of smooth curves of genus g having a theta characteristic L with $h^0(C, L) \geq r + 1$. If $L_0 \to C_0$ is a theta characteristic with

$$\begin{cases} h^0(C_0, L_0) \geq r + 1, \\ h^0(C_0, L_0) \equiv r + 1 \bmod(2) \end{cases}$$

then convince yourself that

$$\operatorname{codim} \mathcal{T}^r_g \leq r(r + 1)/2$$

in a neighborhood of C_0.

(*Hint*: Use Exercises 34–37 and Exercise B-7 of Chapter II.)

Definition. A theta characteristic L is *even* or *odd* according to the parity of $h^0(C, L)$.

From Theorem (**) it follows that the configuration of even and odd theta characteristics is invariant under deformation of the curve, and this configuration will be described in the following exercises. More precisely, keeping our notations, for each theta characteristic $L \to C$ we define

$$q_L: J_2(C) \to \mathbb{Z}/2$$

by

$$q_L(\eta) = h^0(C, L \otimes \eta) - h^0(C, L),$$

and will prove the

Riemann–Mumford Relation. *For* $\omega, \eta \in J_2(C)$

$$q_L(\eta + \omega) + q_L(\eta) + q_L(\omega) = \lambda(\eta, \omega),$$

where $\lambda(\eta, \omega)$ *is the Weil pairing.*

39. Let $\eta \in J_2(C)$ and $\pi: C_\eta \to C$ the double covering associated to η and $\tau: C_\eta \to C_\eta$ the sheet exchange involution. Show that, for any $\omega \in J_2(C)$,

$$h^0(C_\eta, \pi^*L) = h^0(C, L) + h^0(C, L \otimes \eta),$$

$$h^0(C_\eta, \pi^*(L \otimes \omega)) = h^0(C, L \otimes \omega) + h^0(C, L \otimes \eta \otimes \omega).$$

(*Hint*: Consider the representation of τ on $H^0(C_\eta, \pi^*L)$.)

40. Let E be any divisor on C representing ω and write

$$\pi^*E \sim D - \tau D$$

with D, τD disjoint (cf. Exercise 20). Set

$$\begin{aligned} V &= H^0(C_\eta, \pi^*L(D)/\pi^*L) \oplus H^0(C_\eta, \pi^*L/\pi^*L(-\tau D)) \\ &= H^0(C, L(NmD)/L) \oplus H^0(C, L/L(-NmD)). \end{aligned}$$

Note that $V \cong \mathbb{C}^{2d}$, where $\deg D = d$. Using the second expression for V and an isomorphism $L^{\otimes 2} \cong K_C$, define a quadratic form

$$Q: V \times V \to \mathbb{C}$$

by

$$Q(\sigma \oplus \sigma', \lambda \oplus \lambda') = \sum \operatorname{Res}(\sigma\lambda') + \sum \operatorname{Res}(\sigma'\lambda),$$

where the sum is over the points of NmD. Show that Q is well defined and non-degenerate.

41. Denote by Λ_1 and Λ_2 the direct summands of V and set

$$\Lambda_3 = \text{Image}(H^0(C_\eta, \pi^*L(D)) \to V).$$

Show that

(i) $\Lambda_1, \Lambda_2, \Lambda_3$ are d-dimensional isotropic subspaces for Q;
(ii)

$$\dim \Lambda_1 \cap \Lambda_3 = h^0(C_\eta, \pi^*L),$$

$$\dim \Lambda_2 \cap \Lambda_3 = h^0(C_\eta, \pi^*L \otimes \omega);$$

and, using Exercise B-6 of Chapter II, that

$$h^0(C_\eta, \pi^*L) + h^\circ(C_\eta, \pi^*L \otimes \omega) \equiv \deg(D) \bmod(2).$$

By combining this with Exercise 23 you may complete the proof of the Riemann–Mumford relation.

There is another version of the Riemann–Mumford relation due to Thurston.

42. Let $L \cong \mathcal{O}_C(D)$ be a theta characteristic and ϕ a meromorphic 1-form with $(\phi) = 2D$. Let $\eta \in J_2(C) \cong H_1(C, \mathbb{Z}/2)$ and

$$\gamma(t)\colon [0, 1] \to C - \text{support } D$$

be a smooth, non-self-intersecting loop representing η. Show that

$$q_L(\eta) = \frac{1}{\pi\sqrt{-1}} \int_0^1 d \log\langle \gamma'(t), \phi(\gamma(t))\rangle.$$

(*Hint*: Using Exercises 32 and 33 do this explicitly for a hyperelliptic curve, and using Theorem (**) deduce it for all C.)

43. Use Exercises 42 and 30 to deduce the Riemann–Mumford relation.

(**Note**. If γ_1, γ_2 are two smooth loops, then $\gamma_1 + \gamma_2$ may have self-intersections and will have to be smoothed.)

In the next three exercises we will give a smattering of the classical approach to theta characteristics.

44. Let C be a compact Riemann surface of genus g with normalized basis $\lambda_1, \ldots, \lambda_{2g}$ for $H_1(C, \mathbb{Z})$, period matrix Z and associated Riemann theta function $\theta(u)$ as defined in Chapter I. For $a, b \in \mathbb{Z}^g$ show that

$$\theta_{a,b}(u) = e^{-\pi\sqrt{-1}\langle b, u\rangle}\theta\left(u + \frac{a}{2} + \frac{Zb}{2}\right)$$

is odd or even depending on the parity of $\langle a, b\rangle$.

45. Using the preceding exercise and the Riemann singularity theorem show that if L is the line bundle of degree $g - 1$ on C with $u(L) = \kappa$ (=vector of Riemann constants),

$$\begin{aligned} \omega = \sum a_i \frac{\lambda_i}{2} + \sum b_i \frac{\lambda_{g+i}}{2} \in \frac{H_1(C, \frac{1}{2}\mathbb{Z})}{H_1(C, \mathbb{Z})} \\ \cong H_1(C, \tfrac{1}{2}\mathbb{Z}) \\ \cong J_2(C), \end{aligned}$$

then

$$h^0(L \otimes \omega) \equiv \langle a, b\rangle \quad \mathrm{mod}(2).$$

Use this to conclude also the invariance of $h^0(C_t, L_t)$ mod(2).

We return now to the theta characteristics on a hyperelliptic curve, where we will verify directly the Riemann–Mumford relation.

46. Let C and L be as in Exercise 32, and let E_{ij} be the divisor representing the homology class n_{ij} of Exercises 28 and 29. If $I = \{i_1, \ldots, i_{g-1-2m}\}$, show that if $m \geq 0$,

$$h^0(L(E_{ij})) = h^0(L) + \begin{cases} -1 & \text{if } i, j \notin I, \\ +1 & \text{if } i, j \in I, \\ 0 & \text{otherwise.} \end{cases}$$

Derive the corresponding expression in case $m = -1$.

47. Using Exercises 46 and 43 or 44, verify the Riemann–Mumford relation for C.

§4. Quadratic Forms Over $\mathbb{Z}/2$

Let V be a vector space over the field $\mathbb{Z}/2$. By a *bilinear form* on V we will mean a symmetric, bilinear pairing

$$\lambda: V \times V \to \mathbb{Z}/2$$

with the additional requirement that $\lambda(v, v) = 0$ for all $v \in V$ (for a general symmetric bilinear pairing $\lambda(x, y) = \sum a_{ij}x_i y_j$, the restriction of λ to the diagonal is in fact linear:

$$q(x, x) = \sum a_{ij}x_i x_j = \sum a_{ii}x_i^2 = (\sum a_{ii}x_i)^2;$$

here we require that this linear form be zero). Given a quadratic form (i.e., a homogeneous polynomial of degree two)

$$q: V \to \mathbb{Z}/2$$

we have the *associated bilinear form*

$$\lambda(x, y) = q(x + y) + q(x) + q(y).$$

48. Show that if $\dim V = 2n$ there is a unique non-degenerate symmetric bilinear form on V up to isomorphism, which can be written as

$$\lambda(x, y) = \sum_i (x_i y_{n+i} + x_{n+i} y_i).$$

49. Show that there are exactly two quadratic forms having a given non-degenerate associated bilinear form λ, and that if λ is as in the preceding exercise, these can be written as

$$q^+(x) = \sum x_i x_{n+i}$$

and

$$q^-(x) = x_1^2 + x_{n+1}^2 + \sum x_i x_{n+i}.$$

Counting the number of zeros of each of these, show that they are not isomorphic.

50. Interpret the Riemann–Mumford relation above to say that q_L is a quadratic form whose associated bilinear form is the Weil pairing. Show that $q_L \cong q^+$ if L is even, while $q_L \cong q^-$ if L is odd.

In the following exercises we will consider the action of monodromy on the set of theta characteristics, and for this we will use the results of Chapter X together with the additional notations:

S (resp. S^+, S^-) denotes the set of all (resp. even, odd) theta characteristics;

$Sp(J_2)$ is the group $Sp_{2g}(\mathbb{Z}/2)$ of symplectic transformations mod(2), defined by

$$Sp(J_2) = \{A \in GL_{2g}(\mathbb{Z}/2) : \lambda(A\eta, A\omega) = \lambda(\eta, \omega) \text{ for } \eta, \omega \in J_2\},$$

where $J_2 = J_2(C)$ and λ is the Weil pairing. Using the terminology from Chapter X, we will prove the

(***) **Theorem.** *The global monodromy group acts transitively on* S^+ *and* S^-.

51. By arguments similar to those in Chapter X show that $Sp(J_2)$ is generated by Picard–Lefschetz transformations

$$\eta \to \eta + \lambda(\eta, \omega)\omega.$$

Conclude that the monodromy is $Sp(J_2)$ and that the monodromy action is transitive on non-zero elements of J_2.

52. For any theta characteristic L and $A \in Sp(J_2)$, show that

$$l_A(\eta) = q_L(A\eta) + q_L(\eta)$$

is linear in η. Conclude that there exists a unique $\omega \in J_2$ with

$$q_L(A\eta + \omega) = q_L(\eta), \qquad q_L(\omega) = 0, \qquad \eta \in J_2.$$

Denoting by $O(S)$ the group of *affine* linear transformations of J_2 that preserve q_L, this shows that the projection map

$$O(S) \to Sp(J_2)$$

is an isomorphism.

53. Show that every $\omega \in J_2$ with $q_L(\omega) = 0$ occurs for some A in Exercise 52.
(*Hint*: Show that any two quadratic forms mod 2 on $(\mathbb{Z}/2)^{2n}$ with the same number of zeros are conjugate under $Sp_{2n}(\mathbb{Z}/2)$).

From Exercises 51 through 53 you may complete the proof of Theorem (***).

Appendix C

Prym Varieties

In this appendix we will give a brief account of the theory of Prym varieties. Prym varieties are abelian varieties associated to unramified double covers of curves; in studying them we will correspondingly have a number of occasions to refer back to the preceding appendix.

To begin with, our basic objects are as follows: a curve C of genus g and an unramified double cover

$$f: \tilde{C} \to C,$$

with involution

$$\tau: \tilde{C} \to \tilde{C}$$

exchanging sheets of $\tilde{C}$ over C. We will also denote by τ the induced involution on the groups $H_1(\tilde{C}, \mathbb{Z})$, $H^0(\tilde{C}, K)$, and $J(\tilde{C})$; in the first two cases we will denote the $+1$- and -1-eigenspaces of τ by a superscript $+$ or $-$.

To begin with, we represent the map π topologically as in the figure, and choose normalized bases for $H_1(C, \mathbb{Z})$ and $H_1(\tilde{C}, \mathbb{Z})$ as drawn.

Note that we have:

$$f_* \tilde{a}_0 = a_0,$$
$$f_* \tilde{b}_0 = 2b_0,$$

$$\left.\begin{aligned} f^*(a_i) &= \tilde{a}_i + \tilde{a}_{g-1+i} \\ f^*(b_i) &= \tilde{b}_i + \tilde{b}_{g-1+i} \end{aligned}\right\} \quad i = 1, \ldots, g-1.$$

A basis for the lattice $H_1(\tilde{C}, \mathbb{Z})^-$ of the anti-invariant cycles is then given by

$$\tilde{a}_i - \tilde{a}_{i+g-1}, \qquad \tilde{b}_i - \tilde{b}_{i+g-1}, \qquad i = 1, \ldots, g-1.$$

Let $\omega_0, \omega_1, \ldots, \omega_{2g-2}$ be a normalized basis for $H^{1,0}(\tilde{C})$, i.e., one for which

$$\int_{\tilde{a}_i} \omega_j = \delta_{ij}.$$

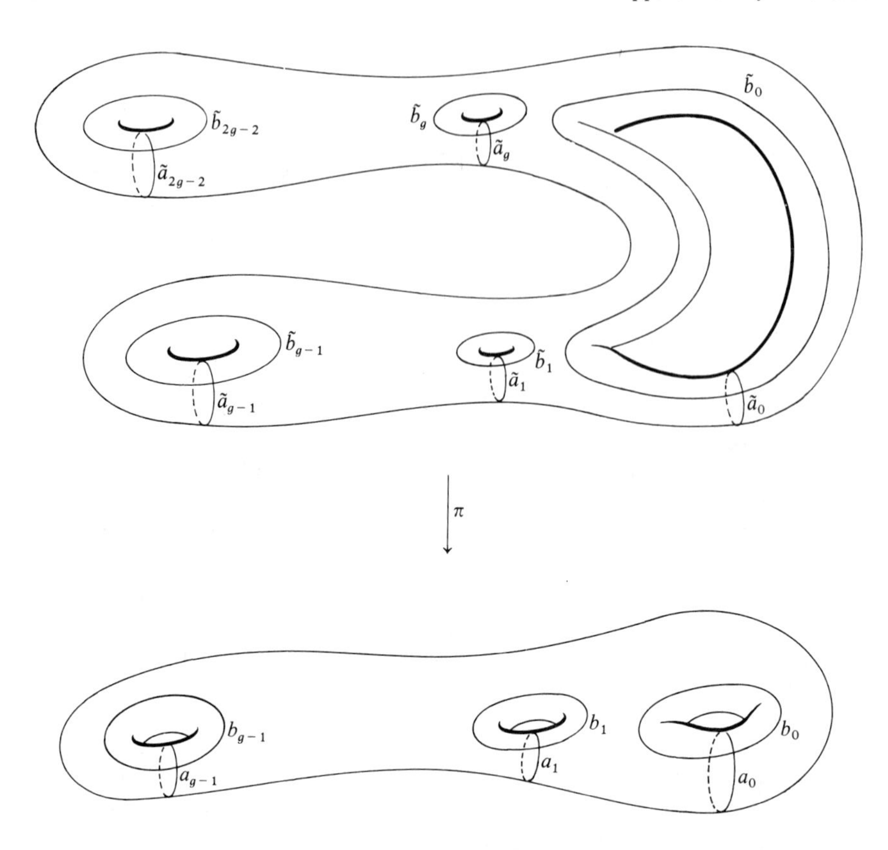

It is then easy to see that a basis for $H^{1,0}(\tilde{C})^+$ is given by

$$\{\omega_0\} \cup \{\omega_i + \omega_{i+g-1} : i = 1, \ldots, g - 1\},$$

while a basis for $H^{1,0}(\tilde{C})^-$ is given by

$$\{\omega_i - \omega_{i+g-1} : i = 1, \ldots, g - 1\}.$$

It is also easy to see that

$$H^1(\tilde{C}, \mathbb{Z})^- \subset (H^{1,0}(\tilde{C})^-)^*$$

is a lattice of maximal rank so that one can define a *complex torus*

$$\operatorname{Prym}(\tilde{C}, \tau) = \frac{(H^{1,0}(\tilde{C})^-)^*}{H_1(\tilde{C}, \mathbb{Z})^-} \subset J(\tilde{C})$$

of dimension $g - 1$. The intersection matrix of the anti-invariant cycles is of the form

$$\left(\begin{array}{c|c} 0 & 2I_{g-1} \\ \hline -2I_{g-1} & 0 \end{array}\right),$$

and this suggests that the *principal polarization on* $J(\tilde{C})$ *restricts to twice a principal polarization* on $\text{Prym}(\tilde{C}, \tau)$. Equipped with this principal polarization the complex torus is called the Prym variety of the two-sheeted covering $f\colon \tilde{C} \to C$. To better understand, let us consider the three basic maps Nm, π^*, and τ,

$$\begin{array}{ccc} J(C) & \xrightarrow{\pi^*} & J(\tilde{C}) \\ \| \wr & & \| \wr \\ \text{Pic}^0(C) & \xleftarrow[Nm]{} & \text{Pic}^0(\tilde{C}), \end{array}$$

where the *norm mapping* is defined as in Exercise 3 of Appendix B by

$$Nm(\sum m_i p_i) = \sum m_i f(p_i).$$

Let Θ and $\tilde{\Theta}$ denote the theta divisors on $J(C)$ and $J(\tilde{C})$ respectively. Recall (Exercise 14 of Appendix B) that

$$\text{Ker } \pi^* = \{0, L_\varepsilon\},$$

where L_ε is a point of order 2 in $J(C)$. Moreover, it is easy to check that

$$\text{Prym}(\tilde{C}, \tau) = \text{Im}(1 - \tau) = \left\{\begin{array}{l}\text{connected component of } Nm^{-1}(0) \\ \text{containing the identity}\end{array}\right\}.$$

As it turns out it is more convenient to look at the norm map Nm as defined in $\text{Pic}^{2g-2}(\tilde{C})$

$$Nm\colon \text{Pic}^{2g-2}(\tilde{C}) \to \text{Pic}^{2g-2}(C).$$

The two basic results in Mumford's approach are the following:

(a) $Nm^{-1}(K_C)$ *is the disjoint union of two translates of* $\text{Prym}(\tilde{C}, \tau)$, P^+, *and* P^-. *Moreover,* $L \in P^+$ *if and only if* $NmL = K_C$ *and* $h^0(\tilde{C}, L)$ *is even.*

(b) $\tilde{\Theta} \cdot P^+ = 2\Xi$ *where* Ξ *is a principal polarization on* P^+.

The first result, and especially the statement about the parity of $h^0(\tilde{C}, L)$, is a crucial one and is based on Exercises 35–37 of Appendix B. Namely one observes that for each $L \in \mathrm{Pic}^{2g-2}(\tilde{C})$ such that $NmL = K_C$, the rank 2 vector bundle $\pi_* L$ is equipped with a quadratic form

$$\pi_* L \otimes \pi_* L \to K_C.$$

Thus, by an argument analogous to the one in Exercises 34–37 of Appendix B,

$$h^0(C, \pi_* L) = h^0(\tilde{C}, L)$$

is constant modulo 2 on P^+ (resp. P^-). From (a) and (b) it follows that

$$\begin{aligned}
\tilde{\Theta} \cap P^+ &= \{L \in \mathrm{Pic}^{2g-2}(\tilde{C})\colon NmL = K_C, h^0(\tilde{C}, L) \text{ is even and at least } 2\} \\
&\subset \tilde{\Theta}_{\mathrm{sing}}, \\
\Xi_{\mathrm{sing}} &= \{L \in P^+ : \mathrm{mult}_L \tilde{\Theta} \text{ is even and at least } 4\} \\
&\cup \{L \in P^+ : \mathrm{mult}_L \tilde{\Theta} = 2 \text{ and } T_L(P^+) \subset \mathscr{T}(\tilde{\Theta})\} \\
&= \Xi'_{\mathrm{sing}} \cup \Xi''_{\mathrm{sing}}.
\end{aligned}$$

We are now going to show that

Claim. *If $g \geq 5$ and $\dim \Xi_{\mathrm{sing}} \geq g - 5$, then almost all points $L \in \Xi_{\mathrm{sing}}$ are of the form $L = (f^*M)(\Delta)$ where $NmL = K_C$, M is a line bundle on C, Δ an effective divisor on $\tilde{C}$, $h^0(C, M) \geq 2$ and $h^0(\tilde{C}, L)$ is even.*

In this assertion "almost all" means a union of Zariski open subsets in each component of maximum dimension in Ξ_{sing}. To prove this we recall the description of Zariski tangent spaces:

$$\begin{aligned}
T_L(\tilde{\Theta}) &= (\mathrm{Im}(\mu_0\colon H^0(\tilde{C}, L) \otimes H^0(\tilde{C}, K_C L^{-1}) \to H^0(\tilde{C}, K_C)))^{\perp}, \\
T_L(P^+) &= (H^{1,0}(\tilde{C})^-)^*.
\end{aligned}$$

Notice that since $NmL = K_C$ we have $K_{\tilde{C}} = \pi^* NmL$. On the other hand, $\pi^* NmL = L \otimes \tau^* L$. It then follows that

$$K_{\tilde{C}} L^{-1} = \tau^* L.$$

Assume now that $L \in \Xi''_{\mathrm{sing}}$ so that

$$\begin{aligned}
NmL &= K_C, \\
h^0(\tilde{C}, L) &= 2, \\
T_L(P^+) &\subset \mathscr{T}_L(\Theta).
\end{aligned}$$

By the Kempf singularity theorem, the equation in $T_L(J(\tilde{C}))$ of the tangent cone $\mathscr{T}_L(\tilde{\Theta})$ is given by

$$\det(s_i t_j) = 0,$$

where $\{s_1, s_2\}$ and $\{t_1, t_2\}$ are bases of $H^0(\tilde{C}, L)$ and $H^0(\tilde{C}, K_{\tilde{C}} L^{-1})$, respectively. By the above, we may assume that

$$t_j = \tau^* s_j, \qquad j = 1, 2,$$

and we decompose the holomorphic differential $s_i t_j$ in its invariant and anti-invariant parts

$$s_i(\tau^* s_j) = \tfrac{1}{2}[s_i(\tau^* s_j) + s_j(\tau^* s_i)] + \tfrac{1}{2}[s_i(\tau^* s_j) - s_j(\tau^* s_i)].$$

It then follows that

$$\begin{aligned} T_L(P^+) \cap \mathscr{T}_L(\tilde{\Theta}) &= T_L(P^+) \cap \{\det(s_i t_j) = 0\} \\ &= \left\{\det\begin{pmatrix} 0 & s_1(\tau^* s_2) - s_2(\tau^* s_1) \\ -(s_1(\tau^* s_2) - s_2(\tau^* s_1)) & 0 \end{pmatrix} = 0\right\}. \end{aligned}$$

Now to say that $T_L(P^+) \subset \mathscr{T}_L(\tilde{\Theta})$ exactly means that

$$(*) \qquad s_1(\tau^* s_2) = s_2(\tau^* s_1).$$

But this in turn means that s_2/s_1 defines a meromorphic function h on C and letting

$$\Delta = \text{l.c.d.}((s_1), (s_2)), \qquad D = (h)_0, \qquad M = \mathcal{O}_C(D),$$

we get $L = (\pi^* M)(\Delta)$, proving our claim in case $L \in \Xi''_{\text{sing}}$. We can now assume that the point $L \in \Xi_{\text{sing}}$ belongs to an irreducible component Z of Ξ'_{sing}, with $\dim Z \geq g - 5$. Suppose now that for a general $L \in Z$

$$h^0(\tilde{C}, L) = r + 1$$

so that

$$L \in (W^r_{2g-2}(\tilde{C}) - W^{r+1}_{2g-2}(\tilde{C})) \cap P^+.$$

It follows that

$$\dim T_L(W^r_{2g-2}(\tilde{C})) \cap T_L(P^+) \geq g - 5.$$

In view of the identifications of $T_L(\tilde{\Theta})$ and $T_L(P^+)$ above, this translates into

$$\dim(\operatorname{Im}(v_0\colon \Lambda^2 H^0(\tilde{C}, L) \to H^{1,0}(\tilde{C})^-)^\perp) \geq g - 5,$$

where $v_0(s_1 \wedge s_2) = s_1(\tau^* s_2) - s_2(\tau^* s_1)$. This means that $\operatorname{codim}(\operatorname{Ker} v_0) \leq 4$. But in $\Lambda^2 H^0(\tilde{C}, L)$ the decomposable vectors form a non-degenerate subvariety of dimension at least five, so that there exists a decomposable vector $s_1 \wedge s_2$ in $\operatorname{Ker} v_0$. This means that $(*)$ holds and therefore we can proceed as before.

We want to apply the above discussion to the double covering

$$f\colon \Theta_{\text{sing}} \to \Gamma$$

described in exercise batch F of Chapter VI, where Θ_{sing} is the singular locus of the theta divisor of a general canonical curve C of genus 5 and Γ is a smooth plane quintic realized as the locus of rank 4 quadrics through C. Let τ be the -1 involution on Θ_{sing}. By Exercise F-7 of Chapter VI and the Riemann singularity theorem, if Q is a general quadric through C, $f^{-1}(Q)$ is the pair of g^1_4's determined by the two distinct rulings of Q.

Before restricting to this situation let us consider any two-sheeted unbranched covering

$$f\colon \tilde{\Gamma} \to \Gamma$$

of a smooth plane quintic, and denote by τ the sheet-interchange involution on $\tilde{\Gamma}$. Let $L_\varepsilon \in J(\Gamma)$ be the point of order 2 determining f, and let $\Xi \subset \operatorname{Prym}(\tilde{\Gamma}, \tau)$ be the theta divisor of the Prym variety of f ($\operatorname{Prym}(\tilde{\Gamma}, \tau)$ is a principally polarized abelian variety of dimension 5). We will show that:

$(**)$ *The following conditions are equivalent*:

(i) $\dim \Xi_{\text{sing}} = 1$;
(ii) $h^0(\Gamma, \mathcal{O}_\Gamma(1) \otimes L_\varepsilon)$ *is even.*

If (i) *or* (ii) *holds then*

(iii) $\Xi_{\text{sing}} = \{(f^*\mathcal{O}_\Gamma(1))(p - \tau p)\colon p \in \tilde{\Gamma}\} \cup \{\textit{finite set}\}$
$\cong \tilde{\Gamma} \cup \{\textit{finite set}\}$.

If (i) *or* (ii) *are not satisfied then the following equivalent conditions are met*

(iv) $\dim \Xi_{\text{sing}} = 0$;
(v) $h^0(\Gamma, \mathcal{O}_\Gamma(1) \otimes L_\varepsilon)$ *is odd.*

If (iv) *or* (v) *holds, then*

(vi) $f^*\mathcal{O}_\Gamma(1) \in \Xi_{\text{sing}}$.

The proof of (**) is a consequence of the claim above. In fact, from the proof of the claim we deduce that, if there exists an irreducible component Z of Ξ_{sing} with $\dim Z \geq g - 5 = 1$ then a general point $L \in Z$ is of the form

$$L = (f^*M)(\Delta),$$

with

$$\begin{cases} NmL = K_\Gamma, \\ h^0(\Gamma, M) \text{ even and at least 2.} \end{cases}$$

Since

$$2 \deg M = \deg Nmf^*M \leq \deg K_\Gamma = 10,$$

we have that $|M|$ is a g_d^r with $d \leq 5$ and $r \geq 1$. As follows from Exercise 18 of Appendix A, only two cases are possible: either $M = \mathcal{O}_\Gamma(1)$ or $M = \mathcal{O}_\Gamma(1)(-p)$. Since $\mathcal{O}_\Gamma(2) = K_\Gamma$, a line bundle L of the form $(f^*M)(\Delta)$ satisfies the condition above only if $L = f^*\mathcal{O}_\Gamma(1)$ or $L = (f^*\mathcal{O}_\Gamma(1))(p - \tau p)$, $p \in \tilde{\Gamma}$. Now for any point $p \in \tilde{\Gamma}$, $h^0(\tilde{\Gamma}, (f^*\mathcal{O}_\Gamma(1))(p - \tau p))$ is even if and only if $h^0(\tilde{\Gamma}, f^*\mathcal{O}_\Gamma(1))$ is odd. Since

$$H^0(\tilde{\Gamma}, f^*\mathcal{O}_\Gamma(1)) \cong H^0(\Gamma, \mathcal{O}_\Gamma(1)) \oplus H^0(\Gamma, \mathcal{O}_\Gamma(1) \otimes L_\varepsilon),$$

we conclude that the following are equivalent:

(a) $\dim Z \geq 1$;
(b) a general point of Z is of the form $f^*\mathcal{O}_\Gamma(p - \tau p)$;
(c) $Z = \{(f^*\mathcal{O}_\Gamma(1))(p - \tau p) : p \in \tilde{\Gamma}\}$;

the equivalence of (a), (b), and (c) establishes (**).

A covering $f: \tilde{\Gamma} \to \Gamma$ is called *even* (resp. *odd*) if $h^0(\Gamma, \mathcal{O}_\Gamma(1) \otimes L_\varepsilon)$ is even (resp. odd). We consider first the even case. Then $\text{Prym}(\tilde{\Gamma}, \tau)$ is a five-dimensional principally polarized abelian variety for which

$$\Xi_{\text{sing}} = \tilde{\Gamma} \cup \{\text{finite set}\};$$

moreover, the -1 involution on Ξ_{sing} corresponds to the τ involution on $\tilde{\Gamma}$. It is then natural to ask if we are in the situation $f: \Theta_{\text{sing}} \to \Gamma$ above.

The following theorem is the result of published and unpublished results of Clemens, Tjurin, Masiewicki, Donagi, and Smith and it says that the answer is yes, and in the strongest sense.

Theorem. *Let C be a general canonical curve of genus* 5 *and Γ the locus of rank* 4 *quadrics containing C. Then the two-sheeted unramified covering*

$$f: \Theta_{\text{sing}} \to \Gamma$$

is even and every unramified two-sheeted covering of a smooth plane quintic arises in this way. Moreover

$$(J(C), \Theta) \cong (\mathrm{Prym}(\Theta_{\mathrm{sing}}, -1), \Xi).$$

(In particular, the "finite set" in Ξ_{sing} *is the empty set.)*

As a corollary of this theorem and the Torelli theorem we infer that the pairs (Γ, L_ε) where

$$\begin{cases} \Gamma \text{ is a smooth plane quintic,} \\ L_\varepsilon^2 \cong \mathcal{O}_\Gamma \text{ and } h^0(\Gamma, \mathcal{O}_\Gamma(1) \otimes L_\varepsilon) \equiv 0 \mod(2), \end{cases}$$

are in 1–1 correspondence with general genus 5 canonical curves C.

The case of odd coverings is not less harmonious. Consider a cubic threefold, i.e., a non-singular cubic hypersurface

$$X \subset \mathbb{P}^4.$$

Such a threefold determines a five-dimensional principally polarized abelian variety (see Clemens–Griffiths [1])

$$(J(X), \Theta),$$

the *intermediate Jacobian* of X. On the other hand, take a line

$$l \subset X$$

and consider all 2-planes π passing through l. Any such 2-plane intersects X along a curve of the form

$$\pi \cap X = l \cup C,$$

where C is a conic. These 2-planes can be viewed as points in a $\mathbb{P}^2$ and the 2-planes π for which

$$C = l_1 \cup l_2 = \{\text{union of 2 lines}\}$$

describe a smooth plane quintic Γ in this $\mathbb{P}^2$. This quintic comes naturally equipped with a two-sheeted unramified covering

$$\{([\lambda], [\pi]) \in G(2, 5) \times \Gamma : \lambda \subset \pi, \lambda \neq l\} = \tilde{\Gamma} \xrightarrow{f_l} \Gamma \subset \mathbb{P}^2,$$

where f_l is the projection. In [1], Tjurin proves

Theorem. *Let X be a cubic threefold and let*

$$f_l : \tilde{\Gamma} \to \Gamma$$

be the two-sheeted covering defined by a line $l \subset X$. Then this is an odd covering and every odd covering arises in this way. Moreover

$$(J(X), \Theta) = (\mathrm{Prym}(\tilde{\Gamma}, \tau), \Xi).$$

This theorem leads to a proof that *the cubic threehold X is not rational*, a result that was first established in Clemens–Griffiths [1]. In fact one first establishes the general result that the rationality of a smooth threehold X implies that the intermediate Jacobian $J(X)$ is a direct sum of Jacobian varieties of smooth curves and then one uses Theorem (**) above to argue that, since $J(X)$ is the Prym variety of an odd two-sheeted covering of a smooth plane quintic, Θ_{sing} is zero-dimensional so that $J(X)$ *cannot be a direct sum of Jacobians of smooth curves.*

Exercises

1. Let Γ be a curve of genus 2, expressed as a double cover of $\mathbb{P}^1$ ramified at $p_1, \ldots, p_6 \in \Gamma$, and let $\eta = p_i - p_j$. If $\tilde{\Gamma} \xrightarrow{f} \Gamma$ is the unramified double cover associated to η (cf. Exercise 13 of Appendix B) and τ the sheet interchange, identify the elliptic curve $\mathrm{Prym}(\tilde{\Gamma}, \tau)$.

2. More generally, suppose Γ is any hyperelliptic curve, expressed as a double cover $\pi: \Gamma \to \mathbb{P}^1$ ramified at $p_1, \ldots, p_{2g+2}$; let $\eta = p_i - p_j$ and let $\tilde{\Gamma} \to \Gamma$ be the corresponding double cover. Show that $\mathrm{Prym}(\tilde{\Gamma}, \tau) = J(\Gamma')$, where Γ' is the hyperelliptic curve given as the double cover of $\mathbb{P}^1$ branched over $\{\pi(p_k): k \neq i, j\}$.

3. In the situation of the preceding exercise, if η is a point of order 2 in $J(\Gamma)$ not of the form $p_i - p_j$, is $\mathrm{Prym}(\tilde{\Gamma}, \tau)$ a Jacobian?

4. Let Γ now be a canonical curve of genus 3, η the point of order 2 in $J(\Gamma)$ given as the difference of the divisors of degree 2 cut by a pair of bitangent lines. Let $\tilde{\Gamma} \to \Gamma$ be the associated double cover and τ the sheet interchange. Of what curve of genus 2 is $\mathrm{Prym}(\tilde{\Gamma}, \tau)$ the Jacobian?

5. Let $\tilde{\Gamma} \to \Gamma$ be a ramified double cover, $\tau: \tilde{\Gamma} \to \tilde{\Gamma}$ the sheet interchange, and let P be the complex torus

$$P = \frac{(H^{1,0}(\tilde{\Gamma})^-)^*}{H_1(\tilde{\Gamma}, \mathbb{Z})^-}$$

(the minus superscripts again refer to the (-1)-eigenspace under the action of τ). Show that P does *not* admit a principal polarization in general.

Note. You may want to use the fact, to be established in the second volume, that if Γ is a general curve of genus g, the standard polarization of $J(\Gamma)$, given by the intersection form on Γ, is the *unique* polarization of $J(\Gamma)$ up to multiplication by scalars.

Chapter VII

The Existence and Connectedness Theorems for $W_d^r(C)$

In this chapter we will prove some of the basic results of Brill–Noether theory, as described in Chapter V. These results fall into two main categories: those which apply to an arbitrary curve, and those which are true only for a general curve. In this chapter we will be concerned with the former, leaving the latter to the second volume.

With the theory developed thus far, the arguments in this chapter are straightforward. In each case, we take the construction of the variety in question ($W_d^r(C)$ or C_d^r) as a determinantal variety, given in Chapter IV, and analyze its particular circumstances. In the case of $W_d^r(C)$, the key feature is the ampleness of one of the two bundles involved in the construction, the direct image of the Poincaré bundle. This allows us to conclude our two qualitative theorems, the existence and connectedness theorems. Our "quantitative" results, the determination of the expected fundamental classes of $W_d^r(C)$ and C_d^r, follow when we apply Porteous' formula (Chapter II) to their determinantal constructions, once we compute the Chern classes of the bundles involved.

It should be remarked that the proofs of the results in this chapter appear nearly in reverse historical order. Thus, the first proofs of the existence theorem were via the computations of the classes of $W_d^r(C)$ and C_d^r in Kleiman–Laksov [1], Kempf [1], and Kleiman–Laksov [2]. Some 10 years later, Fulton and Lazarsfeld [1] proved the connectedness theorem (at the same time proving existence); finally, Lazarsfeld noted that an even simpler proof of the existence theorem could be given.

§1. Ample Vector Bundles

Let us first fix our notation. In what follows, we will denote by $E \to X$ a vector bundle E on a variety X. By the projective bundle $\mathbb{P}E \xrightarrow{\pi} X$ associated to E, we will mean the bundle over X whose fiber over a point $x \in X$ is the projective space of lines in the fiber E_x of E over x; scheme-theoretically,

$$\mathbb{P}E = \operatorname{Proj}\left(\bigoplus_{n=0}^{\infty} \operatorname{Sym}^n E^*\right).$$

The tautological sheaf $\mathcal{O}_{\mathbb{P}E}(1)$ will be the sheaf of sections of the line bundle on $\mathbb{P}E$ whose fiber over a point $(x, \xi) \in \mathbb{P}E$ is the space of linear functionals on the line $\xi \subset E_x$; thus the space of sections of $\mathcal{O}_{\mathbb{P}E}(1)$ over a fiber $\mathbb{P}E_x$ is the dual vector space E_x^*, and the sheaf map $\pi^*E^* \to \mathcal{O}_{\mathbb{P}E}(1)$ induces an isomorphism between $\pi_* \mathcal{O}_{\mathbb{P}E}(1)$ and E^*. More generally, we see that

$$\pi_*(\mathcal{O}_{\mathbb{P}E}(n)) = \mathrm{Sym}^n E^*,$$

and that for any coherent sheaf $\mathcal{F}$ on X,

$$\pi_*(\mathcal{O}_{\mathbb{P}E}(n) \otimes \pi^*\mathcal{F}) = \mathcal{F} \otimes \mathrm{Sym}^n E^*. \tag{1.1}$$

What is the correct notion of ampleness for vector bundles? *A priori* there may be several answers to this, since the direct generalizations of the various definitions or criteria for ampleness in the case of line bundles do not turn out to be equivalent for vector bundles. What turns out to be the correct notion is the following. A vector bundle E is *ample* if the tautological line bundle $\mathcal{O}(1)$ is ample on $\mathbb{P}E^*$. An equivalent definition is that E is ample if tensoring with high symmetric powers of E kills the higher cohomology of coherent sheaves $\mathcal{F}$ on X; i.e., if for any $\mathcal{F}$, there exists an n_0 such that $H^i(X, \mathcal{F} \otimes \mathrm{Sym}^n E)$ vanishes for all $n \geq n_0$ and $i > 0$. That this vanishing result holds for any ample E follows from (1.1) by the Leray spectral sequence and the vanishing of $R^i\pi_* \mathcal{O}_{\mathbb{P}E^*}(n)$ for $i > 0$ and $n \geq 0$. The opposite implication follows from the remark that for any E with the vanishing property above and large enough n we can find a section of $\mathrm{Sym}^n E$ assuming any preassigned values at any two points of X.

We observe that if we have an exact sequence of vector bundles

$$0 \to E \to F \to G \to 0,$$

then F ample implies that G is ample; this follows from the fact that $\mathcal{O}_{\mathbb{P}G^*}(1)$ is just the restriction of $\mathcal{O}_{\mathbb{P}F^*}(1)$ under the inclusion $\mathbb{P}G^* \hookrightarrow \mathbb{P}F^*$ dual to the quotient map $F \to G$. It is also the case that if E and G are both ample, then F must be; this can be proved as follows. Filter $\mathrm{Sym}^k F$ by the images of $\mathrm{Sym}^l E \otimes \mathrm{Sym}^{k-l} F$, $l = 0, \ldots, k$; the successive quotients of this filtration are the bundles $\mathrm{Sym}^l E \otimes \mathrm{Sym}^{k-l} G$. Therefore, for any coherent sheaf $\mathcal{F}$ on X, if we can show that, for high enough k, and any l, the higher cohomology groups of $\mathrm{Sym}^l E \otimes \mathrm{Sym}^{k-l} G \otimes \mathcal{F}$ vanish, this will imply that the higher cohomology groups of $\mathrm{Sym}^k F \otimes \mathcal{F}$ also vanish for large k. By what we just observed, to say that this holds for any $\mathcal{F}$ is the same as saying that F is ample. In case F is the direct sum of E and G, to say that F is ample is actually equivalent to the vanishing of the higher cohomology groups of $\mathrm{Sym}^l E \otimes \mathrm{Sym}^{k-l} G \otimes \mathcal{F}$, for any $\mathcal{F}$, large enough k, and any l. In fact, in this particular case, $\mathrm{Sym}^k F$ is the direct sum of the bundles $\mathrm{Sym}^l E \otimes \mathrm{Sym}^{k-l} G$.

To prove our claim, it is thus sufficient to examine the case when F is the direct sum of E and G. This can be done by direct computation. What has to be shown is that the sections of $\mathrm{Sym}^k(F)$ separate points of $\mathbb{P}F^*$ for some large k. In other terms, given vectors $e + g$, $e' + g'$ in F^* which are not proportional, we must be able to find a section of $\mathrm{Sym}^k F$ that vanishes at $e + g$ but not at $e' + g'$ (here, of course, we are taking e and e' to be in E^* and g, g' to be in G^*). We first remark that, since E, G are ample, there are sections of $\mathrm{Sym}^k F$ that do not vanish at $e + g$ or $e' + g'$. If e is not proportional to e' there are sections of $\mathrm{Sym}^k F$ that vanish at e but not at e', since E is ample, and the same can be said of g, g', and G. The only case to be examined is then the one when e, e' and g, g' are proportional. We can certainly find a section P of $\mathrm{Sym}^k E$ and a section Q of $\mathrm{Sym}^k G$ such that $P(e) = Q(g)$ but $P(e)$, $P(e')$, $Q(g)$, $Q(g')$ are not all zero. Then $P - Q$, viewed as a section of $\mathrm{Sym}^k F$, vanishes at $e + g$, but does not vanish at $e' + g'$ unless e, g are not zero and there are constants α, β such that

$$\alpha \neq \beta, \qquad e' = \alpha e, \qquad g' = \beta g, \qquad \alpha^k = \beta^k.$$

The same can be done for another large exponent k' prime to k. Thus, if every section of $\mathrm{Sym}^{k+k'} F$ vanishing at $e + g$ also vanishes at $e' + g'$, the above must be satisfied, and in addition one must also have that $\alpha^{k'} = \beta^{k'}$. Since k and k' are relatively prime this implies that $\alpha = \beta$, a contradiction. We conclude that the sections of $\mathrm{Sym}^{k+k'} F$ separate points of $\mathbb{P}F^*$, as desired.

Finally, we note that while the condition of ampleness of line bundles on a fixed variety is both open and closed, the condition of ampleness for vector bundles is open but not closed (for example, the ample bundle $\mathcal{O}_{\mathbb{P}^1}(1) \oplus \mathcal{O}_{\mathbb{P}^1}(1)$ on $\mathbb{P}^1$ may specialize to $\mathcal{O}_{\mathbb{P}^1} \oplus \mathcal{O}_{\mathbb{P}^1}(2)$). What is true is that for E a fixed vector bundle and L a line bundle, the condition of ampleness of $E \otimes L$ is both open and closed in L, since $\mathbb{P}E = \mathbb{P}(E \otimes L)$ for all L. In particular, if L is algebraically equivalent to zero, i.e., if the topological first Chern class $c_1(L) \in H^2(X, \mathbb{Z})$ vanishes, then $E \otimes L$ is ample if and only if E is.

Ample vector bundles have some topological properties analogous to those of ample line bundles. For example, the famous hyperplane theorem of Lefschetz states that if Y is an affine variety of dimension n, i.e., if X is any n-dimensional complete variety, L an ample line bundle on X, σ a global section of L with zero divisor Z, and $Y = X - Z$, then

$$H^i(Y, \mathbb{Z}) = 0 \quad \text{for} \quad i > n.$$

The corresponding statement for vector bundles is the

(1.2) Proposition. *Let $E \to X$ be an ample vector bundle of rank r, σ a global section of E, Z the zero locus of σ, and $Y = X - Z$. Then*

$$H^i(Y, \mathbb{Z}) = 0 \quad \textit{for} \quad i \geq \dim X + r.$$

Proof. Let $\tilde{\sigma}$ be the global section of $\mathcal{O}_{\mathbb{P}E^*}(1)$ corresponding to σ via (1.1) and let $\tilde{Z} \subset \mathbb{P}E^*$ be its zero locus. Then via the projection, $\mathbb{P}E^* - \tilde{Z}$ is a bundle of affine spaces over $X - Z = Y$; in particular, for all i

$$H^i(\mathbb{P}E^* - \tilde{Z}, \mathbb{Z}) \cong H^i(Y, \mathbb{Z}).$$

But by the Lefschetz hyperplane theorem applied to $\mathcal{O}_{\mathbb{P}E^*}(1)$,

$$H^i(\mathbb{P}E^* - \tilde{Z}, \mathbb{Z}) = 0 \quad \text{for} \quad i \geq \dim \mathbb{P}E^* + 1 = \dim X + r. \quad \text{Q.E.D.}$$

One immediate consequence of this proposition is that if $E \to X$ is an ample vector bundle with X projective and $\dim X \geq \operatorname{rank} E$, then *every global section of E must have a zero*. To broaden this slightly, we may say that if L is a line bundle on X, E a vector bundle with $L^* \otimes E$ ample, and $\dim X \geq \operatorname{rank} E$, then every homomorphism $\phi: L \to E$ must have rank equal to zero somewhere. This in turn may be generalized to the

(1.3) Proposition. *Suppose X is a projective variety, E, F vector bundles on X of ranks m and n with $E^* \otimes F$ ample, and $\phi: E \to F$ any vector bundle map. If*

$$\dim X \geq (m - k)(n - k),$$

then somewhere ϕ must have rank at most equal to k.

Proof. We will actually prove something more: it will be shown that if $\Sigma_k \subset X$ is any component of the locus X_k of points $x \in X$ such that $\operatorname{rank}(\phi_x) \leq k$, then

$$\operatorname{codim}(\Sigma_k \cap X_{k-1}, \Sigma_k) \leq m + n - 2k + 1$$

(here $\operatorname{codim}(A, B)$ denotes the minimum codimension of any component of A in B); in fact, this bound holds for the codimension of any component of $\Sigma_k \cap X_{k-1}$. In particular, this says that if $\dim \Sigma_k \geq m + n - 2k + 1$ then $X_{k-1} \cap \Sigma_k \neq \emptyset$; the result will of course follow.

To prove this statement, we may as well restrict to $X = \Sigma_k$, and assume $\operatorname{rank}(\phi_x) = k$ everywhere. We now shift attention to $\mathbb{P}E \xrightarrow{\pi} X$, and to the vector bundle map

$$\tilde{\phi}: \mathcal{O}_{\mathbb{P}E}(-1) \to \pi^*F$$

given as the composition of $\pi^*\phi: \pi^*E \to \pi^*F$ with the inclusion $\mathcal{O}_{\mathbb{P}E}(-1) \subset \pi^*E$. Then we observe that the bundle $\mathcal{O}_{\mathbb{P}E}(1) \otimes \pi^*F$ is ample over $\mathbb{P}E$. To see this remark that the tautological bundle on $\mathbb{P}(\mathcal{O}_{\mathbb{P}E}(-1) \otimes \pi^*F^*) = \mathbb{P}E \times_X \mathbb{P}F^*$ is just the restriction of the ample line bundle $\mathcal{O}_{\mathbb{P}(E \otimes F^*)}(1)$ under the Segre inclusion $\mathbb{P}E \times_X \mathbb{P}F^* \to \mathbb{P}(E \otimes F^*)$. It follows that if $Z \subset \mathbb{P}E$ is the zero locus of $\tilde{\phi}$, then

$$H^i(\mathbb{P}E - Z, \mathbb{Z}) = 0 \quad \text{for} \quad i \geq \dim X + n + m - 1.$$

But now if $I \subset F$ is the image bundle of ϕ, then the induced map $\mathbb{P}E - Z \to \mathbb{P}I$ expresses $\mathbb{P}E - Z$ as a bundle of affine spaces over $\mathbb{P}I$. Thus, since the fundamental class of $\mathbb{P}I$ is non-zero,

$$H^{2\dim X + 2k - 2}(\mathbb{P}E - Z, \mathbb{Z}) = H^{2\dim X + 2k - 2}(\mathbb{P}I, \mathbb{Z}) \neq 0,$$

and we conclude that

$$\dim X + n + m - 1 > 2\dim X + 2k - 2;$$

i.e.,

$$\dim X < n + m - 2k + 1. \quad \text{Q.E.D.}$$

§2. The Existence Theorem

The relevance of the above to the study of special linear systems comes from the determinantal description of the varieties $W^r_d(C)$ of linear systems on a curve C given in Chapter IV, which we briefly recall. Let $\mathscr{L}'$ be a Poincaré line bundle on $C \times \operatorname{Pic}^d(C)$, $D = \sum p_i$ a divisor of large degree m on C, Γ' the divisor $D \times \operatorname{Pic}^d(C)$ on $C \times \operatorname{Pic}^d(C)$, and v' the projection from $C \times \operatorname{Pic}^d(C)$ to $\operatorname{Pic}^d(C)$. Then $v'_*(\mathscr{L}'(\Gamma'))$ and $v'_*(\mathscr{L}'(\Gamma')/\mathscr{L}')$ are vector bundles of ranks $m + d - g + 1$ and m over $\operatorname{Pic}^d(C)$, and we defined $W^r_d(C)$ to be the locus where the evaluation map

$$\gamma\colon v'_*(\mathscr{L}'(\Gamma')) \to v'_*(\mathscr{L}'(\Gamma')/\mathscr{L}')$$

has rank $m - g + d - r$ or less. Here we will find it convenient to adopt a slightly different point of view. We begin by recalling that, if we let a be the isomorphism

$$a\colon \operatorname{Pic}^d(C) \to \operatorname{Pic}^{m+d}(C)$$

given by

$$a(L) = L \otimes \mathcal{O}(D)$$

and if $\mathscr{L}$ is a Poincaré line bundle on $\operatorname{Pic}^{m+d}(C)$, then we may choose for $\mathscr{L}'$ the bundle

$$(1_C \times a)^*(\mathscr{L} \otimes \mathcal{O}(-\Gamma')).$$

Thus, if we let Γ be the divisor $D \times \mathrm{Pic}^{m+d}(C)$ in $C \times \mathrm{Pic}^{m+d}(C)$ and denote by v the projection from $C \times \mathrm{Pic}^{m+d}(C)$ to $\mathrm{Pic}^{m+d}(C)$, the image $a(W^r_d(C))$ of $W^r_d(C)$ in $\mathrm{Pic}^{m+d}(C)$ is the locus where the evaluation map

$$\phi: v_* \mathscr{L} \to v_*(\mathscr{L}/\mathscr{L}(-\Gamma))$$

has rank at most $m - g + d - r$. We will study this map, and for this we will write n for $d + m$, and take as our Poincaré line bundle $\mathscr{L}$ on $\mathrm{Pic}^n(C)$ the bundle constructed in Section 2 of Chapter IV, after fixing a point $q \in C$.

Next, we observe that if we choose the points p_i of the divisor $D = \sum p_i$ to be distinct and let Γ_p be the divisor $p \times \mathrm{Pic}^n(C)$ on $C \times \mathrm{Pic}^n(C)$, then $v_*(\mathscr{L}/\mathscr{L}(-\Gamma))$ is a direct sum of line bundles

$$v_*(\mathscr{L}/\mathscr{L}(-\Gamma)) = \bigoplus_i v_*(\mathscr{L}/\mathscr{L}(-\Gamma_{p_i})).$$

Now, $v_*(\mathscr{L}/\mathscr{L}(-\Gamma_p))$ is just the restriction of $\mathscr{L}$ to the section $\{p\} \times \mathrm{Pic}^n(C) \cong \mathrm{Pic}^n(C)$. By the construction of $\mathscr{L}$, if $u: C_n \to \mathrm{Pic}^n(C)$ is the natural map, and we denote by X_p the subset of C_n consisting of all divisors whose support contains the point $p \in C$, we have

$$v_*(\mathscr{L}/\mathscr{L}(-\Gamma_p)) = u_*(\mathcal{O}_{C_n}(X_p - X_q)),$$

since $\mathcal{O}_{C \times C_n}(\Delta) \otimes \mathcal{O}_{\{p\} \times C_n} = \mathcal{O}_{C_n}(X_p)$. If $p = q$, the line bundle $v_*(\mathscr{L}/\mathscr{L}(-\Gamma_p))$ is trivial. If $p \neq q$, $v_*(\mathscr{L}/\mathscr{L}(-\Gamma_p))$ is no longer trivial, but it remains algebraically equivalent to the trivial bundle. Our conclusion, then, is that *$v_*(\mathscr{L}/\mathscr{L}(-\Gamma))$ is a direct sum of line bundles algebraically equivalent to the trivial line bundle.*

We note in passing that if we do not take the points p_i of D to be distinct, for example, if we take $D = m \cdot q$, this is no longer true. What is true in this case, as the reader may verify, is that *$v_*(\mathscr{L}/\mathscr{L}(-\Gamma))$ has a filtration by subbundles whose successive quotients are trivial line bundles.* For the purposes of the succeeding argument this will work just as well.

We turn now to the bundle $v_* \mathscr{L}$, which for brevity we will call E_n or just E if no confusion is likely. This turns out to have a very nice geometric description; namely for $n \geq 2g - 1$, we have the

(2.1) Proposition. (i) *The associated projective bundle $\mathbb{P}E \to \mathrm{Pic}^n(C)$ is isomorphic to the* nth *symmetric product C_n, with projection map corresponding to $u: C_n \to \mathrm{Pic}^n(C)$; and*
(ii) *in terms of this isomorphism,*

$$\mathcal{O}_{\mathbb{P}E}(1) = \mathcal{O}_{C_n}(X_q).$$

Proof. The first statement is basically tautological: since the fiber of $E = v_* \mathscr{L}$ over a point $L \in \mathrm{Pic}^n(C)$ is the vector space $H^0(C, L)$, the points of $\mathbb{P}E$

correspond to pairs (L, ξ) where L is a line bundle of degree n and ξ is a one-dimensional vector space of sections of L; i.e., a divisor $D \in |L|$. As L varies these correspond to all effective divisors of degree n on C.

As for the second statement, to see this we consider the natural map

$$E = \nu_* \mathscr{L} \xrightarrow{\psi} \nu_*(\mathscr{L}/\mathscr{L}(-\Gamma_q))$$

obtained by evaluation and restriction, and the induced map on $\mathbb{P}E$

$$\tilde{\psi}: \mathscr{O}_{\mathbb{P}E}(-1) \to u^*E \to u^*\nu_*(\mathscr{L}/\mathscr{L}(-\Gamma_q)).$$

By definition, $\tilde{\psi}$ is zero on pairs (L, ξ) such that $\sigma(q) = 0$ for all $\sigma \in \xi \subset H^0(C, L)$; that is, in terms of the above isomorphism $\mathbb{P}E \cong C_n$, $\tilde{\psi}$ vanishes on the divisor $X_q \subset C_n$. But we have seen that $\nu_*(\mathscr{L}/\mathscr{L}(-\Gamma_q))$ is the trivial bundle on $\operatorname{Pic}^n(C)$; thus the transpose map ${}^t\tilde{\psi}: \mathscr{O}_{C_n} \to \mathscr{O}_{\mathbb{P}E}(1)$ is a section of $\mathscr{O}_{\mathbb{P}E}(1)$ with zero locus X_q. Finally, since $\mathscr{O}(X_q)$ and $\mathscr{O}_{\mathbb{P}E}(1)$ both restrict to the same bundle $\mathscr{O}(1)$ on the fibers of $E \to X$, ${}^t\tilde{\psi}$ must vanish to order 1 on X_q. Q.E.D.

The identification given in Proposition (2.1) will be used many times. What makes it useful here is that it allows us to establish the

(2.2) Proposition. *E^* is an ample bundle.*

Proof. By the previous proposition this is equivalent to the statement that *X_q is an ample divisor on C_n*. We may see this by using Nakai's criterion (Hartshorne [3]), which states that if X is any projective variety and D a Cartier divisor on X, then D is ample if and only if for every effective algebraic k-cycle Z on X the intersection number $(Z \cdot D^k)$ is positive.

To apply this to the divisor X_q on C_n, since all the divisors $\{X_p\}_{p \in C}$ are algebraically and hence numerically equivalent, it suffices to show that for any irreducible k-dimensional subvariety $Z \subset C_n$, for a general $p \in C$ the intersection $Z \cap X_p$ is non-empty of dimension $k - 1$. But this is clear: if $D = \sum p_i \in Z$ is a general point we may take p to be any of the points p_i not held in common by all $D' \in Z$.

Alternatively, one can use Nakai's criterion to establish the general fact that if $f: X \to Y$ is any finite surjective map and L any line bundle on Y, then f^*L is ample if and only if L is ample. We then apply this to the map $\tau: C^n \to C_n$ from the ordinary to the symmetric product, observing that

$$\tau^*\mathscr{O}_{C_n}(X_q) = \bigotimes_{i=1}^{n} \pi_i^*\mathscr{O}_C(q)$$

is clearly ample (where the $\pi_i: C^n \to C$ are the projection maps). Q.E.D.

We may now combine what we know to conclude the Existence Theorem. We have seen that $W_d^r(C)$ is isomorphic to the locus in $\mathrm{Pic}^n(C)$ where the bundle map

$$\phi: E \to v_*(\mathscr{L}/\mathscr{L}(-\Gamma))$$

has rank at most $m - g + d - r$. Since E^* is ample and $v_*(\mathscr{L}/\mathscr{L}(-\Gamma)) = \bigoplus_{i=1}^m (\mathscr{L}/\mathscr{L}(-\Gamma_{p_i}))$, with each line bundle $v_*(\mathscr{L}/\mathscr{L}(-\Gamma_{p_i}))$ algebraically equivalent to the trivial bundle, we may conclude that $E^* \otimes v_*(\mathscr{L}/\mathscr{L}(-\Gamma_{p_i}))$ is ample and hence that $E^* \otimes v_*(\mathscr{L}/\mathscr{L}(-\Gamma))$ is. Applying Proposition (1.3) we deduce the Existence Theorem of Chapter V, i.e.,

(2.3) Theorem. *If C is any smooth curve of genus g, and d and r non-negative integers such that $\rho = g - (r + 1)(g - d + r) \geq 0$, then $W_d^r(C) \neq \varnothing$.*

§3. The Connectedness Theorem

A fairly elementary examination of the properties of ample vector bundles has yielded the existence theorem for special divisors. We will now go back and establish a stronger statement about ample bundles, which will in particular imply the connectedness theorem for $W_d^r(C)$.

The statement to be proved is the following:

(3.1) Proposition. *Let X be a connected variety, E and F vector bundles of ranks m and n on X with $E^* \otimes F$ ample, and $\phi: E \to F$ any bundle homomorphism. Then for any k such that*

$$\dim X \geq (m - k)(n - k) + 1,$$

the kth *degeneracy locus*

$$Y = \{x \in X : \mathrm{rank}\ \phi_x \leq k\}$$

is connected.

Proof. Replacing ϕ with its transpose if necessary, we can and will require that $n \geq m$. The proof uses one general construction that we shall describe in advance. Let A and B be vector bundles over a space W and let $\mathbb{P} = \mathbb{P}\,\mathrm{Hom}(B, A) \xrightarrow{\pi} W$. Let $\sigma: A \to B$ be a bundle map; by composition σ induces a bundle map

$$\tilde{\sigma}: \mathrm{Hom}(B, A) \to \mathrm{Hom}(A, A).$$

We consider the composition of bundle maps on $\mathbb{P}$

$$\mathcal{O}_{\mathbb{P}}(-1) \stackrel{i}{\hookrightarrow} \pi^* \operatorname{Hom}(B, A) \stackrel{\pi^*\tilde{\sigma}}{\to} \pi^* \operatorname{Hom}(A, A) \stackrel{\operatorname{tr}}{\to} \mathcal{O}_{\mathbb{P}},$$

where i is the standard inclusion, and tr is the trace functional. This map, viewed as a section of the line bundle $\mathcal{O}_{\mathbb{P}}(1)$, will be denoted $\operatorname{tr}(\sigma)$. The key point here is that if $Y \subset W$ is the zero locus of σ and $Y_{\mathbb{P}} \subset \mathbb{P}$ the zero locus of $\operatorname{tr}(\sigma)$, then $Y_{\mathbb{P}}$ meets a fiber $\mathbb{P}_w$ of $\mathbb{P}$ over a point $w \in W$ in a hyperplane if $w \notin Y$, and contains the fiber if $w \in Y$. Thus $\mathbb{P} - Y_{\mathbb{P}}$ is a bundle of affine spaces over $W - Y$, and in particular for all i

$$H^i(W - Y, \mathbb{Z}) = H^i(\mathbb{P} - Y_{\mathbb{P}}, \mathbb{Z}).$$

The proof of Proposition (3.1) is now accomplished by constructing the following configuration of spaces and bundle maps.

(1) Let $\phi: E \to F$ be the given bundle map, $Y \subset X$ the locus $\{\operatorname{rank} \phi_x \leq k\}$, and set $V = X - Y$.

(2) Let $G = G(m - k, E) \stackrel{\alpha}{\to} X$ be the Grassmannian bundle of $(m - k)$-planes in E; and let ϕ_G be the composition

$$\phi_G: S \stackrel{i}{\hookrightarrow} \alpha^*E \stackrel{\alpha^*\phi}{\to} \alpha^*F,$$

where S is the universal subbundle on G and i the standard inclusion. Let Y_G be the zero locus of ϕ_G and set $V_G = G - Y_G$. Y_G is the standard blow-up of Y as described in Chapter IV, Section 4.

(3) Set $\mathbb{P} = \mathbb{P}\operatorname{Hom}(F, E)$ and let $\delta: \mathbb{P} \to X$ be the projection. Let $\phi_{\mathbb{P}} = \operatorname{tr}(\phi)$ be the trace of ϕ, $Y_{\mathbb{P}}$ the zero divisor of $\phi_{\mathbb{P}}$ and set $V_{\mathbb{P}} = \mathbb{P} - Y_{\mathbb{P}}$.

(4) Finally, set $Q = \mathbb{P}\operatorname{Hom}(\alpha^*F, S)$, let $\beta: Q \to G$ be the projection, set $\phi_Q = \operatorname{tr}(\phi_G)$, let Y_Q be the zero divisor of ϕ_Q and set $V_Q = Q - Y_Q$. Note that Q may also be described as a subvariety of the fiber product $\mathbb{P} \times_X G$: if we write

$$\mathbb{P} \times_X G = \{(x, \sigma, \Lambda): x \in X, \sigma: F_x \to E_x, \Lambda \in G(m - k, E_x)\},$$

then we can describe Q by

$$Q = \{(x, \sigma, \Lambda): \operatorname{Im}(\sigma) \subset \Lambda\}.$$

Let $\gamma: Q \to \mathbb{P}$ be the projection.

Summing up, we have the diagram of spaces and bundle maps:

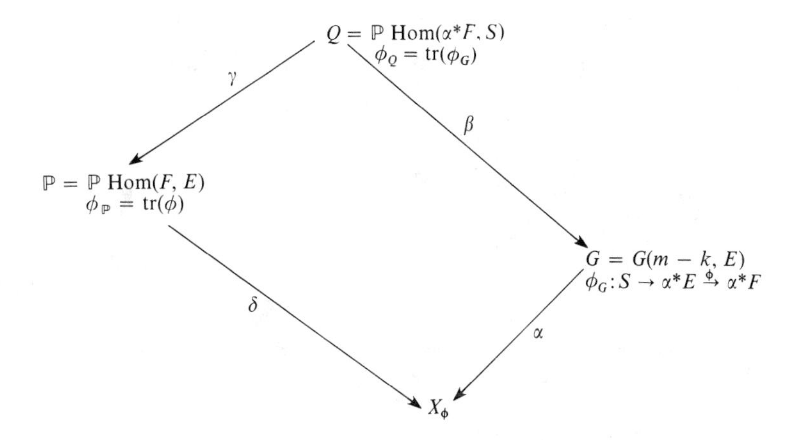

We now proceed clockwise from $\mathbb{P}$. To begin with, our basic hypothesis that $E^* \otimes F$ is ample implies that $\mathcal{O}_{\mathbb{P}}(1)$ is ample, and hence that $V_{\mathbb{P}}$ is an affine variety. Now, since the section $\phi_{\mathbb{P}}$ of $\mathcal{O}_{\mathbb{P}}(1)$ pulls back via γ to the section ϕ_Q of $\mathcal{O}_Q(1)$ (the trace of a map $\phi_x: F_x \to F_x$ whose image lies in Λ is the same as that of $\phi_x|_\Lambda$), we see that $\gamma^{-1}(V_{\mathbb{P}}) = V_Q$; we want to analyze the cohomology of V_Q via this map. For this we observe that the fiber of $\gamma: V_Q \to V_{\mathbb{P}}$ over $(x, \sigma) \in V_{\mathbb{P}}$ is the Grassmannian of $(m - k - \mathrm{rank}(\sigma))$-dimensional subspaces of $E/\mathrm{Im}\,\sigma$; in particular, the fiber dimension of γ is lk over the locus $V_{\mathbb{P}}^l = \{(x, \sigma): \mathrm{rank}(\sigma) = m - k - l\}$, which has codimension $(k + l)(n - m + k + l)$ in $\mathbb{P}$.

The next step requires the notion of constructible sheaf. A sheaf of abelian groups $\mathscr{F}$ on an algebraic variety Z is said to be *constructible* if Z can be expressed as a finite disjoint union of locally closed algebraic subvarieties on each one of which $\mathscr{F}$ is locally constant. It is a fundamental result, to be discussed later, that the direct images of a constructible sheaf under a morphism of algebraic varieties are constructible. This applies, in particular, to the sheaves $R^i\gamma_*\mathbb{Z}$, which, by what we just said, are concentrated on the affine variety $\overline{V}_{\mathbb{P}}^{l+1}$ whenever $i > 2lk$. It is another fundamental property of constructible sheaves, also to be discussed later, that the cohomology of such a sheaf on an affine variety vanishes above the dimension of the variety itself. In our special case, we then conclude that $H^j(\mathbb{P}, R^i\gamma_*\mathbb{Z})$ vanishes whenever there is an l such that

$$\begin{aligned} &i > 2lk, \\ &j > \dim V_{\mathbb{P}}^{l+1} = \dim X + mn - 1 - k(n - m + k) \\ &\qquad - (l + 1)(n - m + 2k + l + 1). \end{aligned}$$

Since we are assuming that $n \geq m$, this is always the case as soon as

$$i + j > \dim X + mn - 1 - k(n - m + k).$$

Using the Leray spectral sequence for γ and $\mathbb{Z}$, we conclude that

$$H^i(V_Q, \mathbb{Z}) = 0 \quad \text{if} \quad i \geq \dim X + mn - k(n - m + k).$$

Moving along, since the map $\beta: V_Q \to V_G$ expresses V_Q as a bundle of affine spaces over V_G, we have

$$H^i(V_G, \mathbb{Z}) = H^i(V_Q, \mathbb{Z}) = 0 \quad \text{for} \quad i \geq \dim X + mn - k(n - m + k).$$

We now apply Lefschetz duality on G. Since G is a real $2(\dim X + (m - k)k)$-dimensional manifold, we have an isomorphism

$$H^1(G, Y_G; \mathbb{Z}) \xrightarrow{\sim} H_{2 \dim X + 2(m-k)k - 1}(V_G, \mathbb{Z});$$

it follows that $H^1(G, Y_G; \mathbb{Z}) = 0$ if

$$2 \dim X + 2(m - k)k - 1 \geq \dim X + mn - k(n - m + k);$$

i.e., that *Y_G is connected if* $\dim X \geq (m - k)(n - k) + 1$. Since the map α carries Y_G onto Y, this concludes the proof. Q.E.D.

As we saw in Section 2 that $W^r_d(C)$ can be described as the $(m - g + d - r)$th degeneracy locus of a bundle map $\phi: E \to F$, where E is of rank $m + d - g + 1$, F is of rank m, and $E^* \otimes F$ is ample. Applying (3.1), we obtain the Connectedness Theorem of Chapter V, that is

(3.2) Theorem. *If C is any smooth curve of genus g, and*

$$\rho = g - (r + 1)(g - d + r) \geq 1,$$

then $W^r_d(C)$ is connected.

We note that in fact (3.1) says something more: it implies that $W^r_d(C)$ is "connected in dimension $\rho - 1$": i.e., that for any $L, M \in W^r_d(C)$ there is a chain of irreducible components $X_0, \ldots, X_n$ of $W^r_d(C)$ with $L \in X_0$, $M \in X_n$, and $\dim X_i \cap X_{i+1} \geq \rho - 1$. This follows if we apply (3.1) to the restriction of the bundle map defining $W^r_d(C)$ to a sufficiently general subvariety $X \subset \mathrm{Pic}^n(C)$ of codimension $\rho - 1$ passing through L and M.

To conclude this section, recall that the proof of Proposition (3.1), and hence of the Connectedness Theorem, are based on two properties of constructible sheaves that we have not proved yet. Here we are going to partially fill this gap.

The first, and basic, property we used is that higher direct images of constructible sheaves are constructible. We are unable to give a self-contained proof of this fact in this book (cf. the bibliographical notes at the end of this chapter); instead, we shall assume it and give a proof of the second property of constructible sheaves that is used. This is expressed by the following result.

(3.3) Proposition. *Let $\mathscr{F}$ be a constructible sheaf on an affine n-dimensional variety X. Then $H^i(X, \mathscr{F})$ vanishes for $i > n$.*

This result is a direct generalization of the Lefschetz hyperplane theorem; this theorem, or an equivalent version of it, is obtained by taking $\mathscr{F}$ equal to the constant sheaf $\mathbb{Z}$.

To prove the proposition we consider X as being embedded in some affine space. By generic projection we may then represent X as a finite covering of $\mathbb{C}^n$; let g stand for the projection. By the Leray spectral sequence,

$$H^i(X, \mathscr{F}) \simeq H^i(\mathbb{C}^n, g_*\mathscr{F})$$

for every i. Since $g_*\mathscr{F}$ is constructible, we are thus reduced to proving the theorem for $X = \mathbb{C}^n$. This will be done by induction on n, the case $n = 1$ being trivial. The statement that needs to be proved is that, if $f: \mathbb{C}^n \to \mathbb{C}^{n-1}$ is a suitable linear projection, then $R^if_*\mathscr{F}$ vanishes for i larger than one. Once we know this, assuming the statement of the proposition to have been proved for $\mathbb{C}^{n-1}$, we get that

$$H^i(\mathbb{C}^{n-1}, R^jf_*\mathscr{F}) = 0 \quad \text{if} \quad i + j > n.$$

The result then follows from the Leray spectral sequence for f.

It remains to find an f with the property that $R^if_*\mathscr{F}$ vanishes for i greater than one. Notice that $\mathscr{F}$ is locally constant off a closed subvariety W of $\mathbb{C}^n$. We choose f in such a way that its restriction to W is a finite map. We shall show that for any x in $\mathbb{C}^{n-1}$ the stalk $(R^if_*\mathscr{F})_x$ is isomorphic to $H^i(f^{-1}(x), \mathscr{F})$; this will be enough for our purposes, since $f^{-1}(x)$ is isomorphic to $\mathbb{C}$ and the restriction of $\mathscr{F}$ to $f^{-1}(x)$ is constructible. Since

$$H^i(f^{-1}(x), \mathscr{F}) = \varinjlim_N H^i(N, \mathscr{F}),$$

where N runs through all neighborhoods of $f^{-1}(x)$, we will be done if we can find, for each such N, a neighborhood U of x and a neighborhood N' of $f^{-1}(x)$ such that $N' \subset N \cap f^{-1}(U)$ and

$$H^i(f^{-1}(U), \mathscr{F}) \cong H^i(N', \mathscr{F}).$$

We view $\mathbb{C}^n$ as the product $\mathbb{C} \times \mathbb{C}^{n-1}$ and f as the projection onto the second factor. If U is a sufficiently small ball centered at x there is an open disc A in $\mathbb{C}$ such that

$$W \cap f^{-1}(U) \subset A \times U,$$
$$\bar{A} \times \bar{U} \subset N.$$

Let t, $z = (z_1, \ldots, z_{n-1})$ be linear coordinates on $\mathbb{C}$ and $\mathbb{C}^{n-1}$ such that U is the unit ball centered at the origin. We may find a continuous function

$$\phi: \mathbb{C} \to \mathbb{R}$$

such that $0 \leq \phi(t) \leq 1$ for every t, ϕ is identically equal to one on A, and (t, z) belongs to N whenever

$$\sum |z_i|^2 \leq \phi(t).$$

We then set

$$N' = \{(t, z): \sum |z_i|^2 \leq \phi(t)\}.$$

To see that the cohomology groups of $\mathscr{F}$ on $f^{-1}(U)$ and N' coincide we compare the two Mayer–Vietoris sequences

$$\begin{array}{ccccccc}
\cdots \to H^{i-1}(A \times U - W, \mathscr{F}) & \longrightarrow & H^i(N', \mathscr{F}) & \longrightarrow & H^i(N' - W, \mathscr{F}) \oplus H^i(A \times U, \mathscr{F}) & \longrightarrow \cdots \\
\wr\| & & \uparrow & & \uparrow \quad\quad\quad\quad \wr\| & \\
\cdots \to H^{i-1}(A \times U - W, \mathscr{F}) & \longrightarrow & H^i(f^{-1}(U), \mathscr{F}) & \longrightarrow & H^i(f^{-1}(U) - W, \mathscr{F}) \oplus H^i(A \times U, \mathscr{F}) & \longrightarrow \cdots
\end{array}$$

and notice that $H^i(f^{-1}(U) - W, \mathscr{F})$ is isomorphic to $H^i(N' - W, \mathscr{F})$ since $\mathscr{F}$ is locally constant off W and $f^{-1}(U) - W$ retracts onto $N' - W$.

§4. The Class of $W^r_d(C)$

We would now like to give a refinement of the existence theorem proved in the second section of this chapter; namely, we will determine the fundamental class of the cycle $W^r_d(C)$ in $\operatorname{Pic}^d(C)$. To do this, we will again use the representation of $W^r_d(C)$ as a determinantal variety associated to the bundle map

$$E \to v_*(\mathscr{L}/\mathscr{L}(-\Gamma))$$

over $\operatorname{Pic}^n(C)$. Given this description of $W^r_d(C)$, we can use Porteous' formula to determine its class; what we need to know are the Chern classes of the two bundles involved. Moreover, since we have seen that $\nu_*(\mathscr{L}/\mathscr{L}(-\Gamma))$ is a direct sum of line bundles with first Chern class 0, it has trivial Chern class; consequently, it only remains to compute the Chern class of E. Porteous' formula will then give the fundamental class w_d^r of $W_d^r(C)$ (in case $W_d^r(C)$ is of the expected dimension ρ) as

$$w_d^r = (-1)^{(r+1)(g-d+r)}\Delta_{r+1,g-d+r}(c_t(E)) \\ = \Delta_{g-d+r,r+1}(c_t(-E)). \tag{4.1}$$

We will compute the Chern class of E by using again the isomorphism $\mathbb{P}E \cong C_n$. We begin by recalling some generalities on the cohomology of projective bundles.

Let E be any complex vector bundle of rank k over a space X. On the projective bundle $\mathbb{P}E \xrightarrow{u} X$, denote by S and Q the universal sub- and quotient bundles. Thus S is the line bundle whose fiber over a point $(x, \xi) \in \mathbb{P}E$ is the one-dimensional subspace $\xi \subset E_x$ and we have the exact sequence

$$0 \to S \to u^*E \to Q \to 0.$$

Notice that, in the analytic case, S is the dual of $\mathscr{O}_{\mathbb{P}E}(1)$. Setting $x = c_1(S^*)$, the Whitney product formula, applied to the above exact sequence, gives

$$(1 - x) \cdot c(Q) = u^*c(E);$$

in other words,

$$c(Q) = u^*c(E) \cdot (1 + x + x^2 + \cdots). \tag{4.2}$$

Taking the homogeneous terms of degree k on both sides, we have (since Q has rank $k - 1$),

$$x^k + u^*c_1(E) \cdot x^{k-1} + \cdots + u^*c_k(E) = 0.$$

Indeed, once one has defined the first Chern class of a line bundle, this relation may be used as the definition of the Chern class of E: an elementary spectral sequence argument shows that as additive groups

$$H^*(\mathbb{P}E, \mathbb{Z}) = \bigoplus_{i=0}^{k-1} x^i \cdot u^*H^*(X, \mathbb{Z});$$

consequently, x^k must be uniquely expressible as a linear combination

$$x^k = \sum_{i=0}^{k-1} u^*\alpha_{k-i} \cdot x^i$$

with $\alpha_i \in H^{2i}(X, \mathbb{Z})$, and we may define $c_i(E) = -\alpha_i$.

There is another way of extracting the Chern classes of E from the multiplicative structure of the cohomology ring $H^*(\mathbb{P}E, \mathbb{Z})$ that is perhaps more direct. Recall that the Gysin homomorphism

$$u_*: H^*(\mathbb{P}E, \mathbb{Z}) \to H^{*-2k+2}(X, \mathbb{Z}),$$

sometimes called integration over the fiber, satisfies the *push–pull formula*: for any classes $\alpha \in H^*(\mathbb{P}E, \mathbb{Z})$, $\beta \in H^*(X, \mathbb{Z})$,

$$u_*(\alpha \cdot u^*\beta) = u_*\alpha \cdot \beta.$$

Applying u_* to (4.2), we obtain

$$u_* c(Q) = c(E) \cdot u_*(1 + x + x^2 + \cdots).$$

Since $u_* c_i(Q) = 0$ for $i < k - 1$, and since $c_{k-1}(Q)$ restricts to the generator of $H^{2k-2}(\mathbb{P}^{k-1}, \mathbb{Z})$ in each fiber, $u_* c(Q) = 1$. Thus we find

$$u_*(1 + x + x^2 + \cdots) = c(-E),$$

where by $c(-E)$ we mean the formal inverse $1/c(E)$; specifically,

$$u_* x^{k-1+i} = c_i(-E). \tag{4.3}$$

We may take this as a definition of the *Segre class* $s(E) = 1/c(E)$, and thereby as a definition of $c(E)$.

In any event, it is the relation (4.3) that we will use to calculate the Chern class of the bundle $E = v_* \mathscr{L}$ on $\operatorname{Pic}^n(C)$. Recall that we have $\mathbb{P}E \cong C_n$, and that via this isomorphism the first Chern class of the tautological bundle $\mathscr{O}_{\mathbb{P}E}(1)$ is the class x of the divisor X_q on C_n. By (4.3), if $u: C_n \to \operatorname{Pic}^n(C)$ is the abelian sum map we have

$$c_i(-E) = u_* x^{d-g+i}.$$

To evaluate the right-hand side, note that all the divisors $\{X_p\}_{p \in C}$ are homologous; so that in cohomology

$$x^\alpha = X_{q_1} \cdot X_{q_2} \cdot \ldots \cdot X_{q_\alpha}.$$

Now, since $X_q = \{D \in C_n : D - q \geq 0\}$, if the q_i are distinct the intersection

$$X_{q_1} \cap \cdots \cap X_{q_\alpha} = \{D \in C_n : D - \textstyle\sum q_i \geq 0\}$$

is just the image of $C_{n-\alpha}$ in C_n via the inclusion $D \to D + \sum q_i$. Moreover, this intersection is transverse: in terms of the identification

$$T_D(C_n) \cong H^0(C, \mathscr{O}(D)/\mathscr{O}),$$

we have, for $D \in X_q$

$$T_D(X_q) = H^0(C, \mathcal{O}(D - q)/\mathcal{O}),$$

so for $D \in \bigcap X_{q_i}$

$$\bigcap_{i=1}^{\alpha} T_D(X_{q_i}) = H^0(C, \mathcal{O}(D - \sum q_i)/\mathcal{O})$$
$$\cong \mathbb{C}^{n-\alpha}.$$

Thus the class x^α is represented by the image of $C_{n-\alpha}$ in C_n, and the Gysin image $u_* x^\alpha$ in $\mathrm{Pic}^n(C)$ is represented by a translate of $W_{n-\alpha}$; i.e.,

$$u_* x^\alpha = w_{n-\alpha}.$$

By Poincaré's formula

$$u_* x^\alpha = \frac{1}{(g - n + \alpha)!} \theta^{g-n+\alpha},$$

and consequently

$$c(-E) = u_*(\sum x^\alpha) = e^\theta.$$

Finally, we arrive at the formula for the Chern class of E:

$$c(E) = e^{-\theta}.$$

To obtain the formula for the class w_d^r of $W_d^r(C)$ in case $W_d^r(C)$ has the right dimension, we apply Porteous' formula:

$$\begin{aligned}
w_d^r &= \Delta_{g-d+r,r+1}(c_t(-E)) \\
&= \Delta_{g-d+r,r+1}(e^{t\theta}) \\
&= \det \begin{pmatrix} \dfrac{\theta^{g-d+r}}{(g-d+r)!} & \dfrac{\theta^{g-d+r+1}}{(g-d+r+1)!} & \cdots \\ \dfrac{\theta^{g-d+r-1}}{(g-d+r-1)!} & \dfrac{\theta^{g-d+r}}{(g-d+r)!} & \\ & & \ddots \\ \vdots & & \\ & & \dfrac{\theta^{g-d+r}}{(g-d+r)!} \end{pmatrix} \\
&= \theta^{(r+1)(g-d+r)} \cdot \Delta(g-d+r, g-d+r-1, \ldots, g-d),
\end{aligned}$$

where in general we set

$$\Delta(a_1, \ldots, a_n) = \det \begin{pmatrix} \frac{1}{a_1!} & \frac{1}{(a_1+1)!} & \cdots & \frac{1}{(a_1+n-1)!} \\ \frac{1}{a_2!} & \frac{1}{(a_2+1)!} & \cdots & \\ \vdots & & & \\ \frac{1}{a_n!} & & \cdots & \frac{1}{(a_n+n-1)!} \end{pmatrix}.$$

It remains now only to evaluate the determinant Δ. This is straightforward: if we multiply the ith row by $(a_i + n - 1)!$, we obtain the determinant

$$\det \begin{pmatrix} \cdots & (a_1+n-1)(a_1+n-2) & (a_1+n-1) & 1 \\ & (a_2+n-1)(a_2+n-2) & (a_2+n-1) & 1 \\ & \vdots & \vdots & \\ \cdots & (a_n+n-1)(a_n+n-2) & (a_n+n-1) & 1 \end{pmatrix}$$

This may in turn be column-reduced to the Vandermonde determinant

$$\det \begin{pmatrix} a_1^{n-1} & \cdots & a_1^2 & a_1 & 1 \\ a_2^{n-1} & & a_2^2 & a_2 & 1 \\ \vdots & & \vdots & \vdots & \vdots \\ a_n^{n-1} & \cdots & a_n^2 & a_n & 1 \end{pmatrix} = \prod_{j>i} (a_i - a_j).$$

Thus,

$$\Delta(a_1, \ldots, a_n) = \frac{\prod_{j>i}(a_i - a_j)}{\prod_{i=1}^n (a_i + n - 1)!}.$$

In the present case, we have $n = r + 1$ and $a_i = g - d + r + 1 - i$, so that

$$\begin{aligned} \prod_{j>i} (a_i - a_j) &= \prod_{1 \le i < j \le r+1} (j - i) \\ &= \prod_{\alpha=0}^{r} \alpha!. \end{aligned}$$

Assembling the pieces, we arrive at the

(4.4) Theorem. *If C is any smooth curve of genus g, and $W^r_d(C)$ is either empty or of the expected dimension ρ, it has fundamental class*

$$w^r_d = \prod_{\alpha=0}^{r} \frac{\alpha!}{(g-d+r+\alpha)!} \cdot \theta^{(r+1)(g-d+r)}.$$

This enumerative formula is the same as Theorem (1.3) in Chapter V. We may remark that the result we just proved implies the Existence Theorem for special divisors, since it shows that the fundamental class of $W_d^r(C)$ is not zero when $\rho \geq 0$.

§5. The Class of C_d^r

In this section we will combine Porteous' formula with the determinantal description of C_d^r given in Section 1 of Chapter IV to compute the class c_d^r of C_d^r. This formula represents a refinement of the formula for the class of $W_d^r(C)$ since, given the class of C_d^r, we can evaluate w_d^r as

$$w_d^r = u_*(x^r c_d^r).$$

To set up our computation of c_d^r, recall that C_d^r is defined to be the $(d-r)$th determinantal variety associated to the differential

$$u_*\colon \Theta_{C_d} \to u_*\Theta_{J(C)}$$

of the map $u\colon C_d \to J(C) \cong \operatorname{Pic}^d(C)$. Since the tangent sheaf $\Theta_{J(C)}$ to the Jacobian of C is trivial, Porteous' formula then tells us that

$$c_d^r = \Delta_{g-d+r,r}(c_t(-\Theta_{C_d})). \tag{5.1}$$

It remains to calculate the Chern classes of C_d. We do this first in case $d \geq 2g-1$, so that

$$C_d \cong \mathbb{P}E$$

is the projective bundle associated to the direct image $E = \nu_*\mathscr{L}$ of the Poincaré line bundle on $\operatorname{Pic}^d(C) = J(C)$. As with any fiber bundle, we have an exact sequence

$$0 \to \Theta_u \to \Theta_{C_d} \to u^*\Theta_{J(C)} \to 0, \tag{5.2}$$

where the "vertical tangent bundle" Θ_u is the bundle whose fiber at any point $D \in C_d$ is the tangent space to the fiber through D of $C_d \to J(C)$. Since $C_d = \mathbb{P}E \to J(C)$ is a projective bundle, we have the Euler exact sequence for Θ_u

$$0 \to \mathscr{O} \to u^*E \otimes \mathscr{O}_{\mathbb{P}E}(1) \to \Theta_u \to 0. \tag{5.3}$$

This sequence may be obtained by observing that for any vector space V, and any elements $v \in V$ and $l \in V^*$, if $\pi: V - \{0\} \to \mathbb{P}V$ is the projection then

$$(\pi_*)_\lambda\left(l(\lambda)\frac{\partial}{\partial v}\right) \in T_{[\lambda]}\mathbb{P}V$$

depends only on the image of λ in $\mathbb{P}V$. This defines a map $V \otimes V^* \to T_{[\lambda]}\mathbb{P}V$ whose kernel at each point $[\lambda]$ of $\mathbb{P}V$ is generated by $V \otimes \lambda^\perp$ and the element "trace" in $\operatorname{Hom}(V, V)^* = V \otimes V^*$ (the Euler relation). As described, this map then globalizes to give the exact sequence (5.3).

Together the two sequences (5.2) and (5.3) yield

$$c_t(\Theta_{C_d}) = c_t(u^*E \otimes \mathcal{O}_{\mathbb{P}E}(1)).$$

To evaluate the right-hand side we write (as before) x for $c_1(\mathcal{O}_{\mathbb{P}E}(1))$ (the class of the divisor $X_q \cong C_{d-1}$ in C_d) and formally factor $c_t(u^*E)$:

$$c_t(u^*E) = \prod_{i=1}^{d-g+1}(1 + t\beta_i).$$

Then

$$\begin{aligned} c_t(u^*E \otimes \mathcal{O}_{\mathbb{P}E}(1)) &= \prod_i (1 + t(x + \beta_i)) \\ &= (1 + tx)^{d-g+1}\prod_i\left(1 + \frac{t\beta_i}{1 + tx}\right) \\ &= (1 + tx)^{d-g+1}c_\tau(u^*E), \end{aligned}$$

where $\tau = t/(1 + tx)$. Since we have seen that $c_s(u^*E) = e^{s\theta}$ (here we are writing θ for $u^*\theta$, a notational convention we will continue to follow for the remainder of the book), we have the formula

$$c_t(\Theta_{C_d}) = (1 + tx)^{d-g+1}e^{-t\theta/(1+tx)} \tag{5.4}$$

for $d \geq 2g - 1$.

Finally, if we embed C_{d-1} in C_d as the divisor X_q, the normal bundle

$$N_{C_{d-1}/C_d} = \mathcal{O}_{C_{d-1}}(X_q)$$

has Chern polynomial $(1 + xt)$. From the sequence

$$0 \to \Theta_{C_{d-1}} \to \Theta_{C_d} \otimes \mathcal{O}_{C_{d-1}} \to N_{C_{d-1}/C_d} \to 0$$

we conclude that

$$\begin{aligned} c_t(\Theta_{C_{d-1}}) &= c_t(\Theta_{C_d})/(1 + xt) \\ &= (1 + xt)^{d-g} e^{-t\theta/(1+tx)}. \end{aligned}$$

Consequently if formula (5.4) is valid for C_d it holds for C_{d-1} as well; and so we conclude that it must hold for all d.

According to (5.1) to compute c_d^r we must evaluate $\Delta_{g-d+r,r}$ for the power series

$$\begin{aligned} c_t(-\Theta_{C_d}) = \frac{1}{c_t(\Theta_{C_d})} &= (1 + xt)^{g-d-1} e^{t\theta/(1+xt)} \\ &= \sum_{k=0}^{\infty} (1 + xt)^{g-d-1-k} \frac{t^k \theta^k}{k!}. \end{aligned} \tag{5.5}$$

Noting that, since $r \geq 1$, the determinant $\Delta_{g-d+r,r}$ involves only terms of degree at least $g - d + 1$ we see that terms in the sum (5.5) corresponding to $k \leq g - d$ do not appear in the determinant. Setting

$$y = -x, \qquad m = g - d,$$

and using

$$(1 + xt)^{-a-1} = \sum_i \binom{a+i}{i} y^i t^i,$$

we have

$$\begin{aligned} \sum_{k \geq g-d} (1 + xt)^{g-d-1-k} \frac{t^k \theta^k}{k!} &= \sum_{\substack{k \geq g-d \\ i \geq 0}} \binom{d-g+k+i}{i} \frac{y^i t^{k+i} \theta^k}{k!} \\ &= \sum_{n \geq 0} \sum_{i=0}^{n} \left[\binom{n}{i} \frac{y^i \theta^{n-i+g-d}}{(n-i+g-d)!} \right] t^{n+g-d}, \end{aligned}$$

and correspondingly

$$\Delta_{g-d+r,r}((1 + xt)^{g-d-1} e^{\tau\theta})$$
$$= \det \begin{pmatrix} \displaystyle\sum_{i=0}^{r} \binom{r}{i} \frac{y^i \theta^{g-d+r-i}}{(g-d+r-i)!} & \cdots & \displaystyle\sum_{i=0}^{2r-1} \binom{2r-1}{i} \frac{y^i \theta^{g-d+2r-1-i}}{(g-d+2r-1-i)!} \\ \vdots & & \\ \displaystyle\sum_{i=0}^{1} \binom{1}{i} \frac{y^i \theta^{g-d+1-i}}{(g-d+1-i)!} & \cdots & \displaystyle\sum_{i=0}^{r} \binom{r}{i} \frac{y^i \theta^{g-d+r-i}}{(g-d+r-i)!} \end{pmatrix}.$$

Rotating the above matrix clockwise through 90°, we get

$$\Delta_{g-d+r,r}((1+tx)^{g-d-1}e^{\tau\theta})$$
$$= (-1)^{r(r-1)/2}\theta^{r(g-d)}$$
$$\times \det\begin{pmatrix} & \dfrac{y}{m!}+\dfrac{\theta}{(m+1)!} & \cdots & \dfrac{y^r}{m!}+\dfrac{ry^{r-1}\theta}{(m+1)!}+\cdots+\dfrac{\theta^r}{(m+r)!} \\ \dfrac{y^2}{m!}+\dfrac{2y\theta}{(m+1)!}+\dfrac{\theta^2}{(m+2)!} & & & \vdots \\ & \vdots & & \\ \dfrac{y^r}{m!}+\dfrac{ry^{r-1}\theta}{(m+1)!}+\cdots+\dfrac{\theta^r}{(m+r)!} & & \cdots & \dfrac{y^{2r-1}}{m!}+\cdots+\dfrac{\theta^{2r-1}}{(m+2r-1)!} \end{pmatrix}$$
$$= (-1)^{r(r+1)/2}\theta^{r(g-d)}$$
$$\times \det\begin{pmatrix} \dfrac{1}{m!} & \cdots & \dfrac{y^{r-1}}{m!}+\cdots+\dfrac{\theta^{r-1}}{(m+r-1)!} & 1 \\ \dfrac{y}{m!}+\dfrac{\theta}{(m+1)!} & \cdots & \dfrac{y^r}{m!}+\cdots+\dfrac{\theta^r}{(m+r)!} & 0 \\ \vdots & & & \vdots \\ \dfrac{y^r}{m!}+\cdots+\dfrac{\theta^r}{(m+r)!} & \cdots & \dfrac{y^{2r-1}}{m!}+\cdots+\dfrac{\theta^{2r-1}}{(m+2r-1)!} & 0 \end{pmatrix}.$$

(5.6)

We will evaluate the determinant in the last expression of (5.6) by a sequence of row and column operations.

First we subtract from each row y times the row above, thereby eliminating the first term of each entry in all but the first row. The $(i+1)$st column is now

$$\frac{y^i}{m!}+\frac{iy^{i-1}\theta}{(m+1)!}+\cdots+\frac{\theta^i}{(m+i)!},$$
$$\frac{y^i\theta}{(m+1)!}+\frac{iy^{i-1}\theta^2}{(m+2)!}+\cdots+\frac{\theta^{i+1}}{(m+i+1)!},$$
$$\frac{y^{i+1}\theta}{(m+1)!}+\frac{(i+1)y^i\theta^2}{(m+2)!}+\cdots+\frac{\theta^{i+2}}{(m+i+2)!},$$
$$\vdots$$
$$\frac{y^{r+i-1}\theta}{(m+1)!}+\frac{(r+i-1)y^{r+i-2}\theta^2}{(m+2)!}+\cdots+\frac{\theta^{r+i}}{(m+r+i)!},$$

for $0 \leq 1 \leq r - 1$; the last column is ${}^t(1, x, 0, \ldots, 0)$. We now repeat the operation, beginning this time with the third row. The $(i + 1)$st column is now

$$\frac{y^i}{m!} + \frac{iy^{i-1}\theta}{(m+1)!} + \cdots + \frac{\theta^i}{(m+i)!},$$

$$\frac{y^i\theta}{(m+1)!} + \frac{iy^{i-1}\theta^2}{(m+2)!} + \cdots + \frac{\theta^{i+1}}{(m+i+1)!},$$

$$\frac{y^i\theta^2}{(m+2)!} + \frac{iy^{i-1}\theta^3}{(m+3)!} + \cdots + \frac{\theta^{i+2}}{(m+i+2)!},$$

$$\vdots$$

$$\frac{y^{r+i-2}\theta^2}{(m+2)!} + \frac{(r+i-2)y^{r+i-3}\theta^3}{(m+3)!} + \cdots + \frac{\theta^{r+i}}{(m+r+i)!},$$

and the last column is ${}^t(1, x, x^2, 0, \ldots, 0)$. Continuing in this fashion the matrix in (5.6) becomes

(5.7)

$$\begin{pmatrix} \dfrac{1}{m!} & \dfrac{y}{m!} + \dfrac{\theta}{(m+1)!} & \cdots & \dfrac{y^{r-1}}{m!} + \cdots + \dfrac{\theta^{r-1}}{(m+r-1)!} & 1 \\ \dfrac{\theta}{(m+1)!} & \dfrac{y\theta}{(m+1)!} + \dfrac{\theta^2}{(m+2)!} & \cdots & \dfrac{y^{r-1}\theta}{(m+1)!} + \cdots + \dfrac{\theta^r}{(m+r)!} & x \\ \vdots & & & & \vdots \\ \dfrac{\theta^r}{(m+r)!} & \dfrac{y\theta^r}{(m+r)!} + \dfrac{\theta^{r+1}}{(m+r+1)!} & \cdots & \dfrac{y^{r-1}\theta^r}{(m+r)!} + \cdots + \dfrac{\theta^{2r-1}}{(m+2r-1)!} & x^r \end{pmatrix}.$$

We now perform the same sequence of operations on the first r columns of the matrix (5.7). When this is done we arrive at the matrix

$$\begin{pmatrix} \dfrac{1}{m!} & \dfrac{\theta}{(m+1)!} & \cdots & \dfrac{\theta^{r-1}}{(m+r-1)!} & 1 \\ \dfrac{\theta}{(m+1)!} & \dfrac{\theta^2}{(m+2)!} & \cdots & \cdots & x \\ \vdots & & & & \vdots \\ \dfrac{\theta^r}{(m+r)!} & \dfrac{\theta^{r+1}}{(m+r+1)!} & \cdots & \dfrac{\theta^{2r-1}}{(m+2r-1)!} & x^r \end{pmatrix}.$$

Expanding this determinant by cofactors of the last column, we obtain

$$c_d^r = \sum_{\alpha=0}^{r} (-1)^{r(r-1)/2} \Delta(m, m+1, \ldots, \widehat{m+\alpha}, \ldots, m+r) y^{\alpha} \theta^{r(g-d+r)-\alpha},$$

where $\Delta(a_1, \ldots, a_k)$ is defined as in the last section; evaluating Δ yields the

Theorem. *If C is any smooth curve of genus g, and C_d^r is either empty or has the expected dimension $\rho + r$, it has fundamental class*

$$c_d^r = \left(\prod_{i=0}^{r} \frac{i!}{(g-d+r+i-1)!}\right) \sum_{\alpha=0}^{r} (-1)^{\alpha} \frac{(g-d+r+\alpha-1)!}{\alpha!\,(r-\alpha)!} \times x^{\alpha} \theta^{r(g-d+r)-\alpha}.$$

Bibliographical Notes

The history of the theorems proved in this chapter is recounted in the bibliographical notes for Chapter V; we will not repeat it here. A more general treatment of ample vector bundles may be found in Hartshorne [1] and in Griffiths [2]. The fact that a section of an ample vector bundle must have a zero if the dimensions so indicates was first proved in Bloch–Gieseker [1]; the stronger connectedness theorem for degeneracy loci of bundle maps was found by Fulton and Lazarsfeld [1], who in the same paper applied it to deduce the connectedness of $W_d^r(C)$. The computations of the expected classes of the loci $W_d^r(C) \subset J(C)$ and $C_d^r \subset C_d$ were first made in Kempf [1] and Kleiman–Laksov [1], and in Kleiman–Laksov [2], respectively. Finally, sources for the material on constructible sheaves are Thom [1], Mather [1], and Artin [1].

Exercises

A. The Connectedness Theorem

In this series of exercises we will derive and then apply the basic connectedness theorem of Fulton and Hansen [1]:

(*) **Theorem.** *Let $X, Y \subset \mathbb{P}^n$ be irreducible varieties with* $\dim X + \dim Y > n$. *Then $X \cap Y$ is connected.*

A-1. Show that Theorem (*) is equivalent to the

(**) **Theorem.** *If $\Delta \subset \mathbb{P}^n \times \mathbb{P}^n$ is the diagonal, and $Z \subset \mathbb{P}^n \times \mathbb{P}^n$ any irreducible subvariety of dimension greater than n, then $Z \cap \Delta$ is connected.*

A-2. Let $Y \subset \mathbb{P}^n$ be any irreducible variety of dimension $k \geq 2$ and $\{X_\lambda\}_{\lambda \in \mathbb{P}^1}$ any pencil of hyperplane sections of Y with base locus of dimension $k - 2$. Show that if there is a point $p \in \bigcap X_\lambda$ of the base of the pencil such that the general X_λ is smooth at p, then the general X_λ is irreducible.

A-3. Let $Y \subset \mathbb{P}^n$ be as in the preceding exercise, and let p, q be any points of Y such that the line $\overline{pq} \not\subset Y$. Let $\Lambda \subset \mathbb{P}^n$ be a general l-plane containing p and q, $l > n - k$. Show that, except possibly in the case $k = 2$, $n = 3$, $\Lambda \cap Y$ is irreducible. (Also, find a counterexample in case Y is a surface in $\mathbb{P}^3$.) Deduce, without any hypotheses on Y, p, and q, that p and q lie in the same connected component of $\Lambda \cap Y$.

A-4. Let Y again be an irreducible, k-dimensional variety in $\mathbb{P}^n$, p, q any points of Y. Suppose that $\{\Lambda_t\}$ is any family of l-planes through p and q. Show that if, for $t \neq 0$, p and q lie on the same connected component of $Y \cap \Lambda_t$, then the same is true for $t = 0$. Use this and the preceding exercise to deduce Theorem (*) in case X is an arbitrary linear subspace of $\mathbb{P}^n$.

A-5. Consider now the birational map

$$\phi : \mathbb{P}^n \times \mathbb{P}^n \dashrightarrow \mathbb{P}^{2n}$$

given by

$$\begin{aligned}\phi([X_0, \ldots, X_n], [Y_0, \ldots, Y_n]) &= \left[1, \frac{X_1}{X_0}, \ldots, \frac{X_n}{X_0}, \frac{Y_1}{Y_0}, \ldots, \frac{Y_n}{Y_0}\right] \\ &= [X_0 Y_0, X_1 Y_0, \ldots, X_n Y_0, \\ &\quad X_0 Y_1, \ldots, X_0 Y_n],\end{aligned}$$

$$\phi^{-1}([Z_0, \ldots, Z_{2n}]) = ([Z_0, Z_1, \ldots, Z_n], [Z_0, Z_{n+1}, \ldots, Z_{2n}]).$$

Let Σ be the graph of ϕ, with projections π_1, π_2 onto $\mathbb{P}^n \times \mathbb{P}^n$ and $\mathbb{P}^{2n}$. Show that:

(a) $\pi_1 : \Sigma \to \mathbb{P}^n \times \mathbb{P}^n$ is simply the blow-up of the locus $X_0 = Y_0 = 0$.
(b) $\pi_2 : \Sigma \to \mathbb{P}^{2n}$ is the blow-up of the union of the planes Λ_1, Λ_2 given by

$$(Z_0 = Z_1 = \cdots = Z_n = 0)$$

and

$$(Z_0 = Z_{n+1} = \cdots = Z_{2n} = 0).$$

(c) The image in $\mathbb{P}^{2n}$ of the proper transform in Σ of the diagonal $\Delta \subset \mathbb{P}^n \times \mathbb{P}^n$ is the plane Λ given by

$$Z_1 = Z_{n+1}, \ldots, Z_n = Z_{2n},$$

(d) Λ is disjoint from Λ_1 and Λ_2.

A-6. Now let $Z \subset \mathbb{P}^n \times \mathbb{P}^n$ be any irreducible subvariety, and set $\tilde{Z} = \pi_2(\pi_1^{-1}(Z)) \subset \mathbb{P}^{2n}$. Using the preceding exercise, show that if $\tilde{Z} \cap \Lambda$ is connected, then $Z \cap \Delta$ is; using Exercise A-3 above, deduce Theorem (**).

A-7. By a similar argument, prove the stronger

Theorem. *Let $Y \subset \mathbb{P}^n$ be any subvariety, X any irreducible variety and $f: X \to \mathbb{P}^n$ any regular map with* $\dim f(X) + \dim Y > n$. *Then $f^{-1}(Y)$ is connected.*

A-8. Let X be any smooth irreducible variety, $f: X \to \mathbb{P}^m$ any map, and assume $m < 2 \cdot \dim X$. Show that if f is an immersion (df is nowhere zero) then f is an embedding.

(*Hint*: Look at $f \times f: X \times X \to \mathbb{P}^m \times \mathbb{P}^m$.)

A-9. Deduce from the preceding exercise that if $X \subset \mathbb{P}^n$ is any smooth, irreducible subvariety and the tangential variety

$$T(X) = \bigcup_{p \in X} \overline{T_p(X)} \subset \mathbb{P}^n$$

has dimension $<2n$, then the secant variety

$$S(X) = \overline{\left(\bigcup_{\substack{p \neq q \\ p, q \in X}} \overline{pq} \right)}$$

equals $T(X)$.

A-10. Using Exercise A-8 above, show that if $X \subset \mathbb{P}^m$ is any smooth, irreducible variety with $\dim X > m/2$, then X has no finite unramified coverings, and hence that $H^1(X, \mathbb{Z}) = 0$

B. Analytic Cohomology of C_d, $d \leq 2g - 2$

We consider here that part of $H^*(C_d, \mathbb{Q})$ generated by x and θ when $d \leq 2g - 2$, and recall (cf. Section 5 of Chapter VIII) that when C is general, this is the entire analytic part $H^*_{an}(C_d, \mathbb{Q})$ of $H^*(C_d, \mathbb{Q})$.

B-1. Let C be a smooth curve of genus g, $d \leq 2g - 2$, and $m \leq d/2$. Show that the $(m + 1) \times (m + 1)$ matrix with entries

$$\Delta_{ij} = (x^{m-i+1}\theta^{i-1} \cdot x^{d-m-j+1} \cdot \theta^{j-1})[C_d]$$

has determinant

$$\det \Delta = (g!)^{m+1} \prod_{\alpha=0}^{m} \frac{\alpha!}{(g - \alpha)!} \neq 0.$$

From this conclude that $x^m, x^{m-1}\theta, \ldots, \theta^m$ are independent in $H^*(C_d, \mathbb{Q})$.

(*Hint*: use the Poincaré formula to show that $(x^{d-\alpha}\theta^\alpha)[C_d] = g!/(g - \alpha)!$)

B-2. Show that when $d \leq 2g - 2$ and $m \geq d/2$, the classes $x^m, x^{m-1}\theta, \ldots, \theta^m$ span a subspace of dimension $d - m + 1$ in $H^{2m}(C_d, \mathbb{Q})$. In this case, show that any $d - m + 1$ of these classes are independent.

B-3. Show that when C is general and $d \leq 2g - 2$

$$\dim H^{2m}_{an}(C_d, \mathbb{Q}) = \begin{cases} m + 1, & m \leq d/2, \\ d - m + 1, & m \geq d/2. \end{cases}$$

B-4. When $d \leq 2g - 2$ and $0 \leq k \leq 2m - d - 1$, show that the relations

$$f_k(x, \theta) = \sum_{\alpha=0}^{d-m+1} (-1)^\alpha \frac{(g - k - \alpha)!}{\alpha!(d - m + 1 - \alpha)!} x^{m-k-\alpha}\theta^{k+\alpha}$$
$$= 0$$

are satisfied. From the preceding exercise, conclude that these constitute a basis for the relations satisfied by the classes $x^\alpha\theta^\beta$ $(\alpha + \beta \leq g - 1)$.

B-5. Show that for $1 \leq \alpha \leq d - 1$

$$x^{d-1-\alpha}\theta^\alpha = \frac{g!}{(g - \alpha)!}(1 - \alpha)x^{d-1} + \frac{(g - 1)!}{(g - \alpha)!}\alpha x^{d-2}\theta.$$

C. Excess Linear Series

C-1. Suppose that E, F, and F' are vector bundles of ranks m, n, and $n + 1$ on a projective variety, and

$$E \xrightarrow{\sigma} F' \xrightarrow{\pi} F$$

a pair of bundle maps with π surjective and $E^* \otimes \operatorname{Ker} \pi$ ample. Using (1.2) show that if Σ' (resp. Σ) is the locus where σ (resp. $\pi \circ \sigma$) has rank k or less,

$$\dim \Sigma' \geq \dim \Sigma - (m - k).$$

C-2. Apply the preceding exercise to conclude that on any curve C,

$$\dim W^r_{d-1}(C) \geq \dim W^r_d(C) - r - 1,$$

and, in particular, that if C possesses ∞^2 g^1_d's then it possesses a g^1_{d-1}.

C-3. Deduce as corollaries of the preceding exercise that

$$\dim W^{r+1}_{d+1}(C) \geq \dim W^r_d(C) - (g - d + r),$$

and

$$\dim W^{r+1}_d(C) \geq \dim W^r_d(C) - (g - d + 2r + 2).$$

Chapter VIII

Enumerative Geometry of Curves

In this chapter we will continue in the direction suggested by Sections 4 and 5 of the previous chapter: that is, we will try to solve some of the enumerative problems that arise in the theory of curves and linear systems. While this is in some sense a quantitative approach, qualitative results may also emerge. For example, the answer to the enumerative question: "How many g_d^r's does a curve C possess" (Theorem (4.4) in Chapter VII) implies the existence theorem (Theorem (2.3) in Chapter VII).

Before we begin, it is necessary to add one major tool to our kit. In Sections 4 and 5 of Chapter VII, the computation of the classes of $W_d^r(C)$ and C_d^r came down to computations of the Chern classes of vector bundles (the push forward $E = v_* \mathscr{L}$ of the Poincaré line bundle, and the tangent bundle Θ_{C_d} to C_d). These were determined on an *ad hoc* basis, using special circumstances in each case. To proceed much further, we need a more general technique for making such computations, and this is the *Grothendieck–Riemann–Roch formula.* In the first section of this chapter, this formula is introduced and, to get started, the Chern classes of E and Θ_{C_d} are re-computed.

Following this, we use the Grothendieck–Riemann–Roch formula to establish the secant plane formula. This may be thought of as an extension of formula (5.6) in Chapter VII giving the class of C_d^r: just as this counts the divisors that fail to impose independent conditions on the canonical series, the secant plane formula counts divisors failing to impose independent conditions on an arbitrary linear series.

Finally, we introduce and compute the classes of the various diagonals in the symmetric product of a curve, and combine this with the secant plane formula to answer some additional enumerative questions.

§1. The Grothendieck–Riemann–Roch Formula

In general, the Grothendieck–Riemann–Roch formula may be used to compute the Chern classes of the alternating sum of the direct images of a coherent sheaf $\mathscr{F}$ on a variety X under a proper morphism $\pi: X \to Y$. To explain this we introduce some terminology. For simplicity, we assume that X and Y are smooth, projective varieties. In addition to the Grothendieck group $K(X)$ constructed from the holomorphic vector bundles on X we may also consider the analogously defined group $\tilde{K}(X)$ constructed from the coherent analytic sheaves on X. There is an obvious map

$$\alpha: K(X) \to \tilde{K}(X), \tag{1.1}$$

and, since any coherent sheaf on a smooth projective variety has a global finite resolution by locally free sheaves, α is surjective; in fact, it is an isomorphism.

Given a proper morphism

$$\pi: X \to Y$$

and a coherent analytic sheaf $\mathscr{F}$ on X, we define

$$\pi_!(\mathscr{F}) = \sum_i (-1)^i R^i \pi_* \mathscr{F} \in K(Y).$$

Next, for $E \to X$ any holomorphic vector bundle with Chern polynomial

$$\begin{aligned} c_t(E) &= \sum_i c_i(E) t^i \\ &= \prod_i (1 + \alpha_i t) \end{aligned} \tag{1.2}$$

we define the *Chern character* to be

$$\operatorname{ch}(E) = \sum_i e^{\alpha_i} \in H^*(X, \mathbb{Q}).$$

This makes sense, since the kth homogeneous term

$$\frac{1}{k!} \sum_i \alpha_i^k$$

on the right is expressible as a polynomial in the elementary symmetric functions of the α_i. Thus, if we write ch_i for the ith homogeneous term of $\operatorname{ch}(E)$,

$$\begin{aligned} \operatorname{ch}_0 &= \operatorname{rank}(E) \\ \operatorname{ch}_1 &= \sum \alpha_i = c_1(E) \\ \operatorname{ch}_2 &= \frac{1}{2} \sum \alpha_i^2 = \frac{1}{2} \left[\left(\sum \alpha_i\right)^2 - 2 \sum_{i<j} \alpha_i \alpha_j \right] \\ &= \frac{c_1(E)^2 - 2c_2(E)}{2}. \end{aligned}$$

The Chern character has the important properties that, for

$$0 \to E \to F \to G \to 0$$

an exact sequence of holomorphic vector bundles,

$$\mathrm{ch}(F) = \mathrm{ch}(E) + \mathrm{ch}(G);$$

and for any bundles E, F

$$\mathrm{ch}(E \otimes F) = \mathrm{ch}(E) \cdot \mathrm{ch}(F).$$

In other words, if we consider the ring structure on $K(X)$ induced by tensor products of bundles, then the Chern character induces a ring homomorphism

$$\mathrm{ch}\colon K(X) \to H^*(X, \mathbb{Q}).$$

Conversely, $\mathrm{ch}(E)$ may be defined to be the value on E on the unique functorial ring homomorphism $K(X) \to H^*(X, \mathbb{Q})$ that maps every line bundle L to $e^{c_1(L)}$.

We observe that, because of the isomorphism (1.1), the definition of the Chern classes and Chern character may be extended to coherent sheaves. If $V \subset X$ is a divisor, then from the exact sequence:

$$0 \to \mathcal{O}_X(-V) \to \mathcal{O}_X \to \mathcal{O}_V \to 0,$$

we infer that

$$\begin{aligned} c_t(\mathcal{O}_V) &= (1 - tv)^{-1} \\ &= 1 + tv + t^2v^2 + \cdots, \end{aligned}$$

where $v \in H^2(X, \mathbb{Q})$ is the fundamental class of V.

The last definition we require is that of the Todd class, $\mathrm{td}(E)$, given for a vector bundle E with Chern polynomial (1.2) by

$$\mathrm{td}(E) = \prod_i \frac{\alpha_i}{1 - e^{-\alpha_i}} \in H^*(X, \mathbb{Q}).$$

As with the Chern character, this is a symmetric power series in the α_i, and so may be expressed in terms of the Chern classes of E. Thus, since

$$\frac{\alpha}{1 - e^{-\alpha}} = 1 + \frac{\alpha}{2} + \frac{\alpha^2}{12} + \cdots,$$

we have

$$\prod_i \frac{\alpha_i}{1-e^{-\alpha_i}} = 1 + \sum \frac{\alpha_i}{2} + \sum \frac{\alpha_i^2}{12} + \sum_{i<j} \frac{\alpha_i \alpha_j}{4} + \cdots$$
$$= 1 + \tfrac{1}{2}c_1(E) + \tfrac{1}{12}(c_1^2(E) + c_2(E)) + \cdots.$$

The notation $\mathrm{td}(X)$ is used when E is the tangent bundle of X. The Todd class is multiplicative in the sense that, for an exact sequence

$$0 \to E \to F \to G \to 0,$$

we have

$$\mathrm{td}(F) = \mathrm{td}(E)\,\mathrm{td}(G).$$

With these explanations, the *Grothendieck–Riemann–Roch formula* is

$$\mathrm{ch}(\pi_!\mathscr{F}) \cdot \mathrm{td}(Y) = \pi_*(\mathrm{ch}(\mathscr{F}) \cdot \mathrm{td}(X)).$$

The simplest case of this is the case in which Y is a point. Here, using the identification $H^0(Y, \mathbb{Q}) = \mathbb{Q}$, we have that $\mathrm{td}(Y) = 1$ and $\pi_!\mathscr{F} = \chi(\mathscr{F}) = \sum(-1)^i h^i(X, \mathscr{F})$, so that the formula reduces to the Hirzebruch–Riemann–Roch formula

$$\chi(\mathscr{F}) = (\mathrm{ch}(\mathscr{F}) \cdot \mathrm{td}(X))[X].$$

Thus, if X is a curve of genus g

$$\chi(\mathscr{F}) = c_1(\mathscr{F}) + \mathrm{rank}(\mathscr{F}) \cdot (1 - g),$$

and if X is a surface

$$\chi(\mathscr{F}) = \frac{c_1^2(\mathscr{F}) - 2c_2(\mathscr{F})}{2} + \frac{c_1(\mathscr{F}) \cdot c_1(X)}{2} + \frac{c_1^2(X) + c_2(X)}{12},$$

which reduces, in case $\mathscr{F}$ is a line bundle, to

$$\chi(\mathscr{F}) = \frac{c_1^2(\mathscr{F}) - c_1(\mathscr{F}) \cdot K_X}{2} + \chi(\mathscr{O}_X).$$

§2. Three Applications of the Grothendieck–Riemann–Roch Formula

Our first application of the Grothendieck–Riemann–Roch theorem will be to rederive the formula of Section 4 of Chapter VII for the Chern classes of the bundle E on $\mathrm{Pic}^n(C)$. To do this, recall that E is defined to be the direct image $\nu_*\mathscr{L}$ of a Poincaré line bundle $\mathscr{L}$ on $C \times \mathrm{Pic}^n(C)$ under the projection $\nu: C \times \mathrm{Pic}^n(C) \to \mathrm{Pic}^n(C)$; we will apply the Riemann–Roch formula to the map ν.

We begin by developing some notation for the cohomology of C and $\mathrm{Pic}^n(C)$, as follows. First, we choose a symplectic basis $\delta_1, \ldots, \delta_{2g}$ for $H^1(C, \mathbb{Z})$ (e.g., the dual of the basis $\gamma_1, \ldots, \gamma_{2g}$ for $H_1(C, \mathbb{Z})$ introduced in Section 5 of Chapter I); we will also denote by $\delta_1, \ldots, \delta_{2g}$ the classes in $\mathrm{Pic}^n(C) = J(C)$ corresponding to the above via the isomorphism

$$u^*: H^1(J(C), \mathbb{Z}) \xrightarrow{\sim} H^1(C, \mathbb{Z}).$$

We will denote by $\delta''_1, \ldots, \delta''_{2g}$ (resp. $\delta'_1, \ldots, \delta'_{2g}$) the pull-backs to $C \times \mathrm{Pic}^n(C)$ of these classes from $\mathrm{Pic}^n(C)$ (resp. C). Finally, we will write θ for the pull-back to $C \times \mathrm{Pic}^n(C)$ of the class $\theta \in H^2(\mathrm{Pic}^n(C), \mathbb{Z})$, and we will denote by η the pull-back of the class of a point on C. Summarizing, we have

$$\delta'_\alpha \delta'_{g+\alpha} = -\delta'_{g+\alpha}\delta'_\alpha = \eta, \qquad \alpha = 1, \ldots, g,$$
$$\delta'_\alpha \delta'_\beta = 0 \quad \text{if} \quad \beta \neq \alpha \pm g,$$

$$\theta = \sum_{\alpha=1}^{g} \delta''_\alpha \delta''_{g+\alpha}.$$

We now make two observations. First, obviously

$$\mathrm{td}(\mathrm{Pic}^n(C)) = 1,$$

and almost as obviously

$$\mathrm{td}(C \times \mathrm{Pic}^n(C)) = 1 + (1 - g)\eta.$$

Secondly, since we have taken $n \geq 2g - 1$, the higher cohomology of the Poincaré bundle $\mathscr{L}$ on the fibers of ν vanishes. Since ν is flat, it follows that

$$R^i\nu_*\mathscr{L} = 0 \quad \text{for} \quad i > 0.$$

Putting these together with the Riemann–Roch formula yields a formula for the Chern character of $E = \nu_*\mathscr{L}$:

$$\text{(2.1)} \qquad \mathrm{ch}(E) = \nu_*((1 + (1 - g)\eta) \cdot \mathrm{ch}(\mathscr{L})).$$

It remains thus to determine the Chern class of $\mathscr{L}$. This may itself be found by applying the Riemann–Roch formula to the original construction of $\mathscr{L}$ (cf. Section 2 of Chapter IV); we leave this as an exercise. It turns out to be substantially simpler in this case to proceed directly. Set

$$c_1(\mathscr{L}) = c^{2,0} + c^{1,1} + c^{0,2}$$

where $c^{i,j}$ is the component of $c_1(\mathscr{L})$ in the (i, j)th term of the Kunneth decomposition (integral coefficients throughout)

$$\begin{aligned} H^2(C \times \operatorname{Pic}^n(C)) = (H^2(C) &\otimes H^0(\operatorname{Pic}^n(C))) \\ &\oplus (H^1(C) \otimes H^1(\operatorname{Pic}^n(C))) \\ &\oplus (H^0(C) \otimes H^2(\operatorname{Pic}^n(C))). \end{aligned}$$

Since $\mathscr{L}$ is trivial on $\{q\} \times \operatorname{Pic}^n(C)$ we infer that $c^{0,2} = 0$, and since $\mathscr{L}$ has degree n on a fiber $C \times \{L\}$ of v it follows that $c^{2,0} = n\eta$. To determine $c^{1,1}$, note that if $\omega\colon C \to \operatorname{Pic}^n(C)$ is the map

$$\omega\colon p \mapsto u(p + (n-1)q),$$

then the universal property of the Poincaré line bundle implies that

$$(1_C \times \omega)^*\mathscr{L} = \mathscr{O}_{C \times C}(\Delta + (n-1)\Gamma_q - \Gamma_q'),$$

where $\Gamma_q = \{q\} \times C$, $\Gamma_q' = C \times \{q\}$ and $\Delta \subset C \times C$ is the diagonal. Since $1_C \times \omega$ induces an isomorphism on the $(1, 1)$ part of H^2, we conclude from the standard formula for the class of the diagonal that

$$c^{1,1} = \Delta^{1,1} = -\sum_{\alpha=1}^{g} (\delta_\alpha''\delta_{g+\alpha}' - \delta_{g+\alpha}''\delta_\alpha').$$

Let us call this class γ. Note that

$$\begin{aligned} \gamma^2 &= -\sum_{\alpha=1}^{g} (\delta_\alpha''\delta_{g+\alpha}'\delta_{g+\alpha}''\delta_\alpha' + \delta_{g+\alpha}''\delta_\alpha'\delta_\alpha''\delta_{g+\alpha}') \\ &= -2\eta\theta, \end{aligned}$$

while evidently

$$\gamma^3 = \eta \cdot \gamma = 0.$$

Summarizing,

$$c_1(\mathscr{L}) = n \cdot \eta + \gamma,$$

and therefore

$$\begin{aligned} \operatorname{ch}(\mathscr{L}) = e^{c_1(L)} &= \sum_{k=0}^{\infty} \frac{(n \cdot \eta + \gamma)^k}{k!} \\ &= 1 + n \cdot \eta + \gamma - \eta\theta. \end{aligned}$$

Plugging this into (2.1), we have

$$\begin{aligned}\mathrm{ch}(E) &= \nu_*((1 + (1 - g)\eta)(1 + n \cdot \eta + \gamma - \eta\theta)) \\ &= \nu_*((n - g + 1)\eta - \eta\theta) \\ &= (n - g + 1) - \theta.\end{aligned}$$

We now solve this for the Chern classes of E. Writing

$$c_t(E) = \sum_{\alpha=1}^{n-g+1} (1 + \lambda_\alpha t)$$

we have

$$\begin{aligned}&\sum \lambda_\alpha = -\theta \\ &\sum \lambda_\alpha^k = 0 \quad \text{for} \quad k > 1.\end{aligned}$$

Then

$$\begin{aligned}c_1(E) &= \sum_\alpha \lambda_\alpha = -\theta \\ c_2(E) &= \sum_{\alpha>\beta} \lambda_\alpha \lambda_\beta \\ &= \frac{(\sum \lambda_\alpha)^2 - \sum_\alpha \lambda_\alpha^2}{2} = \frac{\theta^2}{2}.\end{aligned}$$

More generally, the formulas above for the Newton functions in the λ's show that

$$\begin{aligned}c_t(E) &= e^{\log(\prod(1+\lambda_\alpha t))} \\ &= e^{\sum \log(1+\lambda_\alpha t)} \\ &= e^{-\sum \lambda_\alpha t} \\ &= e^{-t\theta},\end{aligned}$$

or, which is the same, that

$$c_k(E) = (-1)^k \theta^k/k!.$$

Our second application of the Grothendieck–Riemann–Roch formula will be to redo the computations of Section 5 of Chapter VII for the Chern classes of the tangent bundle Θ_{C_d} to the symmetric product C_d. To do this, recall from Chapter IV that if Δ is the universal divisor of degree d on $C \times C_d$, the pointwise identification

$$H^0(C, \mathcal{O}_D(D)) \cong T_D(C_d)$$

of Laurent tails with tangent vectors to C_d globalizes to a sheaf isomorphism

$$(\pi_2)_* \mathcal{O}_\Delta(\Delta) \cong \Theta_{C_d},$$

where π_1, π_2 stand for the projection maps on $C \times C_d$. It is to this situation that we apply the Grothendieck–Riemann–Roch formula, together with the vanishing of the higher direct images of $\mathcal{O}_\Delta(\Delta)$, to obtain

$$\mathrm{td}(C_d) \cdot \mathrm{ch}(\Theta_{C_d}) = (\pi_2)_*(\mathrm{td}(C \times C_d) \cdot \mathrm{ch}(\mathcal{O}_\Delta(\Delta))).$$

Omitting π_1^*, π_2^*, the right-hand side is

$$(\pi_2)_*(\mathrm{td}(C) \cdot \mathrm{td}(C_d) \cdot \mathrm{ch}(\mathcal{O}_\Delta(\Delta))) = \mathrm{td}(C_d)(\pi_2)_*(\mathrm{td}(C) \cdot \mathrm{ch}(\mathcal{O}_\Delta(\Delta))).$$

The $\mathrm{td}(C_d)$ factors cancel out and, recalling that η denotes the pull-back of the class of a point on C, we have

$$\mathrm{ch}(\Theta_{C_d}) = (\pi_2)_*((1 + (1 - g)\eta) \cdot \mathrm{ch}(\mathcal{O}_\Delta(\Delta))).$$

From the usual exact sequence

$$0 \to \mathcal{O}_{C \times C_d} \to \mathcal{O}_{C \times C_d}(\Delta) \to \mathcal{O}_\Delta(\Delta) \to 0$$

we obtain

$$\mathrm{ch}(\mathcal{O}_\Delta(\Delta)) = e^\delta - 1,$$

where $\delta \in H^2(C \times C_d, \mathbb{Z})$ is the class of Δ, and so our problem is simply to determine this class.

To begin with, we note that the inclusion $j: C \to C_d$ sending p to $p + (d - 1)q$ induces an isomorphism

$$j^*: H^1(C_d, \mathbb{Z}) \to H^1(C, \mathbb{Z}). \tag{2.2}$$

To see that j^* is surjective, we note that, as already observed, the composition

$$u^* \circ j^*: H^1(\mathrm{Pic}^d(C), \mathbb{Z}) \to H^1(C, \mathbb{Z})$$

is an isomorphism; to see that j^* is injective, we may apply the Lefschetz hyperplane theorem to the string of smooth ample divisors

$$C \hookrightarrow C_2 \hookrightarrow C_3 \hookrightarrow \cdots \hookrightarrow C_{d-1} \hookrightarrow C_d.$$

As before, we will denote by $\delta_1, \ldots, \delta_{2g}$ a symplectic basis for $H^1(C, \mathbb{Z}) \cong H^1(C_d, \mathbb{Z})$ and by $\delta'_1, \ldots, \delta'_{2g}$ (resp. $\delta''_1, \ldots, \delta''_{2g}$) its pull-back to $C \times C_d$ from C (resp. from C_d); also we will continue to write θ for the pull-back of the class $\theta \in H^2(J(C))$ to C_d, and x for the class of the divisor $X_q = q + C_{d-1} \subset C_d$.

With this understood, we compute the class δ of the universal divisor Δ by breaking it up into its components under the Kunneth decomposition of $H^2(C \times C_d)$:

$$\delta = \delta^{2,0} + \delta^{1,1} + \delta^{0,2},$$

with $\delta^{i,j} \in H^i(C) \otimes H^j(C_d)$. Clearly, since Δ restricted to $\{q\} \times C_d$ is just the divisor X_q,

$$\delta^{0,2} = x.$$

Equally clearly, since Δ restricts on each fiber $C \times \{D\}$ of π_1 to a divisor of degree d on C,

$$\delta^{2,0} = d\eta.$$

Finally, since the map

$$1_C \times j: C \times C \to C \times C_d$$

induces an isomorphism on the (1, 1) part of the cohomology, and since

$$(1_C \times j)^*\Delta = \Delta_C + (d-1)\Gamma_q,$$

where $\Gamma_q = \{q\} \times C$ and Δ_C is the diagonal in $C \times C$, we may calculate as before that

$$\delta^{1,1} = -\sum_{\alpha=1}^{g} (\delta'_\alpha \delta''_{g+\alpha} - \delta'_{g+\alpha}\delta''_\alpha).$$

Calling this class γ, we combine the preceding formulas to obtain

$$\delta = d\eta + \gamma + x;$$

as before we may observe that $\gamma^2 = -2\eta\theta$, while $\eta^2 = \eta\gamma = \gamma^3 = 0$.

Returning to the computation of $c_t(\Theta_{C_d})$, we have

$$\text{(2.3)} \qquad \operatorname{ch}(\Theta_{C_d}) = (\pi_2)_*((1 + (1-g)\eta) \cdot (e^{d\eta+\gamma+x} - 1)).$$

On the right-hand side we have

$$e^{d\eta+\gamma+x} = \sum_k \frac{(d\eta + \gamma + x)^k}{k!}$$

$$= \sum_k \frac{x^k}{k!} + \sum_k \frac{kd\eta x^{k-1}}{k!} + \sum_k \frac{\binom{k}{2} x^{k-2}\gamma^2}{k!} + \sum_k \frac{k\gamma x^{k-1}}{k!}.$$

Using the fact that $\gamma^2 = -2\eta\theta$ this gives

$$e^{d\eta-\gamma+x} = e^x + d\eta e^x - \eta\theta e^x + \gamma e^x.$$

Plugging into (2.3) we find

$$\mathrm{ch}(\Theta_{C_d}) = g - 1 + ((d - g + 1) - \theta)e^x.$$

To see from this what the Chern class of Θ_{C_d} is, note that, by the previous argument, $(d - g + 1) - \theta$ is the Chern character of a vector bundle of rank $d - g + 1$ having Chern polynomial $e^{-t\theta}$. Consequently, $((d - g + 1) - \theta)e^x$ is the Chern character of the tensor product of such a vector bundle with a line bundle having first Chern class x, and by the computation of Section 5 of Chapter VII it follows that

$$c_t(\Theta_{C_d}) = (1 + xt)^{d-g+1}e^{-t\theta/(1+xt)}.$$

Our third application of the Grothendieck–Riemann–Roch theorem will yield a new formula. Recall that C_d^r is the cycle of divisors of degree d on C that fail by r to impose independent conditions on the canonical series $|K|$; equivalently, these correspond to d-secant $(d - r - 1)$-planes to the canonical curve $\phi_K(C) \subset \mathbb{P}^{g-1}$. We would like now to derive a similar formula, but with the complete canonical series replaced by an arbitrary linear series $\mathcal{D} = \mathbb{P}V \subset |L|$. Our approach will be to construct a bundle E_L on C_d whose fiber $(E_L)_D$ over a point $D \in C_d$ is the space of sections of L over D: e.g., if $D = p_1 + \cdots + p_d$, with the p_i distinct,

$$(E_L)_D = \bigoplus_i L_{p_i}.$$

We should then have a natural map from the trivial bundle $V \otimes \mathcal{O}_{C_d}$ on C_d to E_L given by evaluation, and the degeneracy loci of this bundle map will be the cycles in question.

We begin by defining, for any line bundle L on C, a vector bundle E_L on C_d by

$$E_L = (\pi_2)_*(\mathcal{O}_\Delta \otimes \pi_1^*L).$$

Here again π_1, π_2 are the projection maps on $C \times C_d$ and $\Delta \subset C \times C_d$ is the universal divisor. Note that since Δ is flat over C_d and π_1^*L is locally free, the sheaf $\mathcal{O}_\Delta \otimes \pi_1^*L$ is flat over C_d; thus E_L is locally free and we have the identification

$$(E_L)_D \cong H^0(C, L/L(-D)).$$

Moreover, the natural map of sheaves on $C \times C_d$

$$\pi_1^*L \to \mathcal{O}_\Delta \otimes \pi_1^*L$$

given by restriction pushes forward to a map

$$\alpha\colon H^0(C, L) \otimes \mathcal{O}_{C_d} \to E_L$$

from the trivial bundle $H^0(C, L) \otimes \mathcal{O}_{C_d}$ on C_d to E_L. This in turn may be restricted to a map

$$\alpha_V\colon V \otimes \mathcal{O}_{C_d} \to E_L \tag{2.4}$$

for any subseries $V \subset H^0(C, L)$ to give the desired evaluation map.

It remains to determine the Chern class of E_L. The result is expressed in the

(2.5) Lemma. *If L has degree n,*

$$c_t(E_L) = (1 - xt)^{d-n+g-1} e^{t\theta/(1-xt)}.$$

Proof. It will suffice to sketch the derivation, which is essentially the same as in the last application. From the vanishing of the higher direct images and the Grothendieck–Riemann–Roch theorem we obtain

$$\operatorname{td}(C_d)\operatorname{ch}(E_L) = (\pi_2)_*(\operatorname{td}(C \times C_d) \cdot \operatorname{ch}(\mathcal{O}_\Delta(L))).$$

Cancelling out the terms $\operatorname{td}(C_d)$ and using the exact sequence

$$0 \to L \otimes \mathcal{O}_{C \times C_d}(-\Delta) \to L \to L \otimes \mathcal{O}_\Delta \to 0,$$

we have

$$\begin{aligned}\operatorname{ch}(E_L) &= (\pi_2)_*((1 + (1 - g)\eta) \cdot (e^{n\eta} - e^{n\eta - \delta})) \\ &= (\pi_2)_*((1 + (1 - g)\eta) \cdot (1 + n \cdot \eta - (1 + (n - d)\eta - \theta\eta)e^{-x})) \\ &= n - g + 1 + (d - n + g - 1 + \theta)e^{-x},\end{aligned}$$

from which the lemma follows.

§3. The Secant Plane Formula: Special Cases

Before deriving the secant plane formula in general, it is worthwhile to mention one special case in which the calculations are not so cumbersome. In order to facilitate the computations involved, we pause briefly to introduce and illustrate two useful notations.

We denote by

$$(g(t_1, \dots, t_b))_{t_1^{a_1} \dots t_b^{a_b}}$$

the coefficient of $t_1^{a_1} \cdots t_b^{a_b}$ in a formal Laurent series $g(t_1, \dots, t_b)$.

We shall define the binomial coefficients $\binom{n}{i}$, regardless of the sign of the integers n and i, by the following convention

$$\binom{n}{i} = \begin{cases} \dfrac{n(n-1)\cdots(n-i+1)}{i!} & \text{if } i > 0, \\ 1 & \text{if } i = 0, \\ 0 & \text{if } i < 0. \end{cases}$$

For example,

$$\begin{aligned}(1+t)^a &= \sum_i \binom{a}{i} t^i \\ &= \sum_i (-1)^i \binom{-a-1+i}{i} t^i\end{aligned}$$

regardless of the sign of a. Moreover, for any formal series $c(t) = \sum c_j t^j$,

$$\begin{aligned}[(1-xt)^\alpha c(t/(1-xt))]_{t^\beta} &= \left[\sum_j c_j t^j (1-xt)^{\alpha-j}\right]_{t^\beta} \\ &= \left[\sum_{i,j} \binom{-\alpha+j+i-1}{i} c_j x^i t^{i+j}\right]_{t^\beta} \\ &= \left[\sum_j c_j (1+xt)^{\beta-\alpha-1} t^j\right]_{t^\beta},\end{aligned}$$

that is, we have

$$[(1-xt)^\alpha \cdot c(t/(1-xt))]_{t^\beta} = [(1+xt)^{\beta-\alpha-1} c(t)]_{t^\beta}. \tag{3.1}$$

We now consider a linear series $\mathscr{D}$ of degree n and dimension r. This means that we are given a line bundle $L \to C$ of degree n, an $(r+1)$-dimensional linear subspace $V \subset H^0(C, L)$, and then $\mathscr{D} = \mathbb{P}(V)$ consists of all divisors of sections in V. We do not assume that $\mathscr{D}$ is complete or non-special. The cycle $\Gamma_d(V)$ of all divisors of degree d that are subordinate to the linear series V is defined to be

$$\Gamma_d(\mathscr{D}) = \{D \in C_d : E - D \geq 0 \quad \text{for some } E \in \mathscr{D}\}.$$

As will be seen momentarily, there is a natural scheme structure on $\Gamma_d(\mathscr{D})$, and we will prove the

(3.2) Lemma. *Suppose that $n \geq d \geq r$. Then $\Gamma_d(\mathscr{D})$ is r-dimensional and its fundamental class $\gamma_d(\mathscr{D})$ is given by*

$$\gamma_d(\mathscr{D}) = \sum_{k=0}^{d-r} \binom{n-g-r}{k} \frac{x^k \theta^{d-r-k}}{(d-r-k)!}.$$

Proof. As a set, $\Gamma_d(\mathscr{D})$ is the locus of those D such that there exists a section $s \in V$ vanishing in D; i.e., the locus where the bundle map (2.4)

$$\alpha_V : V \otimes \mathcal{O}_{C_d} \to E_L$$

has a kernel. We then *define* $\Gamma_d(\mathscr{D})$ to be the rth determinantal variety of α_V and compute its class accordingly. By a relatively standard formula in the theory of Chern classes (or by Porteous' formula), we have

$$\gamma_d(V) = c_{d-r}(E_L).$$

Using (2.5), $c_{d-r}(E_L)$ is the coefficient of t^{d-r} in the power series $(1 - xt)^{d-n+g-1} e^{\theta t/(1-tx)}$. By (3.1), we have

$$\begin{aligned} c_{d-r}(E_L) &= [(1 - xt)^{d-n+g-1} e^{\theta t/(1-xt)}]_{t^{d-r}} \\ &= [(1 + xt)^{n-g-r} e^{t\theta}]_{t^{d-r}}, \end{aligned}$$

and hence

$$c_{d-r}(E_L) = \sum_{k=0}^{d-r} \binom{n-g-r}{k} \frac{x^k \theta^{d-r-k}}{(d-r-k)!}. \qquad \text{Q.E.D.}$$

Remark. In case $d = n$, we are computing the class v of the linear system V itself, and since $\theta^{g+1} = 0$, (3.2) gives

$$v = \frac{x^\delta \theta^g}{g!} \quad \text{if} \quad \delta = n - g - r \geq 0,$$

where δ is the difference between the "expected dimension" $n - g$ of the complete linear system $|L|$ and the actual dimension r of V. Of course, in case $n \geq 2g - 1$, we knew this already, since V is a linear space of codimension δ in a fiber of $C_d \to J(C)$, and the class of a general fiber is $\theta^g/g!$.

As a second application of (2.5) we will now prove the following result.

Proposition. *Let d, r_α, $\alpha = 1, \ldots, l$ be integers such that $a_\alpha = d - r_\alpha \geq 0$ and $\sum_\alpha a_\alpha = d$. Suppose that there are finitely many effective divisors of degree d subordinate to each of l series $\mathscr{D}_\alpha$ of respective degrees n_α and dimension r_α. Then the number of these divisors, counted with appropriate multiplicity, is*

$$\left[\left(\prod_{\alpha=1}^{l}(1 + t_\alpha)^{n_\alpha - g - r_\alpha}\right)\left(1 + \sum_{\alpha=1}^{l} t_\alpha\right)^g\right]_{t_1^{a_1}\cdots t_l^{a_l}}.$$

Proof. The virtual number of effective divisors of degree d subordinate to $\mathscr{D}_1, \ldots, \mathscr{D}_l$ is the integer

$$\prod_{\alpha=1}^{l} \gamma_d(\mathscr{D}_\alpha) \in H^{2d}(C_d, \mathbb{Z}) \cong \mathbb{Z},$$

where $\gamma_d(\mathscr{D}_\alpha)$ is the class of $\Gamma_d(\mathscr{D}_\alpha)$. By (3.2) this is

$$\left[\prod_{\alpha=1}^{l}(1 + xt_\alpha)^{n_\alpha - g - r_\alpha} \exp\left(\theta \sum_\alpha t_\alpha\right)\right]_{t_1^{a_1}\cdots t_l^{a_l}}. \tag{3.3}$$

To evaluate this we use the Gysin mapping

$$u_*: H^{2d}(C_d, \mathbb{Z}) \to H^{2g}(J(C), \mathbb{Z}).$$

By Poincaré's formula and the push-pull formula we have

$$u_*(x^{d-\beta}\theta^\beta) = \begin{cases} g!/(g-\beta)!\,\xi & \text{if } \beta \leq \min(d, g), \\ 0 & \text{if } \beta > \min(d, g), \end{cases}$$

where ξ is the positively oriented generator of $H^{2g}(J(C), \mathbb{Z}) \cong \mathbb{Z}$. By our convention this can also be written

$$u_*(x^{d-\beta}\theta^\beta) = \begin{cases} \binom{g}{\beta}\beta!\,\xi & \text{if } \beta \leq d, \\ 0 & \text{if } \beta > d. \end{cases}$$

Effectively, then, in (3.3) we may set $x = 1$ and

$$\theta^\beta = \begin{cases} \binom{g}{\beta}\beta! & \text{if } \beta \leq d, \\ 0 & \text{if } \beta > d, \end{cases}$$

to have

$$\begin{aligned} \exp(\textstyle\sum t_\alpha) &= \sum_{\beta=0}^{g} \frac{\theta^\beta}{\beta!}\left(\sum_\alpha t_\alpha\right)^\beta \\ &= \sum_{\beta=0}^{d} \binom{g}{\beta}\left(\sum_\alpha t_\alpha\right)^\beta \\ &= \left(1 + \sum_{\alpha=1}^{l} t_\alpha\right)^g \quad (\text{modulo } (t_1, \ldots, t_l)^{d+1}), \end{aligned}$$

which implies the lemma. Q.E.D.

Example. We shall calculate the number $\gamma_2(\mathscr{D}_1)\gamma_2(\mathscr{D}_2)[C_d]$ of pairs of points common to two pencils $\mathscr{D}_1$ and $\mathscr{D}_2$ of degrees n_1 and n_2. According to the proposition this is the coefficient of $t_1 t_2$ in

$$(1 + t_1)^{n_1 - g - 1}(1 + t_2)^{n_2 - g - 1}(1 + t_1 + t_2)^g,$$

which turns out to be

$$(n_1 - 1)(n_2 - 1) - g.$$

It is interesting to verify this by a geometric argument. The pair of pencils gives a map

$$\phi: C \to \mathbb{P}^1 \times \mathbb{P}^1$$

that has bidegree (n_1, n_2). If we assume that the image curve is birational to C and has δ ordinary double points, then by the genus formula for curves on a quadric

$$\delta = (n_1 - 1)(n_2 - 1) - g,$$

which is the formula above.

Example. The number of $(r+1)$-tuples of points common to a pencil $\mathscr{D}_1$ of degree n_1 and an r-dimensional series $\mathscr{D}_2$ of degree n_2 is

$$\begin{aligned}\gamma_{r+1}(\mathscr{D}_1)\cdot\gamma_{r+1}(\mathscr{D}_2)[C_d] &= [(1+t_1)^{n_1-g-1}(1+t_2)^{n_2-g-r}(1+t_1+t_2)^g]_{t_1^r t_2} \\ &= (n_2-g-r)[(1+t_1)^{n_1-g-1}(1+t_1+t_2)^g]_{t_1^r} \\ &\quad + g[(1+t_1)^{n_1-g-1}(1+t_1+t_2)^{g-1}]_{t_1^r} \\ &= (n_2-g-r)\binom{n_1-1}{r} + g\binom{n_1-2}{r} \\ &= (n_2-r)\binom{n_1-1}{r} - g\binom{n_1-2}{r-1}.\end{aligned}$$

§4. The General Secant Plane Formula

As indicated earlier, we will be concerned with the following problem: given a linear system $\mathscr{D} = \mathbb{P}V \subset |L|$ of degree n and dimension s, find the class of the cycle V_d^r consisting of all divisors of degree d that impose at most $d-r$ conditions on $\mathscr{D}$.

Of course, the formulas in Section 3 are examples of this when $d-r=s$. Another case we have already dealt with is when $\mathscr{D} = |K|$; here

$$V_d^r = C_d^r,$$

(this is clear, at least set-theoretically), and our formula for the class v_d^r of V_d^r will yield yet another proof of the Existence Theorem (2.3) of Chapter VII for special linear systems.

To derive the secant plane formula, we observe that set-theoretically the locus V_d^r is the $(d-r)$th degeneracy locus of the map (2.4); we take this as the definition of V_d^r as an analytic space (note that by relative duality and Lemma (2.3) in Chapter IV the map $\alpha_{|K|}$ is the transpose of the Brill–Noether homomorphism, proving in this case the scheme-theoretic equality $V_d^r = C_d^r$). By Porteous' formula and the formula (2.5) for the Chern class of E_L we have the following preliminary

(4.1) Lemma. *Suppose that $s+1 \geq d-r$, and that V_d^r is either empty or of pure dimension $d - r(s+1-d+r)$. Then the fundamental class v_d^r of V_d^r is equal to*

$$\begin{aligned}v_d^r &= (-1)^{r(s+1-d+r)}\Delta_{r,s+1-d+r}((1+xt)^{d-n+g-1}e^{-\theta t/(1+xt)}) \\ &= \Delta_{s+1-d+r,r}((1+xt)^{n-d-g+1}e^{\theta t/(1+xt)}).\end{aligned}$$

There now arises the major problem of simplifying the determinant in this expression. We shall do this in three steps, the first of which is

Reduction A. *The identity*

$$\Delta_{a,b}((1 + xt)^c e^{\theta t/(1+xt)}) = \Delta_{a,b}((1 - xt)^{a-b-c} e^{t\theta})$$

is valid.

Proof. To begin with, let us write out the determinant $\Delta_{a,b}((1 + xt)^c e^{\theta t/(1+xt)})$ explicitly. We have

$$\begin{aligned}
(1 + xt)^c e^{\theta t/(1+xt)} &= (1 + xt)^c \sum_j \frac{1}{j!} \theta^j t^j (1 + xt)^{-j} \\
&= \sum_j \frac{1}{j!} \theta^j t^j (1 + xt)^{c-j} \\
&= \sum_{i,j} \frac{1}{j!} \binom{c-j}{i} \theta^j x^i t^{i+j} \\
&= \sum_{n=0}^{\infty} \left(\sum_{j=0}^{n} \frac{1}{j!} \binom{c-j}{n-j} \theta^j x^{n-j} \right) t^n,
\end{aligned}$$

so that

$$\Delta_{a,b}((1 + xt)^c e^{\theta t/(1+xt)})$$

$$= \det \begin{pmatrix} \sum_{j=0}^{a} \binom{c-j}{a-j} \frac{\theta^j}{j!} x^{a-j} & \cdots & \sum_{j=0}^{a-b+1} \binom{c-j}{a-b+1-j} \frac{\theta^j}{j!} x^{a-b+1-j} \\ \vdots & & \vdots \\ \sum_{j=0}^{a+b-1} \binom{c-j}{a+b-1-j} \frac{\theta^j}{j!} x^{a+b-1-j} & \cdots & \sum_{j=0}^{a} \binom{c-j}{a-j} \frac{\theta^j}{j!} x^{a-j} \end{pmatrix}.$$

Now, we can try to perform the same reduction on this matrix that we did in the special case $a = g - d + r$, $b = r$, $c = g - d - 1$. The problem is that the success of that sequence of row and column operations in reducing to a matrix of monomials depended on the original matrix having the form above and starting in one corner with a monomial; since in the original case we had $c = g - d - 1$, $a - b + 1 = g - d + 1$, so that the corner entry

$$\begin{aligned}
\sum_{j=0}^{a-b+1} \binom{c-j}{a-b+1-j} \frac{\theta^j}{j!} x^{a-b+1-j} &= \sum_{j=a-b}^{a-b+1} \binom{c-j}{a-b+1-j} \frac{\theta^j}{j!} x^{a-b+1-j} \\
&= -\frac{\theta^{a-b} x}{(a-b)!} + \frac{\theta^{a-b+1}}{(a-b+1)!}
\end{aligned}$$

had only two terms, we had only to add one extra row to the matrix to obtain the desired form. To carry this out more generally, at least for $c \leq a - b$, we would have to add $a - b - c$ extra rows. At the conclusion of the reduction, then, we would be left with an $(a - c) \times (a - c)$ matrix, with b columns consisting of monomials in θ and $a - b - c$ consisting of monomials in x. We would then expand not by cofactors along one column, but by minors along $a - b - c$. This approach will, it may be seen, ultimately yield the secant plane formula below; in the present general context, however, the following reduction seems more straightforward.

The first step is essentially the same as before. The first column of the matrix above is

$$\begin{array}{l}
\binom{c}{a}x^a + \binom{c-1}{a-1}x^{a-1}\theta + \cdots + \binom{c-a}{0}\frac{\theta^2}{a!}, \\[2ex]
\binom{c}{a+1}x^{a+1} + \binom{c-1}{a}x^a\theta + \cdots + \binom{c-a-1}{0}\frac{\theta^{a+1}}{(a+1)!}, \\[2ex]
\binom{c}{a+2}x^{a+2} + \binom{c-1}{a+1}x^{a+1}\theta + \cdots + \binom{c-a-2}{0}\frac{\theta^{a+2}}{(a+2)!}. \\
\qquad\vdots \qquad\qquad\qquad \vdots \qquad\qquad\qquad\qquad\qquad\qquad \vdots
\end{array}$$

Adding to each row from the second to the last x times the row above it thus has the effect of replacing c by $c + 1$ in all but the first row, and similarly in the remaining columns. Performing the same operation from the third row on down then replaces $c + 1$ by $c + 2$ in the third to last rows; and continuing in this fashion we arrive at a matrix with first column

$$\begin{array}{lllll}
\binom{c}{a}x^a & + & \binom{c-1}{a-1}x^{a-1}\theta & + \cdots \\[2ex]
\binom{c+1}{a+1}x^{a+1} & + & \binom{c}{a}x^a\theta & + \cdots \\[2ex]
\binom{c+2}{a+2}x^{a+2} & + & \binom{c+1}{a+1}x^{a+1}\theta & + \cdots \\
\quad\vdots & & \quad\vdots & \\
\binom{c+b-1}{a+b-1}x^{a+b-1} & + & \binom{c+b-2}{a+b-2}x^{a+b-2}\theta & + \cdots
\end{array}$$

and $(i + 1)$st row

$$t\begin{pmatrix} \binom{c+i}{a+i}x^{a+i} & + & \binom{c+i-1}{a+i-1}x^{a+i-1}\theta & + \cdots \\ \binom{c+i}{a+i-1}x^{a+i-1} & + & \binom{c+i-1}{a+i-2}x^{a+i-2}\theta & + \cdots \\ \vdots & & & \\ \binom{c+i}{a-b+i+2}x^{a-b+i+2} & + & \binom{c+i-1}{a-b+i+1}x^{a-b+i+1}\theta & + \cdots \\ \binom{c+i}{a-b+i+1}x^{a-b+i+1} & + & \binom{c+i-1}{a-b+i}x^{a-b+i}\theta & + \cdots \end{pmatrix}.$$

Now adding to each of the first $b - 1$ columns x times the column on its right replaces c by $c + 1$ in all but the right-hand column; repeating this for the first $b - 2$ columns, the first $b - 3$ columns, and so on, replaces c by $c + b - i$ in the ith column; we have then the matrix

$$\begin{pmatrix} \sum_{j=0}^{a}\binom{c+b-1-j}{a-j}\frac{\theta^j}{j!}x^{a-j} & \cdots & \sum_{j=0}^{a-b+1}\binom{c-j}{a-b+1-j}\frac{\theta^j}{j!}x^{a-b+1-j} \\ \vdots & & \vdots \\ \sum_{j=0}^{a+b-1}\binom{c+2b-2-j}{a+b-1-j}\frac{\theta^j}{j!}x^{a+b-1-j} & \cdots & \sum_{j=0}^{a}\binom{c+b-1-j}{a-j}\frac{\theta^j}{j!}x^{a-j} \end{pmatrix}.$$

By the relation $\binom{n}{m} = (-1)^m\binom{-n+m-1}{m}$, we may rewrite this as

$$\begin{pmatrix} \sum_{j=0}^{a}\binom{n}{a-j}\frac{\theta^j}{j!}y^{a-j} & \cdots & \sum_{j=0}^{a-b+1}\binom{n}{a-b+1-j}\frac{\theta^j}{j!}y^{a-b+1-j} \\ \vdots & & \vdots \\ \sum_{j=0}^{a+b-1}\binom{n}{a+b-1-j}\frac{\theta^j}{j!}y^{a+b-1-j} & \cdots & \sum_{j=0}^{a}\binom{n}{a-j}\frac{\theta^j}{j!}y^{a-j} \end{pmatrix},$$

where $n = a - c - b$ and $y = -x$. The point of the reduction so far is that now the binomial coefficient part of each term in the entries depends only on the power of y involved, while the straight factorial part depends only on the power of θ; so that the entries are the coefficients of t in the product of the power series

$$\sum_{j=0}^{\infty}\frac{\theta^j}{j!}t^j = e^{t\theta},$$

and

$$\sum_{i=1}^{\infty} \binom{n}{i} y^i t^i = (1 + yt)^n = (1 - xt)^{a-c-b}.$$

We have accordingly proved the formula

$$\Delta_{a,b}((1 + xt)^c e^{\theta t/(1+xt)}) = \Delta_{a,b}((1 - xt)^{a-c-b} e^{t\theta});$$

more generally, since the coefficients of $e^{\theta t/(1+xt)}$ were irrelevant to the reduction, we have, for any power series $f(t)$,

$$\Delta_{a,b}\left((1 + xt)^c f\left(\frac{t}{1 + xt}\right)\right) = \Delta_{a,b}((1 - xt)^{a-c-b} f(t)).$$

Reduction B. MacDonald's Formula

The last expression for $\Delta_{a,b}((1 + xt)^c e^{\theta t/(1+xt)})$ transposes readily into a (special case of) a formula of MacDonald. To derive this, note that for any power series $f(t) = \sum c_i t^i$ in one variable,

$$\begin{aligned}\Delta_{a,b}(f) &= \det\begin{pmatrix} c_a & c_{a-1} & \cdots & c_{a-b+1} \\ \vdots & & & \vdots \\ c_{a+b-1} & & \cdots & c_a \end{pmatrix} \\ &= \sum_{\sigma \in S_b} \operatorname{sgn}(\sigma) c_{a+1-\sigma(1)} \cdots c_{a+b-\sigma(b)} \\ &= \sum_{\sigma \in S_b} \operatorname{sgn}(\sigma) \left[\prod_{i=1}^{b} f(t_i)\right]_{t_1^{a+1-\sigma(1)} \cdots t_b^{a+b-\sigma(b)}} \\ &= \left[\left(\prod_{i=1}^{b} f(t_i)\right)\Delta(t)\right]_{t_1^a t_2^{a+1} \cdots t_b^{a+b-1}},\end{aligned}$$

where

$$\begin{aligned}\Delta(t) &= \prod_{i>j} (t_i - t_j) \\ &= \sum_{\sigma \in S_b} \operatorname{sgn}(\sigma) t_1^{\sigma(1)-1} t_2^{\sigma(2)-1} \cdots t_b^{\sigma(b)-1}\end{aligned}$$

is the Vandermonde determinant in the variables $t_1, \ldots, t_b$. To obtain an expression more symmetric in the variables t_i, we can reorder the t_i by a permutation $\tau \in S_b$ and write

$$\begin{aligned}\Delta_{a,b}(f) &= \operatorname{sgn}(\tau)\left[\prod f(t_i) \cdot \Delta(t)\right]_{t_1^{a+\tau(1)-1} \cdots t_b^{a+\tau(b)-1}} \\ &= \operatorname{sgn}(\tau)\left[\prod f(t_i) \cdot \Delta(t) \cdot t_1^{b-\tau(1)} \cdots t_b^{b-\tau(b)}\right]_{(\prod t_i)^{a+b-1}}.\end{aligned}$$

Summing this over all $\tau \in S_b$ and noting that if $\tau'(\alpha) = b + 1 - \tau(\alpha)$ then $\mathrm{sgn}(\tau') = \varepsilon_b \cdot \mathrm{sgn}(\tau)$, where $\varepsilon_b = (-1)^{b(b-1)/2}$, we infer that

$$\Delta_{a,b}(f) = \frac{\varepsilon_b}{b!}\left[\prod f(t_i) \cdot \Delta(t)^2\right]_{(\prod t_i)^{a+b-1}}.$$

In our present circumstance, this yields the formula

$$\begin{aligned}\Delta_{a,b}((1+xt)^c e^{t\theta/(1+xt)}) &= \Delta_{a,b}((1-xt)^{a-c-b}e^{t\theta}) \\ &= \frac{\varepsilon_b}{b!}\left[\prod_{i=1}^{b}(1-xt_i)^{a-c-b}e^{\theta\Sigma t_i}\Delta(t)^2\right]_{(t_1\cdot\ldots\cdot t_b)^{a+b-1}}.\end{aligned}$$

Since $\Delta(t)^2$ contains a sum over all permutations, the expression on the right in Reduction B does not represent an essential simplification over the original determinant. However, in some cases the formula is more manageable; for example, let us prove the following result (cf. MacDonald [1]):

(4.2) Proposition. *The virtual number of divisors of degree d imposing only $d - r$ conditions on a given linear series g_n^s and containing $d - r(s + 1 - d + r)$ given points is*

$$\begin{aligned}&\frac{\varepsilon_r}{r!}\prod_{i=1}^{r}\left[(1-t_i)^{g+s-n}\left(1+\sum_i t_i\right)^g\Delta(t)^2\right]_{(t_1\cdots t_r)^{s-d+2r}} \\ &\quad= \frac{\varepsilon_r}{(s+1-d+r)!} \\ &\quad\quad\times\left[\prod_{i=1}^{s+1-d+r}(1+t_i)^{n-g-s}(1+\textstyle\sum t_i)^g\Delta(t)^2\right]_{(t_1\cdots t_{s+1-d+r})^{s+d+2r}}.\end{aligned}$$

Proof. Recalling that $x \in H^2(C_d, \mathbb{Z})$ is the class of $C_{d-1} + p \subset C_d$ for any $p \in C$, we must evaluate the cup-product

$$v_d^r \cdot x^{d-r(s+1-d+r)} \in H^{2d}(C_d, \mathbb{Z}) \cong \mathbb{Z}.$$

Using (4.1) and the two reductions above we find

$$v_d^r = \frac{\varepsilon_r}{r!}\left[\prod_{i=1}^{r}(1-xt_i)^{g-n+s}e^{(\Sigma t_i)\theta}\Delta(t)^2\right]_{(t_1\cdots t_r)^{s-d+2r}}.$$

Since

$$x^{d-\alpha}\theta^\alpha = \begin{cases}\dbinom{g}{\alpha}\alpha!\,\xi & \text{if } \alpha \le d,\\ 0 & \text{if } \alpha > d,\end{cases}$$

where ξ is the positively oriented generator of $H^{2d}(C_d, \mathbb{Z})$, to evaluate the cup-product above we may replace x by 1 and, after expanding $e^{(\Sigma t_i)\theta}$, replace θ^α by $\binom{g}{\alpha}\alpha!$ when $\alpha \leq d$ and by 0 when $\alpha > d$. When this is done

$$\begin{aligned} e^{(\Sigma t_i)\theta} &= \sum_\alpha \frac{(\sum t_i)^\alpha \theta^\alpha}{\alpha!} \\ &= \sum_{\alpha=0}^{d} \binom{g}{\alpha} (\sum t_i)^\alpha \\ &\equiv (1 + \sum t_i)^g \quad (\text{modulo}(t_1, \ldots, t_r)^{d+1}). \end{aligned}$$

Substituting in the expression above gives the first expression of (4.2). The second one is obtained similarly.

Example. The simplest (non-trivial) case of the proposition we just proved is to compute the number of double points of a plane curve of degree n; that is, divisors of degree 2 imposing only one condition on a given two-dimensional linear series. Here we take $d = 2$, $r = 1$, and $s = 2$; trivially $\Delta(t) = 1$, and hence the number of double points is

$$\begin{aligned} [(1-t)^{g+2-n}(1+t)^g]_{t^2} &= \binom{g+2-n}{2} - g(g+2-n) + \binom{g}{2} \\ &= \frac{(n-1)(n-2)}{2} - g, \end{aligned}$$

as expected.

Example. To generalize the above example, we find the number ν_m of $(m+1)$-secant $(m-1)$-planes to a curve $C \subset \mathbb{P}^{2m}$ of degree n. Here $r = 1$, $s = 2m$, and $d = m + 1$, and plugging in we have

$$\begin{aligned} \nu_m &= [(1-t)^{g+2m-1}(1+t)^g]_{t^{m+1}} \\ &= \sum_{\alpha=0}^{m+1} (-1)^\alpha \binom{g+2m-n}{\alpha} \binom{g}{m+1-\alpha}. \end{aligned}$$

It may be an interesting problem to determine when this number is positive, negative or zero.

Example. The next simplest example is the formula for the number of 4-secant lines to a space curve $C \subset \mathbb{P}^3$ of degree n. Here we take $d = 4$, $r = 2$, and $s = 3$; then $\Delta(t)^2 = (t_2 - t_1)^2$, and consequently the number is

$$\begin{aligned} -[((1-t_1)(1-t_2))^{g+3-n}(1+t_1+t_2)^g(t_2^2 - 2t_1t_2 + t_1^2)]_{t_1^3t_2^3} \\ = \frac{(n-2)(n-3)^2(n-4)}{12} - \frac{g(n^2 - 7n + 13 - g)}{2}. \end{aligned}$$

We now return to the derivation of the secant plane formula. For the third stage in the reduction we will use the following additional notational convention

$$\mu(a, b, m, i, \beta) = \binom{m + i - 1}{a + i - \beta} \frac{(a + i - \beta)!}{(a + b - \beta)!(\beta - 1)!}.$$

We have then the

Reduction C.

$$\Delta_{a,b}((1 + xt)^n e^{t\theta}) = \sum_{1 \le \beta_1 < \cdots < \beta_b \le a+b} \Delta(\beta)^2 \prod_{i=1}^{b} \mu(a, b, n, i, \beta_i) \times \theta^{\Sigma(\beta_i - i)} x^{ab - \Sigma(\beta_i - i)}.$$

Proof. To simplify the determinant

$\Delta_{a,b}((1 + xt)^n e^{\theta t})$

$$= \det \begin{pmatrix} \sum_{j=0}^{a} \binom{n}{a-j} \frac{\theta^j}{j!} x^{a-j} & \cdots & \sum_{j=0}^{a+b-1} \binom{n}{a+b-1-j} \frac{\theta^j}{j!} x^{a+b-1-j} \\ \vdots & & \vdots \\ \sum_{j=0}^{a-b+1} \binom{n}{a-b+1-j} \frac{\theta^j}{j!} x^{a-b+1-j} & \cdots & \sum_{j=0}^{a} \binom{n}{a-j} \frac{\theta^j}{j!} x^{a-j} \end{pmatrix}$$

we consider the defining expansion of the determinant

$$\det A = \sum_{\sigma \in S_b} \operatorname{sgn}(\sigma) \prod_{i=1}^{b} a_{i,\sigma(i)}$$

and separate out those terms in the b-fold products $\prod a_{i,\sigma(i)}$ which take from the polynomial $a_{i,\sigma(i)}(x, \theta)$ the term involving θ^{α_i}. The determinant may in this way be expressed as a sum

$$\sum_{\alpha_1, \ldots, \alpha_b} \det A_{\alpha_1, \ldots, \alpha_b},$$

where

$A_{\alpha_1, \ldots, \alpha_b}$

$$= \begin{pmatrix} \binom{n}{a-\alpha_1} \frac{\theta^{\alpha_1}}{\alpha_1!} x^{a-\alpha_1} & \cdots & \binom{n}{a+b-1-\alpha_1} \frac{\theta^{\alpha_1}}{\alpha_1!} x^{a+b-1-\alpha_1} \\ \vdots & & \vdots \\ \binom{n}{a-b+1-\alpha_b} \frac{\theta^{\alpha_b}}{\alpha_b!} x^{a-b+1-\alpha_b} & \cdots & \binom{n}{a-\alpha_b} \frac{\theta^{\alpha_b}}{\alpha_b!} x^{a-\alpha_b} \end{pmatrix}.$$

Setting $\alpha = (\alpha_1, \ldots, \alpha_b)$ we have then

$$\det A_\alpha = \frac{\theta^{\Sigma \alpha_i} x^{ab - \Sigma \alpha_i}}{\prod \alpha_i!} \delta_{n; a-\alpha_1, \ldots, a-b+1-\alpha_b},$$

where

$$\delta_{n;k_1,\ldots,k_b} = \det\begin{pmatrix} \binom{n}{k_1} & \binom{n}{k_1+1} & \cdots & \binom{n}{k_1+b-1} \\ \binom{n}{k_2} & & & \\ \vdots & & & \vdots \\ \binom{n}{k_b} & & \cdots & \binom{n}{k_b+b-1} \end{pmatrix}.$$

To evaluate $\delta_{n;k_1,\ldots,k_b}$, add to each column of the above matrix (except the first) the column on its left; repeat the process for all but the first two columns, and so on. We have then

$$\delta_{n;k_1,\ldots,k_b} = \det\begin{pmatrix} \binom{n}{k_1} & \binom{n}{k_1+1} & \cdots & \binom{n+b-1}{k_1+b-1} \\ \vdots & & & \vdots \\ \binom{n}{k_b} & & \cdots & \binom{n+b-1}{k_b+b-1} \end{pmatrix}$$

$$= \prod_{i=1}^{b} \frac{(n+i-1)!}{(n-k_i)!} \det\begin{pmatrix} \frac{1}{k_1!} & \cdots & \frac{1}{(k_1+b-1)!} \\ \vdots & & \vdots \\ \frac{1}{k_b!} & \cdots & \frac{1}{(k_b+b-1)!} \end{pmatrix}$$

$$= \prod_{i=1}^{b}\left(\frac{(n+i-1)!}{(n-k_i)!}\right)\prod_{i=1}^{b}\left(\frac{1}{(k_i+b-1)!}\right)\Delta(k_b,\ldots,k_1).$$

Here, by the expression $\prod((n+i-1)!/(n-k_i)!)$ we mean the product $\prod((n+i-1)\cdot\ldots\cdot(n-k_{\sigma(i)}+1))$, where $\sigma \in S_b$ and $k_{\sigma(1)} \geq k_{\sigma(2)} \geq \cdots k_{\sigma(b)}$, if $k_{\sigma(i)} + i > 0$ for all i; and 0 if $k_{\sigma(i)} + i < 0$ for any i (the need for this cumbersome convention will disappear shortly). We have then

$$\det A_\alpha = \prod_{i=1}^{b}\left(\frac{(n+i-1)!}{(n-a+\alpha_i+i-1)!}\right)\frac{1}{(a+b-\alpha_i-i)!\,\alpha_i!}$$
$$\times\, \Delta(\alpha_1+1, \alpha_2+2, \ldots, \alpha_b+b)\theta^{\Sigma\alpha_i}x^{ab-\Sigma\alpha_i}$$
$$= \det B_\beta = \prod_{i=1}^{b}\left(\frac{(n-i-1)!}{(n-a+\beta_i-1)!}\,\frac{1}{(a+b-\beta_i)!(\beta_i-i)!}\right)$$
$$\times\, \Delta(\beta_1,\ldots,\beta_b)\cdot\theta^{\Sigma(\beta_i-i)}x^{ab-\Sigma(\beta_i-i)},$$

where $\beta_i = \alpha_i + i$. The reason for the shift from the index set α to the index set β is this: for each choice of $\beta_1 < \beta_2 < \cdots < \beta_b$ and $\sigma \in S_b$, we may set

$$(\beta^\sigma)_i = \beta_{\sigma(i)};$$

noting that

$$\Delta(\beta^\sigma) = \operatorname{sgn}(\sigma)\Delta(\beta)$$

we can sum the quantities $\det B_\beta$ over all rearrangements of β to obtain

$$\sum_{\sigma \in S_b} \det B_{\beta^\sigma} = \prod_{i=1}^{b} \left(\frac{(n+i-1)!}{(n-a+\beta_i-1)!} \frac{1}{(a+b-\beta_i)!} \right) \Delta(\beta) \cdot \theta^{\Sigma(\beta_i - i)} x^{ab - \Sigma(\beta_i - i)}$$
$$\times \sum_{\sigma \in S_b} \operatorname{sgn}(\sigma) \frac{1}{\prod (\beta_{\sigma(i)} - i)!}.$$

The sum in this expression may then be evaluated as

$$= \det \begin{pmatrix} \frac{1}{(\beta_1 - 1)!} & \cdots & \frac{1}{(\beta_1 - b)!} \\ \vdots & & \vdots \\ \frac{1}{(\beta_b - 1)!} & \cdots & \frac{1}{(\beta_b - b)!} \end{pmatrix} = \frac{\Delta(\beta_1, \ldots, \beta_b)}{\prod (\beta_i - 1)!}.$$

and we have

$$\Delta_{a,b}((1+xt)^n e^{\theta t}) = \sum_{1 \le \beta_1 < \beta_2 < \cdots < \beta_b \le a+b} \Delta(\beta)^2$$
$$\times \prod_{i=1}^{b} \frac{(n+i-1)!}{(n-a+\beta_i-1)!(a+b-\beta_i)!(\beta_i-1)!}$$
$$\times \theta^{\Sigma(\beta_i - i)} x^{ab - \Sigma(\beta_i - i)},$$

where again the quotient $(n+i-1)!/(n-a+\beta_i-1)!$ is taken to be the product $(n+i-1)\cdot\ldots\cdot(n-a+\beta_i)$ (note that, now that the β_i's are ordered, we have $i - 1 \ge \beta_i - a$ always). If we wish to avoid this convention, we may rewrite this last formula as

$$\Delta_{a,b}((1+xt)^n e^{t\theta}) = \sum_{1 \le \beta_1 < \beta_2 < \cdots < \beta_b \le a+b} \Delta(\beta)^2 \prod_{i=1}^{b} \binom{n+i-1}{a+i-\beta_i}$$
$$\times \frac{(a+i-\beta_i)!}{(a+b-\beta_i)!(\beta_i-1)!} \theta^{\Sigma(\beta_i - i)} x^{ab - \Sigma(\beta_i - i)}. \quad \text{Q.E.D.}$$

We now apply Reductions A and C to the formula for v_d^r given by Lemma (4.1). Reduction A gives

$$\begin{aligned} v_d^r &= \Delta_{k,r}((1 - xt)^{-\delta} e^{t\theta}) \\ &= (-1)^{rk} \Delta_{r,k}((1 - xt)^{\delta} e^{-t\theta}), \end{aligned}$$

where

$$\delta = n - g - s,$$

$$k = s + 1 - d + r,$$

while Reduction C yields the formulas

$$\text{(4.3)} \quad v_d^r = \sum_{1 \le \beta_1 < \cdots < \beta_k \le k+r} \Delta(\beta)^2 \prod_{i=1}^{k} \mu(r, k, \delta, i, \beta_i) \theta^{\Sigma(\beta_i - i)} x^{rk - \Sigma(\beta_i - i)}$$

$$= \sum_{1 \le \beta_1 < \cdots < \beta_r \le k+r} \Delta(\beta)^2 \prod_{i=1}^{k} \mu(k, r, -\delta, i, \beta_i) \theta^{\Sigma(\beta_i - i)} (-x)^{rk - \Sigma(\beta_i - i)}.$$

This formula is, clearly, a mess. Nonetheless, it does represent a substantial simplification of (4.1). For example it has $\binom{r+k}{r}$ terms rather than the roughly $r!\,k^r$ terms involved in the expansion of (4.1). Most importantly, (4.3) displays much more clearly the qualitative behavior of the class v_d^r: for example, we notice at once that in case $\delta = n - g - s \ge 0$, all terms in the first expression of (4.3) have a positive sign. To motivate this condition, recall that our linear system $\mathscr{D} = \mathbb{P}V$ is a g_n^s corresponding to a subspace $V \subset H^0(C, L)$, and by the Riemann–Roch theorem we have

$$s = \dim \mathbb{P}V \le \dim|L| = n - g + h^1(C, L);$$

thus the following conditions are equivalent:

$$\begin{cases} n - g - s \ge 0, \\ \dim|L| \ge s + h^0(C, KL^{-1}). \end{cases}$$

If we define the *relative deficiency* of a linear series of dimension s and degree n on a genus g curve to be $\delta = n - g - s$, then the positivity of the terms in (4.3) has the following implication:

If the series $\mathscr{D}$ has non-negative relative deficiency and if $r(s + 1 - d + r) \le d$, then the cycle V_d^r is non-empty.

Somewhat more geometrically:

If $C \subset \mathbb{P}^s$ has genus g and degree n, then C possesses d-secant $(d - r - 1)$-planes provided that

$$n \geq s + g \quad \textit{and} \quad r(s + 1 - d + r) \leq d.$$

Unfortunately, the hypothesis $n - g - s \geq 0$ excludes the most interesting case, namely when $\mathscr{D}$ is complete and is a special linear series. The difficulty here is not merely apparent but is real: the conclusion above is not always true, since, as we shall see by example, there are cases in which v_d^r is zero and still others where it is non-zero but is not the class of an effective cycle (what *can* be concluded in these cases is that V_d^r is non-empty and has the wrong dimension). From a technical standpoint the difficulty is that the coefficients of (4.3), viewed as a polynomial in x, may have alternating signs, and since there are relations among x and θ in $H^*(C_d, \mathbb{Q})$ it is not possible to conclude that $v_d^r \neq 0$.

In borderline cases, however, we are able to extract information from (4.3):

Proposition. *If $s + 1 \geq d - r \geq 0$ and $r(s + 1 - d + r) \leq d$, and if $s = n - g + 1$ (i.e., the series $\mathscr{D}$ has relative deficiency -1) then $v_d^r = 0$ if and only if $g < (r + 1)(s - d + r + 1)$. In particular, if $g \geq (r + 1)(s - d + r + 1)$, then $V_d^r \neq \varnothing$; while if $g < (r + 1)(s - d + r + 1)$ then V_d^r is either empty or of the wrong dimension.*

We observe that the condition $\delta = -1$ is satisfied when

$$\mathscr{D} = |K|$$

and that in this case the inequality $g \geq (r + 1)(s - d + r + 1)$ amounts to saying that $\rho \geq 0$, so that this proposition implies the existence theorem for special divisors (Theorem (2.3) in Chapter VII).

Proof of the Proposition. In case $d - r = s + 1$, $V_d^r = C_d$ and there is nothing to prove. Thus we may assume that $s - d + r \geq 0$. In the sum (4.3) we must have

$$1 \leq \beta_1 < \cdots < \beta_r \leq s + 1 - d + 2r,$$

and also, since $\delta = -1$, we must have

$$0 \leq \beta_i - (s + 1 - d + r)$$

for all i. The only index sets yielding non-zero terms are

$$\beta^{(j)} = (s + 1 - d + r, \ldots, \widehat{s + 1 - d + r + j}, \ldots, s + 1 - d + 2r)$$

where $0 \le j \le r$. Then we have

$$\Delta(\beta^{(j)})^2 = \left(\frac{\prod_{i=0}^{r} r!}{j!(r-j)!}\right)^2,$$

$$\prod_{i=1}^{r} \frac{(g+s-n+i-1)!}{(g-n+d-r-2+\beta_i^{(j)})!} = j!,$$

$$\prod_{i=1}^{r} \frac{1}{(s+1-d+2r-\beta_i^{(j)})!} = \frac{(r-j)!}{\prod_{i=0}^{r} i!},$$

$$\prod_{i=1}^{r} \frac{1}{(\beta_i^{(j)}-1)!} = \frac{(s-d+r+j)!}{\prod_{i=1}^{r}(s-d+r+i)!},$$

so that by (4.3)

$$v_d^r = \prod_{i=0}^{r} \frac{i!}{(s-d+r+i)!} \sum_{j=0}^{r} \frac{(s-d+r+j)!}{j!(r-j)!} (-x)^j \theta^{r(s+1-d+r)-j}.$$

Since $\theta^{g+1} = 0$, $v_d^r = 0$ when $g < r(s-d+r)$ (which, incidentally, implies that $g < (r+1)(s+1-d+r)$). Therefore we may assume that $g \ge r(s-d+r)$. To see when the expression for v_d^r is non-zero we multiply by the monomial

$$\theta^{\alpha} x^{d-r(s+1-d+r)-\alpha} \qquad 0 \le \alpha \le d - r(s+1-d+r)$$

of complementary degree. Ignoring the non-zero factor

$$\prod_{i=0}^{r} \frac{i!}{(s-d+r+i)!} \xi,$$

where ξ is the positive generator of $H^{2d}(C_d, \mathbb{Z})$, this gives

$$\sum_{j=0}^{r} \frac{(-1)^j(s-d+r+j)!}{j!(r-j)!} \binom{g}{r(s-d+r)-j+r+\alpha} (r(s-d+r)-j+r+\alpha)!$$

$$= \frac{(s-d+r)!\,g!}{(g-r(s-d+r)-\alpha)!} \sum_{j=0}^{r} (-1)^j \binom{s-d+r+j}{j} \binom{g-r(s-d+r)-\alpha}{r-j}$$

$$= \frac{(s-d+r)!\,g!}{(g-r(s-d+r)-\alpha)!} \sum_{j=0}^{r} \left[\frac{1}{(1+t)^{s-d+r+1}}\right]_{t^j} [(1+t)^{g-r(s-d+r)-\alpha}]_{t^{r-j}}$$

$$= \frac{(s-d+r)!\,g!}{(g-r(s-d+r)-\alpha)!} [(1+t)^{g-\alpha-(r+1)(s-d+r+1)+r}]_{t^r}$$

$$= \frac{(s-d+r)!\,g!}{(g-r(s-d+r)-\alpha)!} \binom{g-(r+1)(s+1-d+r)-\alpha+r}{r}.$$

This is zero for all α such that

$$0 \leq \alpha \leq d - r(s + 1 - d + r)$$

exactly when

$$g < (r + 1)(s - d + r + 1),$$

so that we know that v_d^r is not zero in case $g \geq (r + 1)(s - d + r + 1)$. In fact, this computation establishes the proposition once we realize that the class v_d^r is a polynomial in x and θ and that the intersection pairing is non-degenerate on the subring of $H^*(C_d, \mathbb{Q})$ generated by x and θ.

§5. Diagonals in the Symmetric Product

Our last application of the topology of the symmetric products to enumerative geometry will be concerned with divisors with multiplicities, and to set this up we define the *diagonal mappings* for C_d:

$$\phi_a: C_{n_1} \times \cdots \times C_{n_k} \to C_d$$

to be given by

$$\phi_a(D_1, \ldots, D_k) = a_1 D_1 + \cdots + a_k D_k,$$

where $a = (a_1, \ldots, a_k)$ and $\sum_{i=1}^k a_i n_i = d$.

To state the result we shall retain the notations used previously in this chapter; in addition we set

$$R_a(t) = 1 + \sum_{i=1}^{k} a_i^2 t_i,$$

$$P_a(t) = 1 + \sum_{i=1} a_i t_i,$$

$$e = n_1 + \cdots + n_k.$$

The main result is the

(5.1) Proposition. *The image, via the diagonal mapping ϕ_a, of the fundamental class of $C_{n_1} \times \cdots \times C_{n_k}$ is*

$$(\phi_a)_*[C_{n_1} \times \cdots \times C_{n_k}] = \left[\sum_{\alpha \geq \beta} \frac{(-1)^{\alpha+\beta}}{\beta!(\alpha-\beta)!} P_a(t)^{e-g+\beta} R_a(t)^{g-\beta} x^{d-e-\alpha}\theta^\alpha\right]_{t_1^{n_1}\cdots t_k^{n_k}}.$$

As an application of (5.1) we will be able to derive the classical

De Jonquières' Formula. *Let $a_1, \ldots, a_k, n_1, \ldots, n_k, d$ be positive integers. Let r be a non-negative integer. Suppose the a_i's are distinct, $\sum n_i = d - r$, and $\sum a_i n_i = d$. Set $a = (a_1, \ldots, a_k)$, $n = (n_1, \ldots, n_k)$. Then the virtual number $\mu_{a,n}$ of divisors having n_i points of multiplicity a_i in a given linear series of dimension r and degree d is*

$$\mu_{a,n} = [R_a(t)^g P_a(t)^{d-r-g}]_{t_1^{n_1} \cdots t_k^{n_k}}.$$

The proof of the proposition will be given in several steps, and will be based on the following

Lemma. *Let C be a curve with general moduli. The subring of $H^*(C_d, \mathbb{Q})$ generated by the fundamental classes of algebraic cycles on C_d is generated by x and θ.*

A few words of explanation are in order. The sentence, "If C is a curve (of genus g) with general moduli, then ..." is shorthand for "There are denumerably many proper subvarieties of the moduli space of genus g curves such that, if the point corresponding to the curve C does not belong to any one of them, then ...".

Proof of the Lemma. When $d \geq 2g - 1$ we have seen in Chapter VII (cf. Proposition (2.1)) that there are a vector bundle E_d on $J(C)$ and a biregular morphism $C_d \xrightarrow{\sim} \mathbb{P}(E_d)$ such that the diagram

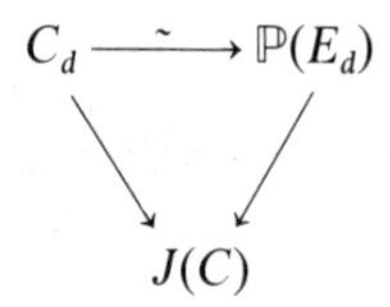

is commutative. Letting $A^*(C_d)$, $A^*(J(C))$ be the Chow rings of $C_d = \mathbb{P}(E_d)$ and $J(C)$, respectively, we have

$$A^*(C_d) = A^*(J(C))[x],$$

where x is the cycle corresponding to $X \subset C_d$ and x satisfies the minimal equation (4.2) of Chapter VII. Of course, one should now read the Chern classes in this formula as having values in the Chow ring of $J(C)$. The lemma follows, when $d \geq 2g - 1$, from this remark together with Poincaré's formula and the assertion that, when C has general moduli, the algebraic part of $H^{2q}(J(C), \mathbb{Q})$ is generated by θ^q (this will be proved in Chapter X).

Assuming the result for C_{d+1} we will deduce it for C_d. Let $Z \subset C_d$ be an analytic subvariety; we want to express the fundamental class z of Z as a polynomial in x and θ. For this we define

$$\psi_d\colon C_d \times C \to C_{d+1}$$

by

$$\psi_d(D, p) = D + p,$$

and we denote by X_q the image of the inclusion

$$j\colon C_d \to C_{d+1}$$

given by

$$j(D) = D + q.$$

If we define the subvariety $Z_q \subset C_{d-1}$ by

$$Z_q = \{D - q\colon D \in Z \text{ and } D - q \geq 0\},$$

then

$$j(Z) = X_q \cdot \psi_d(Z \times C) - j\psi_{d-1}(Z_q \times C)$$

(this means that a point in $X_q \cap \{D + p\colon D \in Z \text{ and } p \in C\}$ arises either when $q \in D$ or when $p = q$). Passing to fundamental classes, this gives

$$z = j^*[\psi_d(Z \times C)] - [\psi_{d-1}(Z_q \times C)].$$

By our induction assumption, $j^*[\psi_d(Z \times C)]$ is a polynomial in x and θ.

Repeating the argument with two points q_1, q_2 where $q = q_1$ and with $\psi_{d-1}(Z_{q_1} \times C)$ replacing Z, we are reduced to establishing the result for $\psi_{d-1}(\psi_{d-2}(Z_{q_1q_2} \times C) \times C)$ where

$$Z_{q_1q_2} = \{D - q_1 - q_2\colon D \in Z \text{ and } D - q_1 - q_2 \geq 0\}.$$

Continuing in this way we are eventually reduced to establishing the result for the image of the standard inclusion

$$C_k \to C_d$$

given by

$$D \to D + q_1 + q_2 + \cdots + q_{d-k}, \qquad D \in C_k.$$

The fundamental class of this cycle is x^{d-k}. Q.E.D.

By the lemma the class on the left-hand side of the formula in (5.1) is a polynomial in x, θ, and to determine the coefficients we will evaluate the pull-backs $\phi_a^*(x^\alpha\theta^\beta)$, $\alpha + \beta = e$, in $H^{2e}(C_{n_1} \times \cdots \times C_{n_k}, \mathbb{Z})$.

We begin by showing that

$$(5.2) \qquad \phi_a^*(\theta^e)[C_{n_1} \times \cdots \times C_{n_k}] = \binom{g}{e} \frac{(e!)^2}{n_1! \cdots n_k!} \prod_i a_i^{2n_i}.$$

When $e > g$, both sides are zero; when $e \le g$, the verification is straightforward but requires introducing some notation. As in Section 2 of this chapter we let $\delta_1, \ldots, \delta_{2g}$ be a symplectic basis for

$$H^1(C, \mathbb{Q}) \cong H^1(C_d, \mathbb{Q})$$

and we denote by $\delta_\alpha^{(i)}, x^{(i)}, \ldots$ the pull-backs of $\delta_\alpha, x, \ldots$ from the ith factor of $C_{n_1} \times \cdots \times C_{n_k}$. We also set

$$\eta_\alpha = \delta_\alpha \delta_{g+\alpha},$$

$$\eta_I = \eta_{i_1} \cdots \eta_{i_l},$$

where $I = (i_1, \ldots, i_l)$, $1 \le i_1 < \cdots < i_l \le g$, and recall that

$$\theta = \sum_{\alpha=1}^{g} \eta_\alpha.$$

Note that

$$(5.3) \qquad \begin{cases} \theta^l = l! \sum_{|I|=l} \eta_I, \\ \eta_I \text{ is the positively oriented generator of } H^{2d}(C_d, \mathbb{Z}) \text{ in case } |I| = d. \end{cases}$$

The reason for the second statement is that on the ordinary cartesian product C^d, η_I is $d!$ times the generator of $H^{2d}(C^d, \mathbb{Z})$, and the mapping $C^d \to C_d$ has degree $d!$. Moreover, any other $2d$-fold product of the δ_α's is zero.

For the multiplication map

$$\mu_{a_i}: J(C) \to J(C)$$

given by

$$\mu_{a_i}(u) = a_i u,$$

we have

$$\mu_{a_i}^*(\delta_\alpha) = a_i \delta_\alpha$$

(think of $J(C)$ as $\mathbb{R}^{2g}/\mathbb{Z}^{2g}$ and of δ_α as dx_α). It follows that

$$\phi_a^*(\delta_\alpha) = \sum_i a_i \delta_\alpha^{(i)},$$

$$\phi_a^*(\eta_\alpha) = \sum_{i,j} a_i a_j \delta_\alpha^{(i)} \delta_{g+\alpha}^{(j)}.$$

Even though this last expression is a little messy, by the last remark in the preceding paragraph in the final evaluation we may discard all the terms such that $i \neq j$ in the above. Thus, when $I = (i_1, \ldots, i_e)$, $e = \sum n_i$, we may write

$$\phi_a^*(\eta_I) = \prod_{j=1}^{e} \left(\sum_h a_h^2 \eta_{i_j}^{(h)} \right).$$

This, together with (5.3), gives

$$\phi_a^*(\eta_I) = \frac{e!}{n_1! \cdots n_k!} \prod_i a_i^{2n_i} \gamma,$$

where γ is the positively oriented generator of $H^{2e}(C_{n_1} \times \cdots \times C_{n_k}, \mathbb{Z})$. When combined with the first equation in (5.3) this proves (5.2).

For the next step, by a similar argument we will show that, when $p + q = e$

$$\text{(5.4)} \qquad \phi_a^*(x^q \theta^p)[C_{n_1} \times \cdots \times C_{n_k}] = p! \binom{g}{p} [R_a(t)^p P_a(t)^q]_{t_1^{n_1} \cdots t_k^{n_k}}.$$

Again the verification is straightforward. We have

$$\phi_a^* x = \sum_i a_i x^{(i)},$$

and then

$$\phi_a^*(x^q) = \sum_{m_1 + \cdots + m_k = q} \frac{q!}{m_1! \cdots m_k!} \prod_i a_i^{m_i} (x^{(i)})^{m_i}.$$

Since $(x^{(1)})^{m_1} \cdots (x^{(k)})^{m_k} \in H^*(\Pi C_{n_i}, \mathbb{Q})$ is dually represented by $C_{n_1 - m_1} \times \cdots \times C_{n_k - m_k}$, we may combine this with (5.3) to have

$$\begin{aligned} &\phi_a^*(x^q \theta^p)[C_{n_1} \times \cdots \times C_{n_k}] \\ &\quad = \sum_{m_1 + \cdots + m_k = q} \frac{q!}{m_1! \cdots m_k!} \prod_i a_i^{m_i} \phi_a^*(\theta^p)[C_{n_1 - m_1} \times \cdots \times C_{n_k - m_k}] \\ &\quad = \sum_{m_1 + \cdots + m_k = q} \binom{g}{p} \frac{(p!)^2 q! \prod_i a_i^{m_i + 2(n_i - m_i)}}{(n_1 - m_1)! \cdots (n_k - m_k)! m_1! \cdots m_k!} \\ &\quad = p! \binom{g}{p} \left[\left(1 + \sum_i a_i^2 t_i\right)^p \left(1 + \sum_i a_i t_i\right)^q \right]_{t_1^{n_1} \cdots t_k^{n_k}}, \end{aligned}$$

where $p + q = \sum_i n_i$. This establishes (5.4).

Completion of the Proof of Proposition (5.1). The expression on the right-hand side of the identity to be proved is

$$\sum_{\alpha} [(-1)^{\alpha} P_a(t)^{e-g} R_a(t)^{g-\alpha} (R_a(t) - P_a(t))^{\alpha} x^{d-e-\alpha} \theta^{\alpha}/\alpha!]_{t_1^{n_1} \cdots t_k^{n_k}}.$$

Multiplying this by $\theta^p x^q$ and evaluating on the fundamental class of C_d gives

$$\left[\sum_{\alpha} (-1)^{\alpha} P_a(t)^{e-g} R_a(t)^{g-\alpha} (R_a(t) - P_a(t))^{\alpha} \frac{(p+\alpha)!}{\alpha!} \binom{g}{p+\alpha}\right]_{t_1^{n_1} \cdots t_k^{n_k}}$$

$$= p!\binom{g}{p}\left[P_a(t)^{e-g} R_a(t)^p \sum_{\alpha} (-1)^{\alpha} \binom{g-p}{\alpha} R_a(t)^{g-p-\alpha} (R_a(t) - P_a(t))^{\alpha}\right]_{t_1^{n_1} \cdots t_k^{n_k}}$$

$$= p!\binom{g}{p}[P_a(t)^{e-p} R_a(t)^p]_{t_1^{n_1} \cdots t_k^{n_k}}.$$

Comparing once more with (5.4) gives the result.

Proof of DeJonquières' Formula. We must derive the formula for the virtual number of divisors

$$a_1 D_1 + \cdots + a_k D_k, \qquad \deg D_i = n_i$$

in a linear series $\mathscr{D}$ of degree d and dimension r, assuming that the dimensions are such that there are a finite number of such divisors to be expected; i.e., that

$$\sum_i n_i = d - r.$$

To do this we evaluate on $\mathscr{D} \cong \mathbb{P}^r$ the formula (5.1) for the fundamental class of $\phi_a(C_{n_1} \times \cdots \times C_{n_k})$. Since

$$\begin{cases} \theta | \mathscr{D} = 0, \\ x^r[\mathscr{D}] = 1, \end{cases}$$

the result is

$$[P_a(t)^{d-r-g} R_a(t)^g]_{t_1^{n_1} \cdots t_k^{n_k}}. \qquad \text{Q.E.D.}$$

Example. We will compute the number $\tau(d, g)$ of tritangent planes to a space curve $C \subset \mathbb{P}^3$ of degree d and genus g. This is obtained by taking

$$n_1 = 3, \qquad a_1 = 2,$$

$$n_2 = d - 6, \qquad a_2 = 1,$$

in DeJonquières' formula, which gives

$$\tau(d, g) = 8\binom{d-3}{3} + 8\binom{g}{2} + 4g(d-5)(d+g-5).$$

Note that for a canonical curve this is

$$\tau(6, 4) = 120,$$

which is the number of odd theta characteristics on a curve of genus 4.

Example. By combining the diagonal formula with the secant-plane formula it is possible to determine the virtual number of k-planes meeting a given curve $C \subset \mathbb{P}^r$ with given multiplicities of intersection. The simplest case of this is finding the number $T(d, g)$ of *tangential trisecants* to a curve $C \subset \mathbb{P}^3$ of degree d and genus g (a tangential trisecant is a tangent line meeting the curve in one additional point). To calculate $T(d, g)$ we intersect, in C_3, the class of the locus of divisors $D = p_1 + p_2 + p_3$ that fail to impose independent conditions on the hyperplane series g_d^3 with the diagonal $\phi_{2,1}(C \times C)$. The answer is

$$T(d, g) = 2(d-2)(d-3) + 2g(d-6).$$

For a canonical curve this is $2 \cdot 12$, in accordance with the Riemann–Hurwitz formula for the number of branch points of each of the g_3^1's.

Bibliographical Notes

The original proof of the Grothendieck–Riemann–Roch formula appears in Borel–Serre [1]; a more recent version may be found in Fulton [2]. Original attempts to derive enumerative formulas for projective curves did not use it, but rather relied on Schubert calculus (as in Giambelli [1]) or on the theory of correspondences, as in MacDonald [1]. The use of Riemann–Roch formulas was first suggested in this setting by Schwarzenberger. Similarly, earlier proofs of formulas such as DeJonquières' (as, for example, in Coolidge [1]) did not use the topology of the symmetric product; the cohomology of C_d and its use in establishing formulas for curves appears first in MacDonald [2] and Mattuck [1].

Exercises

A. Secant Planes to Canonical Curves

In this sequence of exercises $C \subset \mathbb{P}^{g-1}$ will be a smooth canonical curve with d-secant variety

$$\Sigma_d \subset \mathbb{P}^{g-1}$$

defined as follows: In $C_d \times G(d, g)$ we let Γ_d be the closure of the subvariety $\{(D, \Lambda): \overline{\phi_K(D)} = \mathbb{P}\Lambda\}$, $\bar{\Gamma}_d$ the inverse image of Γ_d in the projectivized universal subbundle $\mathbb{P}S$ over $G(d, g)$, and Σ_d the image of $\bar{\Gamma}_d$ under the natural map $\mathbb{P}S \to \mathbb{P}^{g-1}$.

A-1. Let $E_K^* \to C_d$ be the bundle defined in Section 2 of Chapter VIII and $\mathcal{O}(1)$ the tautological line bundle over $\mathbb{P}(E_K^*)$. Show that Σ_d is the image of $\mathbb{P}(E_K^*)$ under the map to $\mathbb{P}^{g-1}$ given by the sections of $\mathcal{O}(1)$ (in this exercise we are only interested in equality of varieties, not of schemes).

A-2. Show that, for $d < g/2$, the map

$$\mathbb{P}(E_K^*) \to \mathbb{P}^{g-1}$$

is birational (use the uniform position lemma).

A-3. Using the preceding two problems, show that

$$\deg \Sigma_d = \sum_{i=0}^{d} \binom{g-d-1-i}{d-i}\binom{g}{i}, \qquad d < g/2.$$

A-4. Alternatively, show that $\deg \Sigma_d$ is the number of divisors of degree d failing to impose independent conditions on a general $(2d-1)$-dimensional subspace $V \subset H^0(C, K)$. Using the secant plane formula show that this number is

$$\deg \Sigma_d = \left[\frac{(1+t)^g}{(1-t)^{g-2d}}\right]_{t_1}.$$

(It may be reassuring to verify that the answers obtained in these two problems do in fact coincide.)

B. Weierstrass Pairs

Definition. Let C be a smooth curve of genus g and α, β a pair of positive integers with $\alpha + \beta = g - 1$. We say that $p, q \in C$ is a *Weierstrass pair of points* (of type (α, β)) if $r(\alpha p + \beta q) \geq 1$.

B-1. Show that not every pair of points $p, q \in C$ is a Weierstrass pair.

B-2. Show that a non-hyperelliptic curve of genus 4 can have at most finitely many Weierstrass pairs of type (1, 2), and use the result at the end of Chapter VIII to count them.

B-3. Show that a bi-elliptic curve of genus 5 may have ∞^1 Weierstrass pairs of type (2, 2). Can a non-trigonal curve of genus 5 have ∞^1 Weierstrass pairs of type (3, 1)?

B-4. By combining (5.3) and the theorem at the end of Chapter VII find an enumerative formula for the number (assumed finite) of Weierstrass pairs of type (α, β).
(*Answer*: $g\alpha\beta((g-1)\alpha\beta - \alpha - \beta)$.)

B-5. Expressing the condition $r(\alpha p + \beta q) \geq 1$ determinantally, find an enumerative formula for the number (assumed finite) of Weierstrass pairs of type (α, β). (*Warning*: Watch out for the diagonal in $C \times C$.)

For $p, q \in C$ the *Weierstrass semigroup* $H(p, q) \subset \mathbb{N} \times \mathbb{N}$ is defined by

$$H(p, q) = \{(m, n)\text{: there exists } f \in \mathcal{M}(C) \text{ with } (f)_\infty = mp + nq\}.$$

B-6. Show that $H(p, q) \subset \mathbb{N} \times \mathbb{N}$ is a sub-semigroup with index

$$(*) \qquad \operatorname{card}(\mathbb{N} \times \mathbb{N} - H(p, q)) \geq \binom{g+2}{2} - 1.$$

(*Hint*: Show that for any fixed α,

$$\operatorname{card}\{\beta : \beta \geq 1, (\alpha, \beta) \notin H(p, q)\} \geq r(\alpha p) - (\alpha - g),$$

and use Exercise E-6 of Chapter I.).

B-7. Show that if p and q are general, then equality holds in $(*)$; that if either p or q is a Weierstrass point for C then equality fails in $(*)$; and that equality may fail in $(*)$ even if neither p nor q is a Weierstrass point of C.

C. Miscellany

C-1. Let C, B_1, B_2 be curves of respective genera g, g_1, g_2. Assume that

$$\phi_i : C \to B_i, \qquad i = 1, 2,$$

is a d_i-sheeted mapping such that

$$\phi = \phi_1 \times \phi_2 : C \to B_1 \times B_2$$

is birational to its image. Show that

$$g \leq (d_1 - 1)(d_2 - 1) + d_1 g_1 + d_2 g_2.$$

(*Suggestion*: Estimate the self-intersection of $\phi(C)$ in $B_1 \times B_2$ by taking the Künneth decomposition of its class and applying the Hodge index theorem (cf. page 126 of Griffiths–Harris [1]).)

C-2. As applications of the preceding exercise observe that a curve of genus $g \geq 3$ cannot be both hyperelliptic and bi-elliptic, that a curve of genus $g \geq 5$ cannot be both trigonal and bi-elliptic; and that a curve of genus $g \geq 6$ cannot be bi-elliptic in more than one way.

C-3. Let C be a smooth curve of genus $g \geq 2$. By a Chern class argument show that the bundle $E_{2g-1} \to J(C)$ introduced in Section 2 of Chapter VII does not have any sub-line bundles. Conclude that there does not exist a section of $u : C_d \to J(C)$ for $d \leq 2g - 1$.

Remark. On the other hand, it is a general fact about projective bundles that, for $d \geq 2g$, there is always a section of $u: C_d \to J(C)$.

C-4. Let V denote the vector space of homogeneous polynomials $f(x_0, x_1)$ of degree d, and for any collection of positive integers $n_1, \ldots, n_k$ with $\sum i \cdot n_i = d$, let $V_n \subset V$ be the variety of polynomials having n_i roots of multiplicity i. Using DeJonquières' formula, show that

$$\deg V_n = \frac{(\sum n_i)!}{n_1! \cdots n_k!} \Pi\, i^{n_i}.$$

C-5. Suppose that $f: C \to B$ is a branched covering of a curve C of genus $g \geq 2$ over a curve B of genus $h \geq 1$. Let $A \subset J(C)$ be the inverse image of $J(B)$ under the pull-back map. Show that the fundamental class of A is *not* a rational multiple of θ^{g-h}.

C-6. From the preceding exercise and the theorem on global monodromy proved in Chapter X conclude that a general curve of genus $g \geq 2$ does not admit a non-constant map to a curve of positive genus.

D. Push–Pull Formulas for Symmetric Products

Let C be a smooth curve of genus g. Given an algebraic cycle Z in C_d we define the following associated cycles

$$\begin{aligned} A(Z) &= \{D + p: D \in Z\} \\ &= \{E \in C_{d+1}: E - D \geq 0 \text{ for some } D \in Z\} \subset C_{d+1}; \\ B(Z) &= \{D - p: D \in Z \text{ and } D - p \geq 0\} \\ &= \{E \in C_{d-1}: D - E \geq 0 \text{ for some } D \in Z\} \subset C_{d-1}. \end{aligned}$$

In both of these we have only described the cycle set-theoretically, and it is understood that multiplicities are to be attached appropriately, e.g., the multiplicity of a component W of $A(Z)$ is card $\{D \in Z: E - D \geq 0\}$ where E is a general point of W. Similarly, we define cycles

$$\begin{aligned} A_k(Z) &= \{E \in C_{d+k}: E - D \geq 0 \text{ for some } D \in Z\}, \\ B_k(Z) &= \{E \in C_{d-k}: D - E \geq 0 \text{ for some } D \in Z\}. \end{aligned}$$

The assignments $Z \to A_k(Z)$, $Z \to B_k(Z)$ induce maps on the analytic part of the rational cohomology

$$\begin{aligned} A_k&: H^{2m}_{an}(C_d, \mathbb{Q}) \to H^{2m+2k}_{an}(C_{d+k}, \mathbb{Q}), \\ B_k&: H^{2m}_{an}(C_d, \mathbb{Q}) \to H^{2m-2k}_{an}(C_{d-k}, \mathbb{Q}). \end{aligned}$$

(Recall that for a curve C with general moduli, $H^*_{an}(C_d, \mathbb{Q})$ is generated by x and θ.)

In this series of exercises we will derive formulas for A_k, B_k and will give some applications of these formulas.

D-1. Define

$$\Psi: C_d \times C \to C_d \times C_{d+1}$$

by

$$\Psi(D, p) = (D, D + p),$$

and let ψ denote the fundamental class of the image. Show that for $z \in H^*_{an}(C_d, \mathbb{Q})$

$$A(z) = (\pi_2)_*(\psi \cup \pi_1^* z),$$

and for $z \in H^*_{an}(C_{d+1}, \mathbb{Q})$

$$B(z) = (\pi_1)_*(\psi \cup \pi_2^* z),$$

where π_1, π_2 are the projections of $C_d \times C_{d+1}$ to C_d, C_{d+1}.

Remark. This involves relating cup-product to intersection via Poincaré duality. An alternate approach to these exercises would be to *define* A, B by the above formulas, and then interpret these maps geometrically by our defining relations.

D-2. For $z \in H^{2d-2m}_{an}(C_d, \mathbb{Q})$ and $w \in H^{2m+2}_{an}(C_{d+1}, \mathbb{Q})$ show that

$$A(z) \cdot w = z \cdot B(w).$$

D-3. Show that A, B take the subring of $\bigoplus_d H^*_{an}(C_d, \mathbb{Q})$ generated by x, θ to itself (cf. the lemma in Section 5 of Chapter VIII).

D-4. Denote again by x, θ the pull-backs to $C_d \times C$ of x, θ on C_d, and by η the pull-back to $C_d \times C$ of the class of a point on C. For Ψ as in D-1 show that

$$\Psi^* \pi_1^* x = x,$$

$$\Psi^* \pi_1^* \theta = \theta,$$

$$\Psi^* \pi_2^* x = x + \eta,$$

$$\Psi^* \pi_2^* \theta = \theta + g \cdot \eta + \gamma,$$

where γ is the class in Section 2 of Chapter VIII.

D-5. Continuing the preceding exercise, if $\alpha + \beta + \lambda + \delta = d + 1$ show that

$$\begin{aligned}(B(x^\alpha\theta^\beta) \cdot x^\lambda\theta^\delta)_{C_d} &= (\pi_1^* x^\lambda\theta^\delta \cdot \pi_2^* x^\alpha\theta^\beta \cdot \psi)_{C_d \times C_{d+1}} \\ &= (\Psi^*\pi_1^* x^\lambda\theta^\delta \cdot \Psi^*\pi_2^* x^\alpha\theta^\beta)_{C_d \times C} \\ &= \alpha(x^{\alpha+\lambda-1} \cdot \theta^{\beta+\delta})_{C_d} + \beta(g - \beta + 1)(x^{\alpha+\lambda} \cdot \theta^{\beta+\delta-1})_{C_d}\end{aligned}$$

(recall that $\gamma^2 = -2\eta\theta$).

D-6. Conclude from D-3 and D-5 that

$$B(x^\alpha\theta^\beta) = \alpha x^{\alpha-1}\theta^\beta + \beta(g - \beta + 1)x^\alpha\theta^{\beta-1}.$$

D-7. Similarly, show that when $\alpha + \beta + \lambda + \delta = d + 1$

$$\begin{aligned}(A(x^\alpha\theta^\beta)\cdot x^\lambda\theta^\delta)_{C_{d-1}} &= \lambda(x^{\alpha+\lambda}\theta^{\beta+\delta})_{C_{d+1}} \\ &\quad + \delta(g - \delta + 1)(x^{\alpha+\lambda+1}\cdot\theta^{\beta+\delta-1})_{C_{d+1}}.\end{aligned}$$

From this conclude that

$$A(x^\alpha\theta^\beta) = (d + 1 - \alpha - 2\beta)x^\alpha\theta^\beta + \beta(g - \beta + 1)x^{\alpha+1}\theta^{\beta-1}.$$

D-8. Show that

$$A_k = \frac{1}{k!}A^k \quad (A^k = k\text{th iterate of } A),$$

$$B_k = \frac{1}{k!}B^k,$$

and using the preceding two exercises conclude that

$$B_k(x^\alpha\theta^\beta) = \sum_{j=0}^{k}\binom{\alpha}{k-j}\binom{\beta}{j}\binom{g-\beta+j}{j}j!\,x^{\alpha-k+j}\theta^{\beta-j},$$

$$A_k(x^\alpha\theta^\beta) = \sum_{i=0}^{k}\binom{\beta}{i}\binom{g-\beta+i}{i}\binom{d+k-\alpha-2\beta}{k-i}i!\,x^{\alpha+i}\theta^{\beta-i}.$$

D-9. Suppose that $n \geq 2g - 1$ and let $\mathcal{D} \subset C_n$ be a g_n^r. Denoting by $\Gamma_d(\mathcal{D}) \subset C_d$ the cycle of divisors of degree d subordinate to the g_n^r, rederive the formula (3.20) of Chapter VIII for the fundamental class $\gamma_d(\mathcal{D})$ of $\Gamma_d(\mathcal{D})$ by noting that

$$[\mathcal{D}] = x^{n-g-r}\frac{\theta^g}{g!},$$

$$\Gamma_d(\mathcal{D}) = A_{n-d}(\mathcal{D}),$$

and applying the preceding exercise.

The remaining exercises will give applications of the above formulas, and for these we will use the following notations:

$C \to H$ is an n-sheeted covering of C over a curve H of genus h;

$\tilde{\Sigma}_0 = \{\pi^{-1}q : q \in H\} \subset C_n$;

$$\begin{aligned}\tilde{\Sigma}_d &= \{\pi^{-1}q + D : q \in H \text{ and } D \in C_d\} \\ &= A_d(\tilde{\Sigma}_0);\end{aligned}$$

$$\begin{aligned}\Sigma_d &= u(\tilde{\Sigma}_d) \\ &= u(\tilde{\Sigma}_0) + W_d(C);\end{aligned}$$

$\tilde{\sigma}_0$, $\tilde{\sigma}_d$, and σ_d are the fundamental classes of $\tilde{\Sigma}_0$, $\tilde{\Sigma}_d$, and Σ_d;
$\Xi \subset C_n$ is the main diagonal, i.e.,

$\Xi = \{p_1 + \cdots + p_{n-2} + 2p : p_1, \ldots, p_{n-2}, p \in C\}$; and
ξ is the fundamental class of Ξ.

D-10. Show that

$$(\tilde{\sigma}_0 \cdot x) = 1,$$
$$(\tilde{\sigma}_0 \cdot \xi) = 2g - 2 - n(2h - 2),$$

and use (5.1) in Chapter VIII to conclude that

$$(\tilde{\sigma}_0 \cdot \theta) = (\sigma_0 \cdot \theta) = nh.$$

D-11. Use D-8 and the relation

$$\begin{aligned}(\tilde{\sigma}_d \cdot \theta^{d+1})_{C_{d+n}} &= (A_d(\tilde{\sigma}_0) \cdot \theta^{d+1})_{C_{d+n}} \\ &= (\tilde{\sigma}_0 \cdot B_d(\theta^{d+1}))_{C_n}\end{aligned}$$

to show that

$$\begin{aligned}(\tilde{\sigma}_d \cdot \theta^{d+1}) &= (\sigma_d \cdot \theta^{d+1}) \\ &= \frac{(g-1)!}{(g-d-1)!}(d+1) \cdot nh.\end{aligned}$$

E. Reducibility of $W_{g-1} \cap (W_{g-1} + u)$ (II)

The following sequence of exercises will establish the following refinement of Theorem (*) in exercise batch C of Chapter VI.

(**) **Theorem.** *Let C be a non-hyperelliptic curve of genus $g \geq 4$ and $v \in J(C)$. Then the intersection*

$$W_{g-1}(C) \cap (W_{g-1}(C) + v)$$

is reducible if and only if either:

(a) $v = u(p - q)$ *where* $p, q \in C$; *or*
(b) C *is a two-sheeted covering* $\pi\colon C \to E$ *of an elliptic curve and* $v \in \pi^*J(E)$

(cf. Weil [1] and Mumford [3]).

These exercises will use the following general result:

Let X be a smooth irreducible variety of dimension at least three and V_1, V_2 ample divisors on X. Then $V_1 \cap V_2$ is connected in codimension one.

Sketch of Proof. For $n \gg 0$ we may assume that $|nV_1|$, $|nV_2|$ are very ample. If $D_1 \in |nV_1|$, $D_2 \in |nV_2|$ are smooth, then $D_1 \cap D_2$ is connected by the Lefschetz theorem. Thus

$$\text{supp}(nV_1) \cap \text{supp}(nV_2) = \text{supp}(V_1 \cap V_2)$$

is the limit of connected varieties, and is therefore connected.

To see that it is connected in codimension 1, we intersect V_1, V_2 with a smooth threefold $X' \subset X$ and use the same argument to conclude that $V_1 \cap V_2 \cap X'$ is connected.

E-1. We have already seen (cf. Exercise C-6 in Chapter VI) that exception (a) in Theorem (**) actually occurs, and this exercise is to show that (b) occurs. Specifically, suppose $\pi: C \to E$ is a two-sheeted cover of an elliptic curve and define

$$\Sigma \subset C_{g-1}$$

by

$$\Sigma = \{D: D - \pi^{-1}p \geq 0 \text{ for some } p \in E\}$$

(thus, Σ consists of all effective divisors of degree $g - 1$ that contain a fiber of $\pi: C \to E$). Show that

$$\dim(u(\Sigma)) = g - 2,$$

and

$$u(\Sigma) \subset W_{g-1}(C) \cap (W_{g-1}(C) + v)$$

for any $v \in \pi^* J(E)$.

E-2. Let $u_*(\sigma)$ be the fundamental class of $u(\Sigma)$. Using Exercise D-11 above show that the intersection number $(u_*(\sigma) \cdot \theta^{g-2})$ equals $(g - 2)(g - 1)!$ and conclude that

$$u(\Sigma) \neq W_{g-1}(C) \cap (W_{g-1}(C) + v)$$

for $v \in \pi^* J(E)$. Hence, the intersection in the statement of Theorem (**) is reducible in case (b).

E-3. Suppose now that C is a non-hyperelliptic curve of genus g and that

$$W_{g-1}(C) \cap (W_{g-1}(C) + v)$$

is reducible for some $v \neq 0$. Using the connectedness result above show that some pair of components $\Xi_1, \Xi_2 \subset W_{g-1}(C) \cap (W_{g-1}(C) + v)$ contains an irreducible variety Γ of dimension $g - 3$.

E-4. Retaining the notations from the preceding exercise, let $u(D)$ be a general point of Γ, and write also

$$u(D) = u(F) + v,$$

where $F \in C_{g-1}$. Using Martens' theorem show that

$$r(D) = r(F) = 0,$$

and using the Riemann singularity theorem together with

$$\Gamma \subset (W_{g-1}(C) \cap (W_{g-1}(C) + v))_{\text{sing}},$$

show that

$$\overline{\phi_K(D)} = \overline{\phi_K(F)};$$

i.e., the divisors D, F span the same hyperplane in $\mathbb{P}^{g-1}$.

E-5. Continuing the notations of the preceding exercise, write

$$D = D' + G,$$
$$F = F' + G,$$

where all divisors are effective and $\operatorname{supp}(D')$ is disjoint from $\operatorname{supp}(F')$. Show that $G \neq 0$.

(*Hint*: if $G = 0$ then by the preceding exercise

$$u(D - F) = v,$$
$$u(D + F) = u(K_C)$$

for all $D \in \Gamma$ where $\dim \Gamma = g - 3$. From this obtain a contradiction when $\dim \Gamma > 0$; i.e., $g \geq 4$.

E-6. Assuming that $\deg D' = \deg F' = d$, on a sufficiently small open set $U \subset \Gamma$ define a map

$$\pi: U \to C_d \times C_d$$

by

$$\pi(D, F) = (D', F').$$

Using Clifford's theorem show that

$$\dim \overline{\phi_K(D' + F')} \geq d,$$

and conclude that a general fiber of π has dimension at most $g - d - 2$.

(*Hint*: Consider hyperplanes in $\mathbb{P}^{g-1}$ containing $\overline{\phi_K(D' + F')}$.)

E-7. Finally, using the preceding exercise show that $\dim(\operatorname{Im} \pi) \geq d - 1$, and using exercise sequence H of Chapter VI conclude the proof of Theorem (**).

F. Every Curve Has a Base-Point-Free g^1_{g-1}

In this sequence of exercises we will prove the following result:

(*) *Let C be a smooth, non-hyperelliptic curve of genus $g \geq 4$. Then there exists a base-point-free g^1_{g-1} on C.*

F-1. Suppose that C is a non-hyperelliptic curve of genus g and Σ is an irreducible component of $W^1_{g-1}(C)$ such that, for a general $u(D) \in \Sigma$, the complete linear series $|D|$ has a base point. Show that one of the following holds (cf. the proof of Mumford's theorem in Chapter IV):

(i) $C \subset \mathbb{P}^2$ is a smooth plane quintic and

$$\Sigma = W^1_5(C) = \{u(D + p - q): p, q \in C\},$$

where $\mathcal{O}_C(D) = \mathcal{O}_C(1)$.

(ii) C is trigonal with $g^1_3 = |F|$, and

$$\Sigma = \{u(F + p_1 + \cdots + p_{g-4}): p_1, \ldots, p_{g-4} \in C\},$$

(iii) $C \xrightarrow{\pi} E$ is bi-elliptic and

$$\Sigma = \{\pi^* J(E) + u(p_1 + \cdots + p_{g-5}): p_1, \ldots, p_{g-5} \in C\}.$$

F-2. Assuming still that C is non-hyperelliptic, show that if $g \geq 6$ then not every component of $W^1_{g-1}(C)$ is of the form (ii) or (iii) in the preceding exercise.

(*Hint*: Use the computation of intersection numbers in exercise batch D above.)

F-3. Using these two exercises conclude the proof of $(*)$.

Bibliography

Accola, R. D. M.

[1] On the number of automorphisms of a closed Riemann surface. *Trans. Amer. Math. Soc.* **131** (1968), 398–408.

[2] Plane models for Riemann surfaces admitting certain half-canonical linear series, part I, in *Riemann Surfaces and Related Topics. Proceedings of the 1978 Stony Brook Conference* (I. Kra and B. Maskit, eds.), Princeton University Press, Princeton, 1981, pp. 7–20.

Andreotti, A.

[1] Recherches sur les surfaces algébriques irrégulières, *Acad. Roy. Belg. Cl. Sci. Mém. Coll. in 8°*, **27** (1952), n° 7.

[2] On a theorem of Torelli, *Amer. J. Math.* **80** (1958), 801–828.

Andreotti, A. and Mayer, A.

[1] On period relations for abelian integrals on algebraic curves, *Ann. Scuola Norm. Sup. Pisa* **21** (1967), 189–238.

Arbarello, E. and Cornalba, M.

[1] Su una congettura di Petri, *Comment. Math. Helv.* **56** (1981), 1–38.

Arbarello, E. and De Concini, C.

[1] On a set of equations characterizing Riemann matrices, *Ann. of Math.* **120** (1984), 119–140.

[2] An analytical translation of a criterion of Welters and its relation with the K–P hierarchy, in *Proceedings of the Conference "Week of Algebraic Geometry"* (Sitges, 1983), to appear in Lecture Notes in Mathematics.

Arbarello, E. and Harris, J.

[1] Canonical curves and quadrics of rank 4, *Compositio Math.* **43** (1981), 145–179.

Arbarello, E. and Sernesi, E.

[1] Petri's approach to the study of the ideal associated to a special divisor, *Invent. Math.* **49** (1978), 99–119.

Artin, M.

[1] Théorème de finitude pour un morphisme propre: dimension cohomologique des schémas algébriques affines, in *Théorie des topos et cohomologie étale des schémas* (SGA IV), tome 3, Lecture Notes in Mathematics 305, Springer-Verlag, Berlin–Heidelberg–New York, 1973, pp. 145–167.

Babbage, D. W.

[1] A note on the quadrics through a canonical curve, *J. London Math. Soc.* **14** (1939), 310–315.

Baily, W.

[1] On the theory of θ-functions, the moduli of abelian varieties, and the moduli of curves, *Ann. of Math.* **75** (1962), 342–381.

Beauville, A.
[1] Prym varieties and the Schottky problem, *Invent. Math.* **41** (1977), 149–196.
[2] L'application canonique pour les surfaces de type général, *Invent. Math.* **55** (1979), 121–140.

Bers, L.
[1] Uniformization, moduli, and Kleinian groups, *Bull. London Math. Soc.* **4** (1972), 257–300.

Bloch, S. and Gieseker, D.
[1] The positivity of the Chern classes of an ample vector bundle, *Invent. Math.* **12** (1971), 112–117.

Borel, A. and Serre, J. P.
[1] Le théorème de Riemann–Roch, *Bull. Soc. Math. France* **86** (1958), 97–136.

Brill, A. and Noether, M.
[1] Über die algebraischen Functionen und ihre Anwendungen in der Geometrie, *Math. Ann.* **7** (1873), 269–310.

Castelnuovo, G.
[1] Ricerche di geometria sulle curve algebriche, *Atti R. Accad. Sci. Torino* **24** (1889), 196–223.
[2] Numero delle involuzioni razionali giacenti sopra una curva di dato genere, *Rendiconti R. Accad. Lincei* (4) **5** (1889), 130–133.

Ciliberto, C.
[1] On a proof of Torelli's theorem, in *Algebraic Geometry–Open Problems* (C. Ciliberto, F. Ghione, F. Orecchia, eds.), Lecture Notes in Mathematics 997, Springer-Verlag, Berlin–Heidelberg–New York–Tokyo, 1983, pp. 113–123.

Clebsch, A.
[1] Zur Theorie der Riemann'schen Flächen, *Math. Ann.* **6** (1873), 216–230.

Clemens, C. H. and Griffiths, P.
[1] The intermediate Jacobian of the cubic threefold, *Ann. of Math.* **95** (1972), 281–356.

Clifford, W. K.
[1] On the classification of loci, in *Mathematical Papers*, Macmillan, London, 1882.

Comessatti, A.
[1] Limiti di variabilità della dimensione e dell'ordine d'una g_n^r sopra una curva di dato genere. *Atti R. Ist. Veneto Sci. Lett. Arti* **74** (1914/15), 1685–1709.
[2] Sulle trasformazioni hermitiane della varietà di Jacobi, *Atti R. Accad. Sci. Torino* **50** (1914–15), 439–455.

Coolidge, J. L.
[1] *Algebraic Plane Curves*, Oxford University Press, Oxford, 1931.

De Concini, C., Eisenbud, D. and Procesi, C.
[1] Hodge algebras, *Astérisque* **91** (1982).

De Concini, C. and Procesi, C.
[1] A characteristic-free approach to invariant theory, *Adv. in Math.* **21** (1976), 330–354.

Deligne, P. and Mumford, D.
[1] The irreducibility of the space of curves of given genus, *Inst. Hautes Études Sci. Publ. Math.* **36** (1969), 75–110.

Donagi, R.
[1] The tetragonal construction, *Bull. Amer. Math. Soc.* (*N.S.*) **4** (1981), 181–185.

Doubilet, P., Rota, G.-C. and Stein, J.
[1] On the foundations of combinatorial theory IX, *Studies in Applied Math.* **53** (1974), 185–216.

Eagon, J. A. and Hochster, M.
[1] Cohen–Macaulay rings, invariant theory, and the generic perfection of determinantal loci, *Amer. J. Math.* **93** (1971), 1020–1058.

Eisenbud, D. and Harris, J.
[1] Divisors on general curves and cuspidal rational curves, *Invent. Math.* **74** (1983), 371–418.
[2] A simpler proof of the Gieseker–Petri theorem on special divisors, *Invent. Math.* **74** (1983), 269–280.
[3] *Curves in Projective Space*, Les Presses de l'Université de Montréal, Montréal, 1982.

Enriques, F.
[1] Sulle curve canoniche di genere p dello spazio a $p-1$ dimensioni, *Rend. Accad. Sci. Ist. Bologna* **23** (1919), 80–82.

Enriques, F. and Chisini, O.
[1] *Teoria Geometrica delle Equazioni e delle Funzioni Algebriche*, 4 vols., Zanichelli, Bologna, 1915, 1918, 1924, 1934.

Farkas, H. M.
[1] Special divisors and analytic subloci of the Teichmüller space, *Amer J. Math.* **88** (1966), 881–901.

Farkas, H. M. and Kra, I.
[1] *Riemann Surfaces*, Springer-Verlag, New York–Heidelberg–Berlin, 1980.

Fay, J. D.
[1] *Theta Functions on Riemann Surfaces*, Lecture Notes in Mathematics 332, Springer-Verlag, Berlin–Heidelberg–New York, 1973.

Fitting, H.
[1] Die Determinantenideale eines Moduls, *Jahresber. Deutsch. Math. Verein.* **46** (1936), 195–228.

Fulton, W.
[1] *Algebraic Curves*, Benjamin, New York, 1969.
[2] *Intersection Theory*, Springer-Verlag, Berlin–Heidelberg–New York–Tokyo, 1984.

Fulton, W. and Hansen, J.
[1] A connectedness theorem for projective varieties, with applications to intersections and singularities of mappings, *Ann. of Math.* **110** (1979), 159–166.

Fulton, W. and Lazarsfeld, R.
[1] On the connectedness of degeneracy loci and special divisors, *Acta Math.* **146** (1981), 271–283.

Giambelli, G. Z.
[1] Ordine di una varietà più ampia di quella rappresentata coll'annullare tutti i minori di dato ordine estratti da una data matrice generica di forme, *Mem. R. Ist. Lombardo* (3) **11** (1904), 101–135.

Gieseker, D.
[1] Stable curves and special divisors, *Invent. Math.* **66** (1982), 251–275.

Grauert, H.
[1] Ein Theorem der analytischen Garbentheorie und die Modulräume komplexer Strukturen, *Inst. Hautes Études Sci. Publ. Math.* **5** (1960), 233–292.

Green, M.
[1] Quadrics of rank four in the ideal of the canonical curve, *Invent. Math.* **75** (1984), 85–104.

Griffiths, P.
[1] The extension problem for compact submanifolds of complex manifolds I, in *Proc. Conf. Complex Analysis (Minneapolis 1964)* (A. Aeppli, E. Calabi, H. Rohrl, eds.), Springer-Verlag, New York, 1965, pp. 113–142.
[2] Hermitian differential geometry, Chern classes, and positive vector bundles, in *Global Analysis* (D. Spencer, S. Iyanaga, eds.), Princeton University Press and University of Tokyo Press, Tokyo, 1969, pp. 185–251.

Griffiths, P. and Harris, J.
[1] *Principles of Algebraic Geometry*, Wiley-Interscience, New York, 1978.
[2] Residues and zero-cycles on algebraic varieties, *Ann. of Math.* **108** (1978), 461–505.
[3] Algebraic geometry and local differential geometry, *Ann. scient. Éc. Norm. Sup.* (4) **12** (1979), 355–432.
[4] The dimension of the variety of special linear systems on a general curve, *Duke Math. J.* **47** (1980), 233–272.

Grothendieck, A.
[1] Technique de descente et théorèmes d'existence en géométrie algébrique V. Les schémas de Picard: théorèmes d'existence, *Séminaire Bourbaki, 14e année*, 1961–62, n° 232.

Gruson, L., Lazarsfeld, R., and Peskine, C.
[1] On a theorem of Castelnuovo, and the equations defining space curves, *Invent. Math.* **72** (1983), 491–506.

Gunning, R. C.
[1] *Lectures on Riemann Surfaces*, Princeton University Press, Princeton, 1966.
[2] Some curves in abelian varieties, *Invent. Math.* **66** (1982), 377–389.

Gunning, R. C. and Rossi, H.
[1] *Analytic Functions of Several Complex Variables*, Prentice-Hall, Englewood Cliffs, N.J., 1965.

Halphen, G. H.
[1] Mémoire sur la classification des courbes gauches algébriques, *J. École Polytechnique* **52** (1882), 1–200.

Harris, J.
[1] Galois groups of enumerative problems, *Duke Math. J.* **46** (1979), 685–724.
[2] A bound on the geometric genus of projective varieties, *Ann. Scuola Norm. Sup. Pisa* (4) **8** (1981), 35–68.
[3] On the Kodaira dimension of the moduli space of curves, II: the even genus case, *Invent. Math.* **75** (1984), 437–466.
[4] Theta characteristics on algebraic curves, *Trans. Amer. Math. Soc.* **271** (1982), 611–638.
[5] The genus of space curves, *Math. Ann.* **249** (1980), 191–204.

Harris, J. and Mumford, D.
[1] On the Kodaira dimension of the moduli space of curves, *Invent. Math.* **67** (1982), 23–86.

Hartshorne, R.
[1] Ample vector bundles, *Inst. Hautes Études Sci. Publ. Math.* **36** (1969), 75–110.
[2] *Algebraic Geometry*, Springer-Verlag, New York–Heidelberg–Berlin, 1977.
[3] *Ample Subvarieties of Algebraic Varieties*, Lecture Notes in Mathematics 156, Springer-Verlag, Berlin–Heidelberg–New York, 1970.

Hochster, M.
[1] Grassmannians and their Schubert subvarieties are arithmetically Cohen–Macaulay, *J. Algebra* **25** (1973), 40–57.

Hodge, W. V. D.
[1] Some enumerative results in the theory of forms, *Proc. Cam. Phil. Soc.* **39** (1943), 22–30.

Hopf, H.
[1] Ein topologischer Beitrag zur reellen Algebra, *Comment. Math. Helv.* **13** (1940/41), 219–239.

Keem, C.
[1] *A remark on the variety of special linear systems on an algebraic curve*, Ph.D. Thesis, Brown University, 1983.

Kempf, G.
[1] Schubert methods with an application to algebraic curves, *Publications of Mathematisch Centrum*, Amsterdam, 1972.
[2] On the geometry of a theorem of Riemann, *Ann. of Math.* **98** (1973), 178–185.

Kempf, G. and Laksov, D.
[1] The determinantal formula of Schubert calculus, *Acta Math.* **132** (1974), 153–172.

Kleiman, S.
[1] r-special subschemes and an argument of Severi's, *Adv. in Math.* **22** (1976), 1–23.

Kleiman, S. and Laksov, D.
[1] On the existence of special divisors, *Amer. J. Math.* **94** (1972), 431–436.
[2] Another proof of the existence of special divisors, *Acta Math.* **132** (1974), 163–176.

Klein, F.
[1] *Über Riemann's Theorie der Algebraischen Functionen und Ihrer Integrale*, Teubner, Leipzig, 1882.

Knudsen, F. F.
[1] The projectivity of the moduli space of stable curves III, *Math. Scand.* **52** (1983), 200–212.

Kodaira, K. and Spencer, D. C.
[1] On deformations of complex analytic structures I, II, *Ann. of Math.* **67** (1958), 328–466.

Laksov, D.
[1] The arithmetic Cohen–Macaulay character of Schubert schemes, *Acta Math.* **129** (1972), 1–9.

Lang, S.
[1] *Abelian Varieties*, Interscience, New York, 1959.

Lax, R. F.
[1] On the dimension of varieties of special divisors, *Trans. Amer. Math. Soc.* **203** (1975), 141–159.

Lewittes, J.
[1] Riemann surfaces and the theta function, *Acta Math.* **111** (1964), 37–61.

Little, J.
[1] Translation manifolds and the converse of Abel's theorem, *Compositio Math.* **49** (1983), 147–171.

Lüroth, J.

[1] Note über Verzweigungsschnitte und Querschnitte in einer Riemann'schen Fläche, *Math. Ann.* **4** (1871), 181–184.

MacDonald, I. G.

[1] Some enumerative formulae for algebraic curves, *Proc. Camb. Phil. Soc.* **54** (1958), 399–416.

[2] Symmetric products of an algebraic curve, *Topology* **1** (1962), 319–343.

Martens, H. H.

[1] A new proof of Torelli's theorem, *Ann. of Math.* **78** (1963), 107–111.

[2] On the varieties of special divisors on a curve, *J. Reine Angew. Math.* **227** (1967), 111–120.

[3] Varieties of special divisors on a curve II, *J. Reine Angew. Math.* **233** (1968), 89–100.

Mather, J.

[1] Stratifications and mappings, in *Dynamical Systems* (M. M. Peixoto, ed.), Academic Press, New York, 1973.

Matsumura, H.

[1] *Commutative Algebra*, second edition, Benjamin, Reading, Mass., 1980.

Matsusaka, T.

[1] On a theorem of Torelli, *Amer. J. Math*, **80** (1958), 784–800.

Mattuck, A.

[1] Symmetric products and Jacobians, *Amer. J. Math.* **83** (1961), 189–206.

Mayer, A. L.

[1] Special divisors and the Jacobian variety, *Math. Ann.* **153** (1964), 163–167.

Meis, T.

[1] Die minimale Blätterzahl der Konkretisierungen einer kompakten Riemannschen Fläche, *Schr. Math. Inst. Univ. Munster No. 16* (1960).

Mulase, M.

[1] Cohomological structure of solutions of soliton equations; isospectral deformations of ordinary differential operators and a characterization of Jacobian varieties. Preprint, 1984.

Mumford, D.

[1] *Abelian Varieties*, Oxford University Press, Bombay, 1970.

[2] Prym varieties I, in *Contributions to Analysis* (L. V. Ahlfors, I. Kra, B. Maskit, L. Niremberg, eds.), Academic Press, New York, 1974, pp. 325–350.

[3] *Curves and Their Jacobians*, The University of Michigan Press, Ann Arbor, 1975.

[4] Stability of projective varieties, *L'Enseignement Mathém.* **23** (1977), 39–110.

[5] *Tata Lectures on Theta I*, Birkhäuser, Boston–Basel–Stuttgart, 1983.

[6] *Tata Lectures on Theta II*, Birkhäuser, Boston–Basel–Stuttgart, 1984.

Mumford, D. and Fogarty, J.

[1] *Geometric Invariant Theory*, second enlarged edition, Springer-Verlag, Berlin–Heidelberg–New York, 1982.

Musili, C.

[1] Postulation formula for Schubert varieties, *J. Ind. Math. Soc.* **36** (1972), 143–171.

Northcott, D. G.

[1] *Finite Free Resolutions*, Cambridge University Press, Cambridge, 1976.

Petri, K.

[1] Über die invariante Darstellung algebraischer Funktionen einer Veränderlichen, *Math. Ann.* **88** (1922), 242–289.

[2] Über Spezialkurven I, *Math. Ann.* **93** (1925), 182–209.

Porteous, I. R.

[1] Simple singularities of maps, in *Proceedings of Liverpool Singularities-Symposium* 1 (C. T. C. Wall, ed.), Lecture Notes in Mathematics 192, Springer-Verlag, Berlin–Heidelberg–New York, 1971, pp. 286–307.

Riemann, B.

[1] Theorie der Abel'schen Functionen, in *Mathematische Werke*, Teubner, Leipzig, 1876.

Saint-Donat, B.

[1] On Petri's analysis of the linear system of quadrics through a canonical curve, *Math. Ann.* **206** (1973), 157–175.

[2] Variétés de translation et théorème de Torelli, *C.R. Acad. Sci. Paris Ser. A* **280** (1975), 1611–1612.

Sernesi, E.

[1] L'unirazionalità dei moduli delle curve di genere dodici, *Ann. Scuola Norm. Sup. Pisa* (4) **8** (1981), 405–439.

Serre, J. P.

[1] *Algèbre Locale-Multiplicités*, Lecture Notes in Mathematics 11, Springer-Verlag, Berlin–Heidelberg, 1965.

Severi, F.

[1] Sulla classificazione delle curve algebriche e sul teorema di esistenza di Riemann II, *Rendiconti R. Accad. Lincei* (5) **241** (1915), 1011–1020.

[2] *Vorlesungen über Algebraische Geometrie*, Teubner, Leipzig, 1921.

Shafarevich, I. R.

[1] *Basic Algebraic Geometry*, Springer-Verlag, Berlin–Heidelberg–New York, 1977.

Shiota, T.

[1] *Characterization of Jacobian varieties in terms of soliton equations*, Ph.D. Thesis, Harvard University, 1984.

Shokurov, V. V.

[1] The Noether–Enriques theorem on canonical curves, *Math. USSR Sbornik* **15** (1971), 361–403.

Siegel, C. L.

[1] *Topics in Complex Function Theory*, 3 vols., Wiley-Interscience, New York, 1969, 1971, 1973.

Smith, R. and Varley, R.

[1] On the geometry of $\mathcal{N}_0$, to appear in *Rend. Sem. Mat. Univers. Politecn. Torino*.

Springer, G.

[1] *Introduction to Riemann Surfaces*, Addison-Wesley, Reading, Mass., 1957.

Thom, R.

[1] Ensembles et morphismes stratifiés, *Bull. Amer. Math. Soc.* **75** (1969), 240–284.

Tjurin, A. N.

[1] The geometry of the Poincaré theta-divisor of a Prym variety, *Math. USSR Izvestija* **9** (1975), 951–986.

Torelli, R.
[1] Sulle varietà di Jacobi, *Rendiconti R. Accad. Lincei. Cl. Sci. Fis. Mat. Nat.* (5) **22** (1913), 98–103.

Walker, R. J.
[1] *Algebraic Curves*, Princeton University Press, Princeton, 1950.

Weil, A.
[1] Zum Beweis des Torellischen Satzes, *Nachr. Akad. Wiss. Gottingen. Math. Phys. Kl.* **IIa** (1957) 33–53.

Welters, G. E.
[1] A criterion for Jacobi varieties, to appear.

Weyl, H.
[1] *Die Idee des Riemannschen Fläche*, Teubner, Leipzig, 1913.
[2] *The Classical Groups*, Princeton University Press, Princeton, 1946.

Index

Abel's Theorem 18
Accola's theorem 275
Abelian
  differentials 7
  sums 17
  variety 21
Adjoint
  conditions 58
  divisor 57
  ideal 57
Adjunction formula
  in the plane 53
  on a general surface 138
Ample vector bundle 305
Andreotti–Mayer Theorem 253
Associated map 37
Automorphisms 44

Base locus 5
Base-point-free
  linear series 5
  pencil trick 126
Bi-elliptic curves 198, 210, 269, 276, 280,
  of genus 5 272
  of genus 6 218
  embeddings of 221
Brill–Noether
  matrix 154, 159
  number 159, 205
Canonical
  curve 12, 117–118
  line bundle 7
  map 10, 15
Castelnuouvo's
  bound 116
  lemmas 120, 151
  number 116
  theorem 151
  theorem on k-normality 141

Chern
  character 331
  polynomial 85
Chern classes
  in $G(2,4)$ 104
  of kernel bundles 105
  of a symmetric product 336
  of a tensor product 89
Clifford's theorem 107
  refinements of 137
Complete interjections 138
Complete linear series 4
Completeness of the adjoint series 54, 56, 60
Complex torus 20
Connectedness theorem
  of Fulton–Hansen 271, 326
  for degeneracy loci 311
  for $W^r_d(C)$ 212, 314
Constructible sheaf 313
C^r_d
  definition of 154
  dimension of 159
  in genus 4 196
  non-emptiness 206
  tangent space to 162, 163, 197
Cubic threefold 302
Curve 1
Curves of genus 2 33, 40, 46, 48, 278
Curves of genus 3 35, 40, 46, 248, 279, 303
Curves of genus 4 34, 40, 47, 196, 206, 276, 279
Curves of genus 5 206, 270, 279, 300
Curves of genus 6 209, 218, 279

Deck transformation 47, 281
Degeneracy locus 83
  existence theorem 307
  connectedness theorem 311

Degree
  of a divisor 2
  of a line bundle 7
  of a map 9
De Jonquières' formula 359
Determinantal varieties 67, 83
Diagonal mappings 358
Difference map 223, 262, 276
Dimension theorem 214
Divisor 2
Duality theorem 7

Effective divisor 3
Embedding
  criteria for 40
  of degree $g+3$ 40
  theorem 216
Enriques–Babbage theorem 124
Euler exact sequence 321
Exceptional divisor 9, 63, 205
Excess linear series 329
Existence theorem 206, 311
Extremal curves 117, 147, 263

Families
  of curves 288
  of linear systems 182
  of line bundles 169
  of theta characteristics 288
Fermat curve 44
First fundamental theorem of invariant theory 77
First-order deformation
  of a linear series 185
  of a line bundle 169
Fitting ideal 179
Fundamental theorems of invariant theory 71, 77

Galois
  covering 47
  extension 281
Gap sequence 41
General position theorem 109
Generic determinantal variety 67
Genus 1
  Castelnuovo's bound on 116
  formula for plane curves 53, 58, 60
  of space curves 143
Gherardelli's theorem 147
Gieseker's lemma 145
Gorenstein relation 59
G^r_d 153, 177, 187, 188
Grothendieck
  decomposition 145
  group 85, 331
  Riemann–Roch formula 333

Heat equation 24, 250
Hodge theorem for curves 54
Hurwitz formula, *see* Riemann–Hurwitz formula
Hyperelliptic curve 10, 41, 43
  embeddings of 221
  in genus 6 218
  Prym varieties of 303
  theta-characteristics on 287–292
  Torelli theorem for 247, 269
  $W_2(C)$ 279

Ideal of initial forms 62
Index of speciality 8
Inflectionary point 37
Isogeny 49

Jacobian variety 17
Jacobi inversion theorem 19, 27

Keem's theorems 200
Kempf's singularity theorem 241
Klein canonical curve 123
k-normal 140, 141
Kodaira–Spencer class 171
Koszul complex 139

Laurent tail 14
Lefschetz hyperplane theorem 306, 315
Linear equivalence 4
Linearly normal 140
Linear series 4
Linear system 4

MacDonald's formula 350
Martens' theorem 191
  for birational maps 198
Mittag–Leffler problem 14
Mittag–Leffler sequence 13
Moduli space 28
Monodromy group 111

Multiplicity
of a divisor 2
of a point in a variety 62
of a point on a curve 36
Mumford's theorem 193
for birational maps 198
for g_d^2's 198

Nearly extremal curves 149
Net 4
of quadrics in $\mathbb{P}^4$ 270
Newton functions 156
Noether, Max
"AF + BG" theorem 139
theorem 117, 151
Non-degenerate 6
Normal covering 47
Normalized period matrix 22
Normalization 31
Normal Weierstrass point 42
Norm map 281

Order 2
Oscnode 220
Osculating k-plane 37

Pencil 4
Period matrix 16, 21
normalized 22
Petri's construction 127
Petri's theorem 131
Picard group 18
Plane curves 31
genus formula for 53, 59
linear series on 56
Plane sextic curve (*see* Sextic plane curve)
Plane quintic curve (*see* Quintic plane curve)
Plücker
formula 39
relations 71
Poincaré
complete reducibility theorem 49
formula 25
line bundle 166
residue 53
Postulation 56, 115, 120, 142, 144, 149, 199, 220
Prill's problem 268
Principally polarized abelian variety 21
Principal part (of a meromorphic differential) 14
Projections 35, 51
Projective bundle 304, 317
Projectively normal 140, 221
Projectivized tangent cone 62
Prym varieties 297
Push-pull formula 91, 318
for symmetric products 367

Quadrics 102
containing a curve 142
canonical 255
of genus 5 270, 275
in $\mathbb{P}^3$ 140
linear spaces on 102
spaces of 100
Quasi-algebraically closed 282
Quintic plane curve 244, 264, 270–274, 300

Ramification
divisor 9
index 9
sequence, of a map 37
Rational normal curve 6, 39
Rational normal scroll 95, 121, 147
Relative divisor 165
Residual 10
Residue 2
theorem 3
Riemann
bilinear relations 16, 21
constant 27
inequality 55
singularity theorem 226
theorem 27
theta function 23
Riemann–Hurwitz formula 8, 39
Riemann–Mumford relation 290
Riemann–Roch theorem 4, 7
for an abelian variety 21
geometric version 12
proof via adjoint series 55–56
for surfaces 148
Riemann surface associated to an algebraic function 31

Schottky problem 249
Schubert varieties 74

Secant plane 12, 152
  formula for 340, 345
Secant variety 231
  of a canonical curve 364
Second fundamental theorem of invariant theory 71
Segre class 318
Semicanonical divisors (*see* Theta characteristics)
Serre's criterion 98
Sextic plane curves 209, 220, 264, 275
Sheet number 9
Siegel modular group 23
Siegel upper half plane 22
Simple abelian variety 49
Smoothness theorem 214, 215
Special divisor 8
Stable curve 29
Standard monomials 72
Superabundance 56, 115, 120, 142, 144, 149, 199, 220
Sylvester's determinant 87
Symmetric product 18, 136, 309
  Chern classes of 322, 336
  cohomology of 328
  tangent spaces to 171, 197

Tangent cone 61
  to the theta divisor 226, 255
  to $W^r_d(C)$ 240
Theta characteristics
  definition of 287
  in genus 5 270–274, 275
Theta divisor 22
  intersection of translates of 267
  singular locus of 250
  tangent cones to 226, 255, 270
  tangent planes to 268
Theta function 23
Todd class 332
Torelli's theorem 27, 245, 263, 265, 268
Trigonal curves 118, 198
  embeddings of 219, 221
  of genus 5 270, 279
  of genus 6 218
  linear series on 138

Uniform position theorem 112
Universal divisor 164
  cohomology class of 336
Universal family
  of abelian varieties 251
  of curves 29

Veronese map 6
Veronese surface 245

Web 4
Weierstrass
  gap sequence 41
  pairs 365
  points 41
  semigroup 41
Weight (of a Weierstrass point) 42
$W^r_d(C)$
  connectedness of 212
  definition of 176, 177
  dimension of 189, 192, 193, 198, 200
  examples of 196, 199, 206–211, 218
  existence of 206
  irreducibility of 214
  multiplicity of 240–241
  tangent spaces to 189, 197, 269, 240
Weil
  pairing 284
  reciprocity 283

Grundlehren der mathematischen Wissenschaften

Continued from page ii

225. Schütte: Proof Theory
226. Karoubi: K-Theory, An Introduction
227. Grauert/Remmert: Theorie der Steinschen Räume
228. Segal/Kunze: Integrals and Operators
229. Hasse: Number Theory
230. Klingenberg: Lectures on Closed Geodesics
231. Lang: Elliptic Curves: Diophantine Analysis
232. Gihman/Skorohod: The Theory of Stochastic Processes III
233. Stroock/Varadhan: Multi-dimensional Diffusion Processes
234. Aigner: Combinatorial Theory
235. Dynkin/Yushkevich: Markov Control Processes and Their Applications
236. Grauert/Remmert: Theory of Stein Spaces
237. Köthe: Topological Vector-Spaces II
238. Graham/McGehee: Essays in Commutative Harmonic Analysis
239. Elliott: Probabilistic Number Theory I
240. Elliott: Probabilistic Number Theory II
241. Rudin: Function Theory in the Unit Ball of C^n
242. Blackburn/Huppert: Finite Groups I
243. Blackburn/Huppert: Finite Groups II
244. Kubert/Lang: Modular Units
245. Cornfeld/Fomin/Sinai: Ergodic Theory
246. Naimark: Theory of Group Representations
247. Suzuki: Group Theory I
248. Suzuki: Group Theory II
249. Chung: Lectures from Markov Processes to Brownian Motion
250. Arnold: Geometrical Methods in the Theory of Ordinary Differential Equations
251. Chow/Hale: Methods of Bifurcation Theory
252. Aubin: Nonlinear Analysis on Manifolds, Monge—Ampère Equations
253. Dwork: Lectures on p-adic Differential Equations
254. Freitag: Siegelsche Modulfunktionen
255. Lang: Complex Multiplication
256. Hormander: The Analysis of Linear Partial Differential Operators I
257. Hormander: The Analysis of Linear Partial Differential Operators II
258. Smoller: Shock Waves and Reaction-Diffusion Equations
259. Duren: Univalent Functions
260. Freidlin/Wentzell: Random Perturbations of Dynamical Systems
261. Remmert/Bosch/Güntzer: Non Archemedian Analysis—A Systematic Approach to Rigid Analytic Geometry
262. Doob: Classical Potential Theory & Its Probabilistic Counterpart
263. Krasnoselščskiĭ/Zabreĭko: Geometrical Methods of Nonlinear Analysis
264. Aubin/Cellina: Differential Inclusions
265. Grauert/Remmert: Coherent Analytic Sheaves
266. de Rham: Differentiable Manifolds
267. Arbarello/Cornalba/Griffiths: Geometry of Algebraic Curves, Vol. I